Laboratory Animals
Regulations and Recommendations for Global Collaborative Research

Laboratory Animals
Regulations and Recommendations for Global Collaborative Research

Javier Guillén
Senior Director and Director of European Activities
AAALAC International, Pamplona, Spain

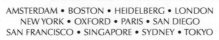

AMSTERDAM • BOSTON • HEIDELBERG • LONDON
NEW YORK • OXFORD • PARIS • SAN DIEGO
SAN FRANCISCO • SINGAPORE • SYDNEY • TOKYO

Academic Press is an Imprint of Elsevier

Academic Press is an imprint of Elsevier
525 B Street, Suite 1800, San Diego, CA 92101-4495, USA
32 Jamestown Road, London NW1 7BY, UK
225 Wyman Street, Waltham, MA 02451, USA

British Library Cataloguing-in-Publication Data
A catalogue record for this book is available from the British Library

Library of Congress Cataloging-in-Publication Data
A catalog record for this book is available from the Library of Congress

ISBN: 978-0-12-397856-1

For information on all Academic Press publications
visit our website at elsevierdirect.com

Typeset by TNQ Books and Journals
www.tnq.co.in

Printed and bound in United States of America

14 15 16 17 18   10 9 8 7 6 5 4 3 2 1

Contents

5.   The European Framework on Research Animal Welfare
      Regulations and Guidelines

*Javier Guillén, Jan-Bas Prins, David Smith, and
Anne-Dominique Degryse*

6.   Israeli Legislation and Regulation on the Use of Animals in Biological and Medical Research

Rony Kalman, Alon Harmelin, Ehud Ziv, and Yacov Fischer

7. **Animal Experimentation in Africa: Legislation and Guidelines**

Amanda R. Hau, Faisal A. Guhad, Margaret E. Cooper, Idle O. Farah, Ouajdi Souilem, and Jann Hau

8. **Laws, Regulations, and Guidelines Shaping Research Animal Care and Use in India**

Syed SYH Qadri and Christian E. Newcomer

9. Oversight of Animal Research in China

Kathryn Bayne and Jianfei Wang

10. Laws, Regulations, Guidelines, and Principles Pertaining to Laboratory Animals in Far East Asia

Tsutomu Miki-Kurosawa, Jae Hak Park, and Chou-Chu Hong

11. Laws, Regulations, Guidelines, and Principles Pertaining to Laboratory Animals in Southeast Asia

*Montip Gettayacamin, Richard Grant, Imelda Liunanita Winoto,
Dondin Sajuthi, Yasmina Arditi Paramastri, Joanna Debby Khoo,
Pradon Chatikavanij, Jason Villano, and Abdul Rahim Mutalib*

12. Laboratory Animals Regulations and Recommendations for Global Collaborative Research: Australia and New Zealand

John Schofield, Denise Noonan, Yvette Chen, and Peter Penson

Contributors

Juan Manuel Baamonde Centro de Estudios Científicos (CECs), Valdivia, Chile

Michael Baar Canadian Council on Animal Care, Ottawa, ON, Canada

Kathryn Bayne AAALAC International, Frederick, MD, USA

B. Taylor Bennett National Association for Biomedical Research, Washington, DC, USA

Ekaterina Akimovna Botovchenco Rivera Federal University of Goias, Brazil

John F. Bradfield AAALAC International, Frederick, MD, USA

Cecilia Carbone Facultad de Ciencias Veterinarias, Universidad Nacional de La Plata, La Plata, Buenos Aires, Argentina

Pradon Chatikavanij National Research Council of Thailand (NRCT), Bangkok, Thailand

Yvette Chen The University of Melbourne, Australia

Margaret E. Cooper Faculty of Veterinary Medicine, University of Nairobi, Kenya; and DICE, The University of Kent; and Wildlife Health Services, UK

Anne-Dominique Degryse Pierre-Fabre, Service de Zootechnie, Institute de Recherche Pierre-Fabre, Castres, France

Idle O. Farah National Museums of Kenya, Nairobi, Kenya

Yacov Fischer Israeli Council for Animal Experimentation; and Ministry of Health, Jerusalem, Israel

Montip Gettayacamin AAALAC International, Samutprakarn, Thailand

Cynthia S. Gillett University of Minnesota, Minneapolis, MN, USA

Rafael Hernandez Gonzalez National Institute for Medical Science and Nutrition-SZ, Animal Resources Department, Mexico City, Mexico

Richard Grant SNBL (Cambodia) Ltd, Khan Daun Penh, Phnom Penh, Cambodia

Gilly Griffin Canadian Council on Animal Care, Ottawa, ON, Canada

Faisal A. Guhad Jigjiga Export Slaughter House (JESH) PLC, Somali Regional State, Federal Democratic Republic of Ethiopia

Javier Guillén AAALAC International, Pamplona, Spain

Alon Harmelin Department of Veterinary Resources, The Weizmann Institute of Science, Rehovot, Israel

Amanda R. Hau Faculty of Law, University of Lund, Lund, Sweden

Jann Hau Faculty of Health and Medical Sciences, University of Copenhagen; and Department of Experimental Medicine, University Hospital, Copenhagen, Denmark

Chou-Chu Hong   AAALAC International, New Taipei City, Taiwan

Rony Kalman   Hebrew University, Jerusalem, Israel

Joanna Debby Khoo   Agri-Food and Veterinary Authority of Singapore, Singapore

Tsutomu Miki-Kurosawa   The Institute of Experimental Animal Sciences, Osaka University Medical School, Suita-shi, Osaka, Japan

Abdul Rahim Mutalib   Department of Veterinary Pathology and Microbiology, Faculty of Veterinary Medicine, University Putra Malaysia, Serdang, Malaysia

Christian E. Newcomer   Association for Assessment and Accreditation of Laboratory Animal Care International, Frederick, MD, USA

Denise Noonan   The University of Adelaide, SA, Australia

Yasmina Arditi Paramastri   Department of Pathology, Microbiology and Immunology, Vanderbilt University Medical Center, Nashville, TN, USA

Jae Hak Park   Laboratory Animal Medicine, College of Veterinary Medicine, Seoul National University, Seoul, Korea

Peter Penson   Rosanna, Victoria, Australia

Jan-Bas Prins   Leiden University Medical Centre, Leiden, The Netherlands

Syed SYH. Qadri   National Institute of Nutrition (ICMR), Hyderabad, Andhra Pradesh, India

Dondin Sajuthi   Primate Research Center, Veterinary Teaching Hospital, Bogor Agricultural University (IPB), Bogor, Indonesia

John Schofield   J & L Consulting, New Zealand

David Smith   FELASA, Chester, UK

Ouajdi Souilem   Laboratory of Physiology and Pharmacology, National School of Veterinary Medicine, Sidi Thabet, Tunisia

Patri Vergara   International Council for Laboratory Animal Science (ICLAS)

Jason Villano   Unit for Laboratory Animal Medicine, University of Michigan Medical School, Ann Arbor, MI, USA

Jianfei Wang   GlaxoSmithKline R&D China, Pudong, Shanghai, P.R. China

Imelda Liunanita Winoto   Primate Research Center, Bogor Agricultural University, Bogor, Indonesia

Ehud Ziv   Diabetes Research Unit, Hadassah University Hospital, Jerusalem, Israel

Chapter 1

Same Principles, Some Differences

Javier Guillén[1] and Patri Vergara[2]

[1]AAALAC International, [2]International Council for Laboratory Animal Science (ICLAS)

Animals play an essential role in the development of many areas of science. Science, as in many other activities of human life, is being globalized, and the care and use of animals in science is being subjected to this transformation. Globalization of science implies an increase in the level of collaborative research, which may be affected by different cultural and legal requirements. Societal ethical concerns on animal use for scientific purposes have led to the development and implementation of regulatory frameworks in many geographical areas. And even in those areas where specific regulations do not exist, practices in the field are greatly influenced by them. Communication and the spread of science, as well as increasing interinstitutional collaboration and outsourcing, are also important factors.

When reviewing the existing regulations described in this book, one easily realizes that although there are differences in the way they are implemented, they share the same ethical core principles. The same principles also underlie most nonstatutory guidelines, or recommendations, on the care and use of laboratory animals.

However, there are still differences in the way these principles are defined—both within the regulations and in their implementation. When principles are enshrined in legislation, legislators inevitably feel the need to provide definition and to establish clear boundaries between what is legal and what is not. This is the reason for the variation in standards that we so frequently find across countries or geopolitical areas. Why do they differ? We can identify a number of reasons including tradition, politics, financial implications, pragmatism, culture, etc. The outcome is the large variation in requirements and standards for such things as regulatory bodies, personal responsibilities, cage and enclosure dimensions, etc. This creates the current situation where we all speak about the same aims and follow the same principles, but where animals are treated differently in practice; differences that can have an adverse effect on collaborative research.

However, there is also a very positive side to this development. The increased use of laboratory animal welfare legislation has created new regulations in areas

Laboratory Animals. http://dx.doi.org/10.1016/B978-0-12-397856-1.00001-5

1

where they did not exist before and has prompted the revision and improvement of existing laws and guidelines. While it is true that the same outcomes can be achieved by different means, the use of performance standards is spreading, thus providing a very powerful tool for harmonization. And this is mainly thanks to the interest, energy, and enthusiasm of the laboratory animal professionals who are continuously seeking to spread knowledge, ethical principles, and good practices across the world. International organizations are the key to achieving these, and there are very good examples. The International Council for Laboratory Animal Science (ICLAS), with its national, scientific, institutional, associate, and affiliate members, is playing an important role. For example, ICLAS has collaborated with Council for International Organizations for Medical Sciences to revise the International Guiding Principles for Biomedical Research Involving Animals.[1] These principles, which are in accordance with the principles of the regulations described in this book, can facilitate the advancement of international collaboration in biomedical sciences. The Association for the Assessment and Accreditation of Laboratory Animal Care (AAALAC International) has accredited around 900 animal care and use programs in almost 40 countries around the globe, serving as a unique tool for the improvement and harmonization of animal care and use standards and practices. Several AAALAC members have contributed to this book due to their international expertise. The World Organization for Animal Health has focused on animal welfare and the use of animals in research and education in different chapters of the Terrestrial Code.[2] The principles described in Chapter 6.1 are perfectly aligned with the principles in all laboratory animal regulations and are developed for animals used in research and education in Chapter 6.8.

At a more regional level, national associations have established umbrella organizations that contribute to the same aim. The recommendations of the Federation of European Laboratory Animal Science Associations (FELASA) are well known and recognized,[3] and similar organizations such as the Asian Federation of Laboratory Animal Science Associations and the Federation of South American Laboratory Animal Science Associations exist in Asia and South America. It is noteworthy to mention the initiative of FELASA and the American Association for Laboratory Animal Science (AALAS) to create the AALAS-FELASA Liaison Body, which is working on the development of common sets of recommendations.

This book includes the main regulations and guidelines used in 11 geographical areas of the world, from the most regulated environments (e.g. Europe) to others where it is hardly possible to identify specific regulations (e.g. Africa). Some information will not be included in this book; in some instances, it was not possible to detail as in the case of multinational areas; in other cases, it was not possible to get specific information, as for example, from Arabian countries. Although no such regulations are included in this book, there are an increasing number of institutions in several Arabian countries establishing quality animal care and use programs in modern facilities.

If science is to advance faster and for the benefit of humans, animals, and the environment, then ideally, laboratory animal regulations should have no political frontiers. Over time, reference books may become outdated. We hope this book becomes outdated soon, provided the reason is that there is no need to describe multiple regulations—only one.

REFERENCES

1. Council for International Organizations for Medical Sciences; and International Council for Laboratory Animal Science 2012. International Guiding Principles for Biomedical Research Involving Animals. Available at: www.iclas.org.
2. World Health Organization. Terrestrial Animal Health Code 2012. Available at: http://www.oie.int/international-standard-setting/terrestrial-code/access-online/.
3. Guillen J. FELASA guidelines and recommendations. *J Am Assoc Lab Anim Sci* 2012;**51**(3):311–21.

Chapter 2

Oversight of Research Animal Welfare in the United States

John F. Bradfield[1], B. Taylor Bennett[2], and Cynthia S. Gillett[3]

[1]AAALAC International, Frederick, MD, USA, [2]National Association for Biomedical Research, Washington, DC, USA, [3]University of Minnesota, Minneapolis, MN, USA

Chapter Outline

Laboratory Animals. http://dx.doi.org/10.1016/B978-0-12-397856-1.00002-7

INTRODUCTION

The United States (US) has many regulations and guidelines regarding the care and use of animals. A comprehensive review of all laws, regulations, and guidelines in the US regarding the welfare, care, and use of animals is beyond the scope of this chapter, but the key regulations and guidelines that impact the care and use of animals in research and teaching will be covered. In many ways the oversight framework for animal use is a reflection of societal values and priorities regarding the overall standing of animals in society. The regulatory landscape in the US, as in many other countries, is continuously evolving and as one reviews the current standards, it must be kept in mind that the implementation of animal use oversight is a rather dynamic process which depends on our evolving understanding of animals, their needs and our obligation to be good stewards when they are in our care.

The careful consideration of animal welfare guidelines and policies involves an international perspective. The US has animal welfare laws, guidelines, and policies that are by their very nature, US-based. However, there is a global view of animal welfare that impacts the US landscape in tangible ways. While non-US-based laws and guidelines do not carry a statutory requirement in the US, they are often used as appropriate resources that impact animal care and use in this country.

The key regulations and guidelines that will be the focus of this chapter include the Animal Welfare Regulations as mandated in US law, the Animal Welfare Act (PL-89-544),[1] the Public Health Service Policy on Humane Care and Use of Laboratory Animals[2] (PHS Policy), which is a statutory mandate of the Health Research Extension Act of 1985, "Animals in Research" (PL-99-158),[3] and the US Government Principles for the Utilization and Care of Vertebrate Animals Used in Testing, Research, and Training.[4] Apart from the Animal Welfare Regulations, the *Guide for the Care and Use of Laboratory Animals* (*Guide*, NRC, 2011),[5] further details the requirements of the PHS Policy and is used by institutions to comply with the PHS Policy. For agricultural animals used in research and teaching, a similar set of guidelines has been published by the Federation of Animal Science Societies, the "*Guide for the Care and Use of Agricultural Animals in Research and Teaching*" (*Ag Guide*, FASS, 2010).[6] Another set of guidelines with universal use are the Guidelines on Euthanasia (2013) published by the American Veterinary Medical Association (AVMA).[7]

The focus of these regulations and guidelines are not meant to imply that the many others that impact animal use and welfare are not important, but simply to direct attention to the core of animal welfare standards in the US which focus on animals in research, teaching, and testing. The US has many relevant guidelines and regulations about animals and welfare, several of which are listed below. The breadth and scope of these laws and guidelines are simply too great to include in one chapter, but these help provide the overall framework in the US regarding the consideration of animal welfare.

A partial list of additional regulations and guidelines in the US include: the Humane Methods of Slaughter Act,[8] Horse Protection Act,[9] Twenty-Eight Hour Law,[10] Guidelines for the Humane Transportation of Research Animals,[11] Guidelines of the American Society of Mammalogists for the Use of Wild Mammals in Research,[12] Guidelines to the Use of Wild Birds in Research,[13] Psychological Well-Being of Nonhuman Primates,[14] Guidelines for the Welfare and Use of Animals in Cancer Research,[15] Guidelines for the Care and Use of Mammals in Neuroscience and Behavioral Research,[16] Guidelines for Use of Live Amphibians and Reptiles in Field and Laboratory Research,[17] and Environmental Enrichment for Nonhuman Primates Resource Guide.[18]

These and many other guidelines are used throughout the US as they apply to a particular species, or discipline of research and are often developed by the various research societies engaged in animal-related research that consider the humane care and use of animals paramount in the conduct of high quality science.

Brief overviews of the core documents that comprise the focus of this chapter are described.

The Animal Welfare Act (Public Law 89-544)[1]

The Animal Welfare Act (AWA) originally enacted in 1966 and was entitled the "Laboratory Animal Welfare Act". Its title was changed to the current title when the Act was amended in 1970 (PL-91-579). The purposes of the original act were to protect against theft of pet dogs and cats, prevent the sale or use of dogs and cats that had been stolen, and ensure that research facilities provided humane animal care and use. The AWA covers all warm-blooded animals used, or intended for use in research, teaching, testing, experimentation, or exhibition purposes, or as a pet. This does not include birds, rats of the genus *Rattus*, and mice of the genus *Mus*, bred for use in research; horses not used for research purposes; and other farm animals used or intended for use for food or fiber, for improving animal nutrition, breeding, management, or production efficiency, or for improving the quality of food or fiber.

The AWA authorized the Secretary of Agriculture, United States Department of Agriculture (USDA), and the Animal and Plant Inspection Service (APHIS) to develop regulations to implement the Act. These Animal Welfare Regulations (AWRs) are set forth in Title 9 of the Code of Federal Regulations (CFR) Chapter, 1, Subchapter A—Animal Welfare, Part 1—Definitions, Part 2—Regulations, and Part 3—Standards.[19] Enforcement duties are the responsibility of the APHIS Deputy Administrator for Animal Care. To comply with the AWA, the USDA-APHIS requires all registered research facilities to adhere to the AWRs. As part of ensuring that institutions follow the regulations, USDA-APHIS conducts unannounced inspections of research facilities by Veterinary Medical Officers (VMO) who inspect at least annually and more often if deemed necessary by the Agency. The USDA uses a risk-based inspection system to support and focus its inspection strategy, allowing more frequent and in-depth inspections at facilities experiencing problems and fewer at those that are consistently in compliance. The process uses several objective criteria including past compliance history to determine inspection frequency.

The AWA has been amended six times since 1966 with the most significant changes made in 1985 in which an amendment entitled the "Improved Standards for Laboratory Animals Act" strengthened the standards for providing laboratory animal care, increased enforcement of the AWA, provided for collection and dissemination of information to reduce unintended duplication of experiments using animals, and mandated training for those who handle animals. The 1985 amendment also included standards for exercise of dogs, psychological well-being of primates, limitation of multiple survival surgeries, and a requirement that investigators consult with a veterinarian in the design of experiments which have the potential for causing pain, to ensure the proper use of anesthetics, analgesics, and tranquilizers. Investigators must also consider alternatives to procedures that may cause pain or distress. Each research facility must demonstrate upon inspection and annually report, that professional standards of animal care, treatment, and use are employed during research, teaching, or testing.

Perhaps most significant, the 1985 amendment required the Chief Executive Officer (CEO) of each research facility to appoint an Institutional Animal Care and Use Committee (IACUC) consisting of at least three members including a doctor of veterinary medicine and one member who is not affiliated with the institution. The IACUC is charged to act as an agent of the research facility in assuring compliance with the AWA. Central to this charge is the authority of the IACUC mandated in the AWR's. The role and authority of the IACUC in the care, use, and oversight of research animals cannot be overstated and it is somewhat a unique premise that is at the core of animal use oversight in the US. The AWRs stipulate that the IACUC is required to conduct a variety of activities to ensure compliance with the regulations. These responsibilities include inspection of all animal facilities and study areas at least once every six months, and review the research facility's program of animal care and use once every six months. The IACUC must file a report of its inspection with the Institutional Official (IO) of the research facility. If significant deficiencies or deviations are not corrected in accordance with the specific plan approved by the IACUC, the USDA and any federal funding agencies must be notified in writing. The IACUC must also review and approve all proposed activities involving the care and use of animals in research, testing, or teaching procedures and all subsequent significant changes of ongoing activities. As part of this review, the IACUC must evaluate procedures which minimize discomfort, distress, and pain, and when an activity is likely to cause pain, and a veterinarian has been consulted in planning for the administration of anesthetics, analgesics, tranquilizers, and that paralytic agents are not employed except in the anesthetized animal. The IACUC must also determine that animals which experience severe or chronic pain are euthanatized consistent with the design of study, that the living conditions meet the species' needs, that necessary medical care will be provided, that all procedures will be performed by qualified individuals, that survival surgery will be performed aseptically, and that no animal will undergo more than one operative procedure unless justified and approved by the IACUC. Methods of euthanasia must be consistent with the definition contained in the regulations.

On behalf of the research facility, the IACUC must also assure that the principal investigator (PI) considered alternatives to painful procedures and that the work being proposed does not unnecessarily duplicate previous experiments. To provide this assurance, the IACUC must review the written narrative description provided by the investigator and the description must include the methods and sources used in determining that appropriate alternatives were not available. The IACUC can grant exceptions to the regulations and standards, if they have been justified in writing by the principal investigator.

The research facility is required to provide training to scientists, animal technicians, and other personnel involved with the care and use of animals and training must include humane practices of experimentation, methods that minimize pain and distress, the proper use of anesthetics, analgesics and tranquilizers, the use of appropriate information services such as the National Agricultural Library, and methods to report deficiencies in animal care and treatment.

Public Health Service Policy (PHS Policy)[2]

The National Institutes of Health (NIH) are a major component of the Public Health Service of the US, Department of Health and Human Services (the largest funding source for biomedical research in the US). In 1971, the NIH issued its first policy on the care and treatment of laboratory animals (NIH Policy, "Care and Treatment of Laboratory Animals." [NIH 4206], NIH Guide for Grants and Contracts, No.7, June 14, 1971). The Policy required that institutions receiving NIH grant awards provide written assurance that a committee of at least three members, one of which was a veterinarian, had been established or that the institution was accredited by a "professional accrediting body." The policy was revised in 1973 and 1979 with a significant revision in 1985 as a result of the passage of the Health Research Extension Act (HREA, 1985) (Public Law 99-158).[3] The HREA provided statutory authority of the PHS Policy for PHS-funded research involving live vertebrate animals used (or intended for use) in biomedical and behavioral research. An additional revision of the policy was released in September, 1986 which reflected the changes required by the HREA. In 2002, another revision occurred to reflect changes in the process of submitting IACUC approval and the information that must be included in the Animal Welfare Assurance.

Under the PHS Policy, each institution using animals in PHS-sponsored projects must provide acceptable written assurance of its compliance with the policy. In this Animal Welfare Assurance the institutions must describe:

1. The Institutional Program for the Care and Use of Animals
2. The Institutional Status
3. The Institutional Animal Care and Use Committee.

The institutional program must include a list of every branch and major component, the lines of authority for administering the program, the qualifications, authority, and responsibility of the veterinarian(s), the membership of the Institutional Animal Care and Use Committee and the procedures which they follow must be stated. The employee health program must be described for those who have frequent animal contact. A training or instruction program in the humane practices of animal care and use must be available to scientists, animal technicians, and other personnel involved in animal care, treatment, and use.

The institutional status must be stated as either Category one (1) (The American Association for Accreditation of Laboratory Animal Care (AAALAC) accredited) or Category two (2) (nonaccredited). Institutions in Category two (2) must establish a reasonable plan with a specific timetable for correcting any departures from the recommendations in the *Guide for the Care and Use of Laboratory Animals* (NRC, 2011).

The IACUC must be appointed by the CEO, who can delegate that authority to the Institutional Official. The IACUC must consist of at least five members; one of whom is a veterinarian with program responsibility, a practicing scientist, an individual whose expertise is in a nonbiological science, and an individual who is not affiliated with the institution. The functions of this committee are analogous to those required under the AWA with the exception that it must use the *Guide* to review the animal facilities and the institutional program for humane care and use of animals. The other major difference is that the PHS Policy covers all vertebrate animals.

The institution is responsible for maintaining all the necessary records to document compliance with the PHS Policy and for filing annual reports developed by the IACUC which detail any changes in the program and indicate the dates of the semiannual inspections and programmatic reviews. Institutions must also report any serious or continuing noncompliance with the policy, any serious deviations from the provisions of the *Guide*, and any IACUC suspension of an activity.

The PHS Policy is intended to implement and supplement the "U.S. Government Principles for the Utilization and Care of Vertebrate Animals in Testing, Research and Training" (US Government Principles). The nine principles are published in the PHS Policy and in the Appendix of the *Guide*.

The National Institutes of Health, Office of Laboratory Animal Welfare (OLAW) is tasked with oversight of animal care and use for PHS-funded research. OLAW requires that institutions base their animal care and use programs on the *Guide* and as previously described, each institution receiving PHS funds for animal research must maintain a PHS Assurance Statement that has been approved by OLAW. OLAW endorses the performance-based concepts of the *Guide* and provides guidance as to how institutions can best implement and comply with the PHS Policy and the *Guide*. Institutions with PHS-assurances must self-report instances of noncompliance with the *Guide* and PHS Policy and the means to correct them. In addition to this self-reporting to ensure compliance, OLAW sometimes conducts on-site evaluations of programs to ensure they comply with the *Guide*, PHS Policy, and the details of the institution's assurance statement.

The Guide for the Care and Use of Laboratory Animals (NRC 2011)[5]

The original "*Guide*" was first published in 1963 by the US Department of Health, Education, and Welfare, Public Health Service and was titled the, *Guide for Laboratory Animal Facilities and Care*. It has been revised or updated seven times since in 1965, 1968, 1972, 1978, 1985, 1996, and 2011 with more recent editions published by the Institute for Laboratory Animal Resources (ILAR). Eventually it became the *"Guide for the Care and Use of Laboratory Animals"*, commonly referred to as the *Guide*. The PHS Policy requires institutions that receive PHS funding adhere to the standards of the *Guide*. The *Guide* states on page XIII that it is intended to *"assist institutions in caring for and using animals in ways judged to be scientifically, technically, and humanely appropriate and is also intended to assist investigators in fulfilling their obligation to plan and conduct animal experiments in accord with the highest scientific, humane, and ethical principles."* The 1996 version (7th edition) stressed the importance of performance standards in developing acceptable standards for the care and use of laboratory animals. The performance standards approach requires clearly defined appropriate outcomes; the criteria for determining how outcomes are achieved; and the methods for evaluating the outcomes, without specifying exactly how the outcome is accomplished. The performance-based approach described in the 1996 *Guide* is in contrast to the engineering approach in which both the outcome and methods to achieve them are specified. The performance-based approach recognizes the wide array of scientific disciplines and animal models employed. It allows institutions to best determine how to achieve the standards described in the *Guide* based on the needs of the animals and the type of research conducted. The current, 2011 edition of the *Guide* (8th edition) extends the performance-based approach and underscores the need for the IO, IACUC, Attending Veterinarian and principal investigator to collaborate when implementing the standards of the *Guide*. The 8th edition contains five chapters: Key Concepts; Animal Care and Use Program; Environment, Housing, and Management; Veterinary Care; and Physical Plant. The purpose of the *Guide* is to assist institutions in laying the foundation for a comprehensive animal care and use program that relies on the use of performance standards and professional judgment to assure that such use is in accordance with the highest scientific, humane, and ethical principles. The *Guide* strongly affirms the principle that those who care for and use laboratory animal must assume responsibility for their well-being. Laboratory animals are defined as any vertebrate animal used in research, testing, and teaching. Key terms in the *Guide* that help the reader prioritize the importance of the recommendations are: *"must"* which denotes an imperative or mandatory requirement; *"should"* is a string recommendation for achieving a goal; and *"may,"* which is a suggestion for consideration.

Guide for the Care and Use of Agricultural Animals in Research and Teaching (FASS, 2010)[6]

The *Guide for the Care and Use of Agricultural Animals in Agricultural Research and Teaching* was published originally in 1988 and revised in 1999 and 2010 and is commonly referred to as the *Ag Guide*. The current 3rd edition, published by the Federation of Animal Science Societies, is intended to provide standards for agricultural animals involved in research and teaching with the viewpoint stated in the preface that, *"Farm animals have certain needs and requirements and these needs and requirements do not necessarily change because of the objectives of the research or teaching activity. Therefore, regardless of the teaching or research objectives, the FASS Ag Guide should serve as a primary reference document for the needs and requirements of agricultural animals."* Similar to the *Guide*, the *Ag Guide* emphasizes the use of performance standards and professional judgment when implementing institutional standards described in the *Ag Guide*, to be applicable among diverse institutions and their agricultural programs.

The 3rd edition of the *Ag Guide* contains 11 chapters: Institutional Policies, Agricultural Animal Health Care, Husbandry, Housing and Biosecurity, Environmental Enrichment, Animal Handling and Transport, Beef Cattle, Dairy Cattle, Horses, Poultry, Sheep and Goats, and Swine. There are many similarities between the standards of the *Ag Guide* and the *Guide* with the exception that standards unique to individual species of farm animals are described in separate chapters. Many aspects of animal care and use are described in this comprehensive document devoted to the care and use of agricultural animals.

US Government Principles for the Utilization and Care of Vertebrate Animals Used in Testing, Research, and Training (Federal Register, Vol. 50, No. 97, 1985)[21] (US Government Principles)

The US Government Principles were drafted by the Interagency Research Animal Committee in response to public interest in the care and use of laboratory animals. This document stresses compliance with federal laws, policies, and guidelines and establishes overarching principles that should be applied when using animals, or sponsoring the use of animals, in research, teaching, and testing. There are nine principles:

I   The transportation, care, and use of animals should be in accordance with the Animal Welfare Act (7 U.S.C. 2131 et. seq.) and other applicable Federal Laws, guidelines, and policies.[1]

II Procedures involving animals should be designed and performed with due consideration of their relevance to human or animal health, the advancement of knowledge, or the good of society.

III The animals selected for a procedure should be of an appropriate species and quality and the minimum number required to obtain valid results. Methods such as mathematical models, computer stimulation, and in vitro biological systems should be considered.

IV Proper use of animals, including the avoidance or minimization of discomfort, distress, and pain when consistent with sound scientific practices, is imperative. Unless the contrary is established, investigators should consider that procedures that cause pain or distress in human beings may cause pain or distress in other animals.

V Procedures with animals that may cause more than momentary or slight pain or in distress should be performed with appropriate sedation, analgesia, or anesthesia. Surgical or other painful procedures should not be performed on unanesthetized animals paralyzed by chemical agents.

VI Animals that would otherwise suffer severe or chronic pain or distress that cannot be relieved should be painlessly killed at the end of the procedure or, if appropriate, during the procedure.

VII The living conditions of animals should be appropriate for their species and contribute to their health and comfort. Normally, the housing, feeding, and care of all animals used for biomedical purposes must be directed by a veterinarian or other scientist trained and experienced in the proper care, handling, and use of the species being maintained or studied. In any case, veterinary care shall be provided as indicated.

VIII Investigators and other personnel shall be appropriately qualified and experienced for conducting procedures on living animals. Adequate arrangements shall be made for them in service training, including the proper and humane care and use of laboratory animals.

IX Where exceptions are required in relation to the provisions of these principles, the decisions should not rest with the investigators directly concerned, but should be made, with due regard to Principle II, by an appropriate review group such as an institutional animal research committee. Such exceptions should not be made solely for the purposes of teaching or demonstration.

The US Government Principles are featured prominently in the PHS Policy, the *Guide*, the *Ag Guide* and many other animal care and use guidelines and have become the foundation for laboratory animal care and use in the US.

The American Veterinary Medical Association Guidelines on Euthanasia

The first report of the AVMA Panel on Euthanasia was published in the AVMA Journal in 1963. The Panel report was revised in 1972, 1978, 1986, 1993, 2000, and 2013. In 2007, the guidelines were updated and renamed the *AVMA Guidelines on Euthanasia*. The guidelines are intended for veterinarians who carry out or oversee euthanasia of animals and as a result have become widely used by research institutions in the US. Many IACUCs have developed institutional policies that require euthanasia procedures that conform to the recommendations of the AVMA Guidelines on Euthanasia. The 2013 guidelines titled *AVMA Guidelines for the Euthanasia of Animals: 2013 Edition*, provide a comprehensive discussion of many aspects of euthanasia for a variety of species.[7] The 2013 updated guidelines include detailed, expanded explanation of methods, techniques, and agents of euthanasia to assist veterinarians in applying their professional judgment.

The Association for the Assessment and Accreditation of Laboratory Animal Care, International

The Association for the Assessment and Accreditation of Laboratory Animal Care (AAALAC) was first established in 1965 as a private nonprofit organization that accredits animal care and use programs. The founders of AAALAC envisioned a voluntary, collaborative, peer-reviewed evaluation of animal care and use programs that engaged both the scientific and laboratory animal care communities to promote high standards of animal care and use. AAALAC is governed by a Board of Trustees comprised of scientific organizations and relies upon members of the Council on Accreditation and Ad Hoc consultant specialists to conduct site visits to participating institutions. This triennial site visit process is the basis for rigorous, peer-reviewed assessments to ensure high standards of animal care and use are met by the participating institutions. In 1996, AAALAC was renamed the Association for the Assessment and Accreditation of Laboratory Animal Care, International to reflect its growth worldwide. Until recently, AAALAC used the *Guide* as the primary standard by which programs were evaluated and in 2011 also added two additional documents as primary standards of accreditation, the *Ag Guide* for research programs that involve agricultural animals and the European Convention for the Protection of Vertebrate Animals Used for Experimental and Other Scientific Purposes, Council of Europe, European Treaty Series 123 (ETS 123),[20] for accredited programs in Europe. In addition to these primary standards of accreditation, there are many other documents used by AAALAC International as reference resources to aid in the accreditation process. Accreditation by AAALAC International is voluntary but over the years, accreditation has come to represent a standard of excellence in proactive, continuous self-improvement in animal care. The peer review accrediting process fosters collaboration among stakeholders to ensure high quality science through the promotion of high quality animal welfare. The impact of the AAALAC International accreditation process in the US has been the establishment of a system of self-assessment and improvement that has become the benchmark of quality animal research programs and includes many institutions across the spectrum of animal care and use; industry, academia, hospitals, nonprofit, and governmental organizations.

The Cornerstones of Research Animal Oversight in the US

The Animal Welfare Act and Regulations, the PHS Policy, the *Guide*, and the *Ag Guide*, all describe the central role of the IACUC in providing the necessary institutional oversight of animals used in research. It is the IACUC that is responsible for oversight and monitoring to ensure that the regulations and guidelines are fully implemented. The concept and importance of this committee is vital in animal use oversight in the US. The AWR's, PHS Policy, *Guide*, and *Ag Guide* also stipulate the key roles of the IO and Attending Veterinarian (AV). The prominent role and authority given to the AV in the US may be considered somewhat unique by those in other countries. The AV and the program of veterinary care comprise key aspects of animal health, welfare, and oversight in the US. The *Guide* also describes the requirement for the IO, the IACUC, and the AV to collaborate in the implementation and oversight of the entire animal care and use program and it is this concept that provides the basis for effective animal care and use oversight. Although these three entities are crucial to the effective institutional oversight of animal care and use, one must not overlook the critical role of the investigator and research team. The *Guide* places a high degree of responsibility on those engaged in research to be good stewards of the animals with which they work. It is these four principal entities that form the fabric of research animal compliance in the US.

THE PRINCIPLES

AWA/AWR's

Section 2131 of the AWA titled, "Congressional statement of policy" on page one states, congress finds that—"(3) *measures which eliminate or minimize the unnecessary duplication of experiments on animals can result in more productive use of Federal funds and* (4) *measures which help meet the public concern for laboratory animal care and treatment are important in assuring that research will continue to progress.*" The animal welfare regulations further state that the IACUC review of proposed animal activities will include consideration of the following principles: procedures will avoid or minimize discomfort, distress, and pain to the animals; ensure that the principal investigator has considered alternatives to procedures that may cause more than momentary or slight pain or distress to the animals and has provided a written narrative description used to determine that alternatives were not available; and further that a written assurance be provided that activities do not unnecessarily duplicate previous experiments; procedures that may cause more than momentary or slight pain of distress will be performed with appropriate sedative, analgesics, or anesthetics unless withholding such agents is justified for scientific reasons and will continue for only the necessary period of time; that paralytics be used only with anesthetics; that a veterinarian be involved in the planning of the experiment; and that timely euthanasia be performed.

PHS Policy and OLAW

The Health Research Extension Act of 1985, Public Law 99-158, "Animals In Research" provides the statutory mandate for the PHS Policy. The introduction of the policy states, "*It is the Policy of the Public Health Service (PHS) to require institutions to establish and maintain proper measures to ensure the appropriate care and use of all animals involved in research, research training, and biological testing activities (hereinafter referred to as activities) conducted or supported by the PHS. The PHS endorses the "US Government Principles for the Utilization and Care of Vertebrate Animals Used in Testing, Research, and Training" developed by the Interagency Research Animal Committee. This Policy is intended to implement and supplement those Principles."*

The introductory paragraph of the US Government Principles states, in part, *"Whenever U.S. Government agencies develop requirements for testing, research, or training procedures involving the use of vertebrate animals, the following principles shall be considered; and whenever these agencies actually perform or sponsor such procedures, the responsible Institutional Official shall ensure that these principles are adhered to."* The Principles were incorporated into the PHS Policy in 1986 and continue to provide a framework for conducting research in accordance with the Policy."

The 3Rs are not specifically mentioned in the PHS Policy or the US Government Principles. However, several of the US Government Principles directly relate to the concept of the 3Rs. Principle III relates to using only the minimum number of animals necessary and Principle IV indicates that animal pain and distress should be minimized or avoided and procedures that cause pain or distress in humans may cause pain or distress in other animals. Principle V specifies that appropriate use of sedatives, anesthetics, and analgesics should be employed to minimize pain or distress. Principle VI states that humane euthanasia should be employed to relieve chronic or severe pain or distress. Also, the PHS Policy does require adherence to the USDA Animal Welfare Act which requires that *"The principal investigator has considered alternatives to procedures that may cause more than momentary or slight pain or distress to the animals, and has provided a written narrative description of the methods and sources, e.g. the Animal Welfare Information Center, used to determine that alternatives were not available."* OLAW has provided further clarification that the US Government Principles are a federal mandate and that Principles III and IV embody key aspects of the 3-Rs. OLAW further states that *"consideration of the three "Rs" should be incorporated into IACUC review, as well as other aspects of the institution's program."*

Guide

The *Guide* is an internationally accepted primary reference on animal care and use, and its use is required in the United States by the PHS Policy; the principles supporting the *Guide* are the same as for the PHS Policy. The *Guide*, on page one, *"strongly affirms the principle that all who care for, use or produce animals for research, testing or teaching must assume responsibility for their well-being."* And *"The Guide plays an important role in decision making regarding the use of vertebrate laboratory animals because it establishes the minimum ethical, practice, and care standards for researchers and their institutions."* With reference to the US Animal Welfare Act and Regulations and the PHS Policy, the *Guide* states, *"Taken together, the practical effect of these laws, regulations, and policies is to establish a system of self-regulation and regulatory oversight that binds researchers and institutions using animals."* Ethics of animal use is of primary concern when making the decision to use animals in research and requires *"critical thought, judgment and analysis."* The *Guide* specifies that using animals in research is, *"a privilege granted by society"* with the expectation that such use will lead to significant new knowledge or the improvement of human and/or animal well-being. The *Guide* endorses the US Government Principles and places the responsibility for their implementation on the research community. The concepts of replacement, refinement, and reduction (3Rs) and their relevance to the ethical use of animals are discussed in detail in the first chapter of the *Guide* on "Key Concepts". In regard to the 3Rs, the *Guide* states on page five that *"the Three Rs have become an internationally accepted approach for researchers to apply when deciding to use animals in research and in designing humane animal research studies."*

Ag Guide

While the *Ag Guide* does not reference the 3-Rs directly, like the PHS Policy and the *Guide* it affirms the fundamental importance of the US Government Principles in the use of agricultural animals. The preface to the *Ag Guide* states, *"The US Government Principles for the Utilization and Care of Vertebrate Animals Used in Testing, Research, and Training of the IRAC (1985; Appendix 1) are endorsed in this guide as a basis for professional judgments about the appropriate treatment and use of agricultural animals in research and teaching activities."*

SCOPE/APPLICABILITY

AWA/AWR's

The AWR's apply to animal activities including *"research, teaching, testing, experimentation or exhibition purposes, or as a pet."* The AWRs further defines the species of animals that are covered and those not covered by the regulations. Those covered include, *"any live or dead dog, cat, nonhuman primate, guinea pig, hamster, rabbit, or any other warm-blooded animal...."* being used for the purposes stated above. Animals excluded are: birds, rats of the genus *Rattus*, mice of the genus *Mus* bred for use in research, horses not used for research purposes, and other farm animals, such as, but not limited to, livestock or poultry, used or intended for use as food or fiber, livestock or poultry used or intended for use for improving management, or production efficiency, or for improving the quality of food or fiber.

PHS Policy and OLAW

In Section II-Applicability, the PHS Policy states *"The PHS Policy is applicable to all PHS-conducted or supported activities involving animals, whether the activities are performed at a PHS agency, an awardee institution, or any other institution and conducted in the United States, the Commonwealth of Puerto Rico, or any territory or possession of the United States. Institutions in foreign countries receiving PHS support for activities involving animals shall comply with this Policy, or provide evidence to the PHS that acceptable standards for the humane care and use of the animals in PHS-conducted or supported activities will be met.*" It further states that, "*All institutions are required to comply, as applicable, with the Animal Welfare Act, and other Federal statutes and regulations relating to animals.*" OLAW provides further guidance that the PHS Policy is applicable to the generation of custom antibodies in vertebrate species and to larval forms of fish and amphibians, but is not applicable to avian and other egg-laying vertebrate species prior to hatching. The policy defines "animal" as any live, vertebrate animal used or intended for use in research, research training, experimentation, or biological testing or for related purposes.

Guide

The PHS requires institutions to use the *Guide* as a basis for developing and implementing an institutional program for activities involving animals. In the section titled Applicability and Goals, the *Guide* states that it applies to all vertebrate animals, but "*does not address in detail agricultural animals used in production, agricultural research or teaching, wildlife and aquatic species studied in natural settings, or invertebrate animals (e.g. cephalopods) used in research, but establishes general principles and ethical considerations that are also applicable to these species and situations.*" It further states that "*The goal of the Guide is to promote the humane care and use of laboratory animals by providing information that will enhance animal well-being, the quality of research and the advancement of scientific knowledge that is relevant to both humans and animals.*" The intended audience of the *Guide* includes: the scientific community; administrators; IACUC's; veterinarians; educators and trainers; producers of laboratory animals; accreditation bodies; regulators; and the public.

Ag Guide

The *Ag Guide* states that "*Farm animals have certain needs and requirements and these needs and requirements do not necessarily change because of the objectives of the research or teaching activity. Therefore, regardless of the teaching or research objective, the FASS Ag Guide should serve as a primary reference document for the needs and requirements of agricultural animals.*" There are detailed recommendations for specific species including beef and dairy cattle, horses, poultry, sheep, goats, and swine.

AUTHORIZATION OF USER-BREEDING INSTITUTIONS/ INSPECTIONS/PENALTIES BY COMPETENT AUTHORITIES

AWA/AWR's

The Animal Welfare Regulations require dealers/breeders to be licensed and research facilities and carriers to be registered. The AWR define three types of licensees. Type A licensees are breeders that only sell animals bred and raised on their premises. Type B licensees include businesses that purchase and resell animals. Type C licensees are involved in showing or displaying animals to the public. The type of license that an organization has is dependent on the type of activity constituting the majority of their business. Retail pet stores are not required to be licensed in most cases, as are individuals who have sales of less than the amount defined in the regulations or who do not maintain more breeding animals or sell more animals than defined in the regulations. Prior to obtaining a license, each applicant must demonstrate their compliance with the standards and regulations and pay a small application fee. Licenses must be renewed annually, along with a licensing fee based upon their annual sales for dealer and number of animals exhibited for exhibitors.

For research facilities to be registered, they must have programs in place which meet the requirement contained in the regulations relative to composition and function of the Institutional Animal Care and Use Committee, a program that assures that personnel who care for and use animals are properly qualified to do so, a program of adequate veterinary care, meeting the record keeping requirements described in the regulations, filing an annual report, and making the premises and records available for unannounced inspections by the officials from the United States Department of Agriculture's Animal Plant Health Inspection Service (APHIS). By law, inspection of research facilities must take place at least annually. Registration must be renewed every 3 years.

Responsibility for administering the AWA was delegated within the USDA to the administrator of the Animal and Plant Health Inspection Service. Enforcement duties are the responsibility of the APHIS Deputy Administrator for Animal Care (AC). Inspections of research facilities are conducted by VMO working under one of the AC regional supervisors. They identify and report Non-Compliant Items (NCIs). Depending upon the nature of the action, subsequent enforcement actions are handled by the Investigative and Enforcement Services (IES). The IES investigators follow up on the NCIs and prepare case reports. If a case warrants prosecution, APHIS takes legal action, usually through an administrative law process. Many cases are closed with an official warning, but sometimes IES issues stipulations which may include a civil fine or other penalty. Fines may be as much as $10,000 per animal per day that the facility is found to be in noncompliance with the regulations and or standards. The more serious cases or those involving repeat violations are submitted to an administrative law judge who can suspend or revoke the violator's USDA license and impose a fine. If a violation is serious enough, IES will work with the Department of Justice to build a criminal case.

The first step in the enforcement process is a 90 day re-inspection, which takes place if a facility is making clear progress toward compliance and the inspector found only a few minor NCIs with only a few repeat NCIs demonstrating no signs of animal health or welfare being in jeopardy. The next step is the issuance of an Official Warning Letter (7060). A 7060 is a notification to an individual or company regarding an alleged violation. It may be issued with or without an IES investigation. The reasons for issuing a 7060 include an inspector finding that a facility: is out of compliance after a 90 day re-inspection, has multiple repeat NCIs, has a direct NCI (one that has a high potential to have a serious, adverse impact on the animals health and well-being), has incomplete documentation of a serious NCI, or is making slow progress toward compliance. The next step in the process is the issuance of a stipulation. A stipulation is an agreement in which the USDA gives notice of an alleged violation and agrees to accept a specified penalty to settle the matter. The settlement agreement form used by IES requires that the penalty be paid within a designated time frame and states that the payment constitutes a waiver of the alleged violator's right to a hearing and a finding that violations of the law have occurred. The fourth and final step in the process is an Office of General Counsel (OGC) Prosecution. An OGC complaint gives notice to a facility of a formal allegation of possible violations of the AWA. The complaint does not mean the facility is guilty of these violations, but serves as a notice that they must respond and either agree to the allegations in the complaint, or seek a hearing date before a USDA Administrative Law Judge (ALJ). The ALJ will issue a Decision and Order based upon the evidence presented by APHIS and the facility. The facility has the right to appeal this decision. A copy of a Decision and Order is made available on the USDA website. This process is initiated following an IES investigation when there are serious NCIs, and/or repeat direct or multiple direct NCIs with no progress toward compliance and where animal health and welfare have been impacted when the facility has usually had previous enforcement actions.

In addition to the enforcement actions described above, the USDA has the authority to temporarily suspend the license of a dealer and to confiscate and destroy animals being held by a licensee or registered facility, if the animals are suffering due to failure to comply with the regulations and/or standards.

PHS Policy and OLAW

The PHS Policy states on page 19 that *"Each awardee institution is subject to review at any time by PHS staff and advisors, which may include a site visit, in order to assess the adequacy or accuracy of the institution's compliance or expressed compliance with this Policy."* The OLAW is the office charged with general administration and coordination of the PHS Policy. OLAW will evaluate allegations of noncompliance with the policy which may include conducting site visits to selected institutions. OLAW does not have established penalties for noncompliance nor a schedule of regularly conducted inspections; however, OLAW may periodically conduct random site visits to institutions to assess compliance with PHS Policy and the *Guide*. Institutions with a PHS Assurance are required to report to OLAW, any activities that are judged by the IACUC to be non-compliant with approved protocols, protocol suspensions, and deviations from the *Guide* and PHS Policy. If warranted, OLAW can rescind an institution's PHS assurance which may have direct consequences on the institution's eligibility to receive NIH grant awards.

Guide

The *Guide* itself does not authorize inspections or require reporting to competent authorities; OLAW is authorized to monitor and oversee all institutions that hold a PHS assurance for adherence to the PHS Policy and the *Guide* as outlined above. Given the prominent role of the IACUC and the level of institutional authority placed upon it, the *Guide's* importance in assisting the IACUC in protocol review, assessment, and oversight of the program is central to the animal oversight/regulatory process in the United States.

Ag Guide

There are no statutory mandates or penalties associated with the *Ag Guide*. The standards of the *Ag Guide* are recommendations by the agricultural animal research community and are accepted as such.

AAALAC International

Although participation in AAALAC accreditation is voluntary and confidential, there are standards that must be met in order for an institution to attain or maintain accreditation. Apart from a requirement to adhere to the AWR's, AAALAC uses the *Guide* and the *Ag Guide* as primary standards for accreditation for US institutions. AAALAC conducts site visits at least once every three years to assess animal care and use programs and determines whether the standards of the *Guide* and/or *Ag Guide* are met. Animal care and use programs must specify which standard they have implemented, the *Guide*, the *Ag Guide*, or both (note—AAALAC International also uses ETS 123 as a primary standard of accreditation, but this standard is used mainly for institutions in Europe). In the US, accredited programs that do not meet one or more of the standards of the *Guide* or *Ag Guide* are placed on Deferred or Probationary accreditation, depending on the nature of the concern, until the concern is resolved. For programs that are unable to meet the standards of the *Guide* or *Ag Guide*, AAALAC International will withhold accreditation or revoke accreditation (if the program is currently accredited). Because accreditation by AAALAC International is widely recognized as a high standard, it has become valued by many institutions in the US and maintaining accreditation is respected.

NONHUMAN PRIMATES: SPECIAL CONSIDERATIONS AND RESTRICTIONS

AWA/AWR's

Within the standards promulgated under the AWA is a section on nonhuman primates entitled "*Environment enhancement to promote psychological well-being.*" In order to comply with these requirements, dealers, exhibitors, and research facilities must develop, implement, and document a plan that promotes the psychological well-being of these animals. This plan must be in accordance with currently accepted standards in the field and directed by the AV. The plan must include provisions to address social grouping of nonhuman primates known to exist in social groups, environmental enrichment to allow animals to express noninjurious species–typical behavior, special consideration such as occurs with infants and juveniles, animals showing signs of psychological distress, animals on research projects requiring IACUC approved restricted activity, individually housed animals, and great apes weighing over 50 kg. In addition, the plan must address the use of restraint devices. The component of a plan that addresses social grouping must include provisions for dealing with primates exhibiting vicious or overly aggressive behavior or are debilitated because of age or other conditions; animals suspected of having contagious diseases; and animals that are not compatible with other animals. The IACUC can exempt animals from the program for scientific reasons and these exemptions must be reviewed and approved annually. The AV can exempt animals from the plan for reasons of health or well-being. These exemptions must be reviewed every 30 days unless the basis for the exemption is a permanent condition.

PHS Policy and OLAW

The PHS Policy does not make specific mention of nonhuman primates. OLAW has guidance regarding nonhuman primates: one regarding social housing expectations and the other encouraging positive reinforcement training (OLAW, frequently asked questions, http://grants.nih.gov/grants/olaw/faqs. htm#useandmgmt_14).

Guide

Information regarding nonhuman primate environment, husbandry, minimum space requirements for cages and other primary enclosures, environmental enrichment, social housing, veterinary care, and occupational health, and safety issues are discussed in the *Guide*. The use of chair restraint for nonhuman primate studies is specifically discouraged. Emphasis is placed on careful consideration for establishing pair or group housing of nonhuman primates and the need to consider important species-specific factors that can affect compatibility: age, behavioral repertoire, sex, natural social organization, breeding, and health status. When considering the cage space requirements for nonhuman primates, professional judgment is paramount and decisions should not be based on body weight alone. Further, the *Guide* states, *"Because of the many physical and behavioral characteristics of nonhuman primate species and the many factors to consider when using these animals in a biomedical research setting, species-specific plans for housing and management should be developed. Such plans should include strategies for environmental and psychological enrichment."*

Ag Guide

The *Ag Guide* does not apply to research, teaching or testing with nonhuman primates.

GENETICALLY ALTERED ANIMALS: SPECIAL CONSIDERATION

AWA/AWR's

The animal welfare regulations and standards make no special provisions for genetically altered animals.

PHS Policy and OLAW

Neither the PHS Policy nor OLAW make specific mention of genetically altered animals.

Guide

Because genetically modified animals (GMAs), particularly mice and fish, are important animal models and the results of genetic manipulation can be unpredictable, there may be unexpected consequences that impact animal well-being. The *Guide* recommends that the first offspring of a newly generated GMA line be carefully observed from birth into early adulthood for signs of disease, pain, or distress and that when the initial characterization of a GMA reveals a condition that negatively affects animal well-being, this should be reported to the IACUC for review. Careful consideration of the phenotype is important during IACUC review of proposed studies to ensure that *"proactive measures can circumvent or alleviate the impact of the genetic modification on the animal's well-being and to establish humane endpoints specific to the GMA line."*

Ag Guide

The *Ag Guide* provides careful consideration and recommendations for genetically altered farm animals and defines specific terms. A transgenic animal is one that carries a foreign gene that has been deliberately inserted into the genome. Genetic engineering is defined as the direct manipulation of an organism's genes including heritable and nonheritable recombinant DNA constructs. The genetic engineering of agricultural animals has been extensively reviewed (National Research Council, 2002; Council on Agricultural Science and Technology, 2003, 2007, 2009).[21] In many instances, research with genetically engineered farm animals is regulated by the USDA and NIH as most of these studies are considered biomedical activities. Also, the US Food and Drug Administration has guidelines for the conduct of research with genetically modified animals (http://www.fda.gov/AnimalVeterinary/DevelopmentApprovalProcess/Genetic Engineering/GeneticallyEngineeredAnimals/default.htm).[22] The *Ag Guide* describes cloning as one method of genetic engineering involving somatic cell nuclear transfer. Cloning is considered an assisted reproductive technology similar to artificial insemination, embryo transfer, and in vitro fertilization. The welfare of cloned animals is similar to that of other conventional agricultural animals. Because of concerns related to the disposition of cloned agricultural animals in the food chain, it is recommended that "*institutions and researchers participate in the Livestock Industry Clone Registry whereby animal clones are registered in the database or registry* (www. livestockcloneregistry.com)." The *Ag Guide* indicates that health and welfare concerns which may arise with transgenic and genetically engineered animals are important to understand so that issues related to animal care and well-being are thoroughly considered. The *Ag Guide* states on page 14 that, "*The scientist is responsible for identifying physiologic and phenotypic changes and must have a plan to address changes that affect animal health to facilitate and ensure animal welfare.*" It is further acknowledged that the IACUC, animal care staff, and veterinarians as well as researchers should be involved in the monitoring and care of genetically engineered animals because of unexpected phenotypes that can result. Protocols involving genetically engineered animals should receive careful review by the IACUC with consideration of such issues as proper animal identification, public safety, animal welfare, and final disposition of the animals. In several of the species-specific chapters there are recommendations regarding known welfare concerns associated with genetically altered animals, from dystocia and large offspring to retained placenta, hydrops and multiple births and the emphasis to provide careful IACUC review, monitoring and oversight, with recognition of potential welfare concerns that are at the forefront of these guidelines.

INSTITUTIONAL AND DESIGNATED PERSONNEL RESPONSIBILITIES

AWA/AWR's

The AWR define the responsibilities of five key individuals or groups: the CEO, the IO, the AV, the PI, and the IACUC. The regulations require that members of the IACUC be appointed by the CEO. At a minimum, this committee must include a chair, a veterinarian with training and experience in laboratory science and medicine and with direct or delegated programmatic responsibilities for activities involving animals, and at least one person who is not affiliated with the research facility who is expected to provide representation for the general community interests in the proper care and treatment of animals. If the committee consists of more than the mandated three members, no more than three can come from the same administrative unit.

The IACUC is charged to act as an agent of the research facility in assuring compliance with the AWA. It is required to inspect all animal facilities and study areas at least once every 6 months, and to review the condition of the animals and the practices involving pain and distress to the animals to assure compliance with the regulations and standards promulgated under the AWA. Once every 6 months, the IACUC is also required to review the research facility's program to assure the care and use of the animals conforms to the regulations and standards. The IACUC must file a report of its inspections with the IO of the research facility. The IO is an individual within the research facility who is legally authorized to commit on behalf of the facility that the regulations and standards will be followed. Failure to adhere to the plan and schedule that results in a deficiency, which may be or is a threat to animal health or safety, remaining uncorrected shall be reported in writing within 15 business days by the IACUC, through the institutional official, to APHIS and any federal agency funding that activity.

The IACUC must also review and approve all proposed activities involving the care and use of animals in research, testing, or teaching procedures and all subsequent significant changes of ongoing activities. It is the responsibility of the PI to submit these proposals and for the design and implementation of the research involving animals.

Research facilities are required to file an annual report in which either the CEO or the IO must assure that professionally accepted standards governing the care, treatment, and use of animals were in place prior to, during, and following the actual use of the animals; the PI considered alternatives to painful procedures; and the facility is adhering to the regulations and standards.

PHS Policy and OLAW

PHS Policy assigns each awardee institution the responsibility of providing verification of approval by the IACUC of those components of the PHS application or proposal related to the care and use of animals. Institutions must have on file with OLAW an approved written Animal Welfare Assurance document which describes the institution's compliance with the PHS Policy. That Assurance must describe the lines of authority and responsibility for administering the program and ensuring compliance. There must be an IO named who has the authority to sign the institution's Assurance, making a commitment on behalf of the institution that the requirements of the PHS Policy will be met. The CEO acting as the IO, or a designated delegate must appoint an IACUC that consists of at least five members including: the veterinarian(s) with direct or delegated program authority and responsibility for animal activities; one practicing scientist experienced in research involving animals; one member whose primary concerns are in a nonscientific area; and one individual who is not affiliated with the institution in any way other than as a member of the IACUC.

OLAW further specifies that each institution must provide: personnel training necessary to comply with the PHS Policy; an occupational health and safety program for personnel who have frequent contact with animals; and an animal facility disaster plan.

Guide

The *Guide* reflects the PHS Policy in regard to institutional responsibilities and required specific personnel. The *Guide,* on pages 13 and 14, states that the primary oversight responsibilities for the institutional animal care and use program rest with the IO, the AV, and the IACUC. Together they *"establish policies and procedures, ensure regulatory compliance, monitor Program performance, and support high-quality science and humane animal use."* The IO bears ultimate responsibility for the Program and must have *"the authority to allocate the resources needed to ensure the Program's overall effectiveness."* The AV is responsible for the health and well-being of all laboratory animals used at the institution. The AV must have sufficient institutional authority and resources to manage the program of veterinary care. The IACUC (or institutional equivalent) is responsible for assessment and oversight of the institution's Program components and facilities, and review and approval of proposed animal use and of proposed significant changes to animal use. The IACUC should have appropriate institutional authority and resources to fulfill its responsibilities. The *"IO is responsible for resource planning and ensuring alignment of Program goals of quality animal care and use with the institution's mission."* Further, the IO ensures that the Program's overall effectiveness and the needs are met. The needs of the Program *"should be clearly and regularly communicated to the IO by the AV, the IACUC, and others associated with the program* (e.g., *facilities management staff, occupational health and safety personnel, scientists)."*

Ag Guide

The *Ag Guide* states on page one that there are critical components in an effective agricultural animal care and use program and include "(1) *clearly established lines of authority and responsibility; (2) an active Institutional Animal Care and Use Committee (IACUC); (3) procedures for self monitoring of the IACUC through semi-annual review of programs and facility oversight by the institutional officer; (4) appropriately maintained facilities for proper management, housing and support of animals; (5) an adequate program of veterinary care; and (6) training and occupational health programs for individuals who work with animals."* From these critical components, three relate to the important roles of the IO, the IACUC, and the AV. The *Ag Guide* furthers states that there should be clearly designated lines of authority within the animal care and use program. The CEO should appoint the IACUC which is tasked with monitoring the program of animal care and use and specific responsibilities of the IACUC are outlined. The IACUC should be comprised of at least five members, each fulfilling a specific role. In addition to describing the responsibilities of the IACUC, the role of the AV is also provided on page nine. The attending veterinarian *"must have authority to ensure that the provisions of the program are met. The attending veterinarian must be provided access to all research and teaching animals and to any related documents including health care records. The attending veterinarian also must be involved in the development of and oversight of the veterinary care program, as well as other aspects of animal care and use such as protocol review, establishment of anesthetic and analgesic guidelines, study removal criteria, training of animal users, and responsible conduct of research activities."*

OVERSIGHT AND ETHICAL REVIEW PROCESS

AWA/AWR's

The IACUC is responsible for reviewing and approving and requiring modifications of proposed activities involving the use of animals in research, teaching, testing, and experimentation. As part of this review, the IACUC must evaluate procedures which minimize discomfort, distress, and pain. In addition, the IACUC must determine that a veterinarian has been consulted in planning for the administration of anesthetics, analgesics, and tranquilizers, and that paralytic agents are used only in anesthetized animals. The IACUC must also determine that animals which experience severe or chronic pain are euthanatized consistent with the design of the study, that the living conditions meet the species' needs, that necessary medical care will be provided, that all procedures will be performed by qualified individuals, that survival surgery will be performed aseptically, and that no animal will undergo more than one major operative procedure that is not justified and approved. On behalf of the research facility, the IACUC must also assure that the principal investigator considered alternatives to painful procedures and that the work being proposed does not unnecessarily duplicate previous experiments. To provide assurance of the former, the IACUC must review the written narrative description provided by the investigator. This description must include the methods and sources used in determining that alternatives were not available.

In reviewing proposed activities and modifications, the IACUC can grant exceptions to the regulations and standards, if they have been justified in writing by the principal investigator.

The IACUC must conduct continuous reviews of activities at least annually and may suspend an activity if it was previously approved for failure to adhere to the description of the activity approved by the committee.

PHS Policy and OLAW

PHS Policy requirements for IACUC oversight and ethical review of animal activities are similar to the AWRs previously described. The PHS Policy does not explicitly require post approval monitoring procedures to compare the practices described in approved protocols against the manner in which they are actually conducted. IACUCs are charged, however, with program oversight and as such are responsible for conducting semiannual facility inspections and programmatic review, review of protocols and amendments, reporting noncompliance, ensuring that individuals who work with animals are appropriately trained and qualified, and addressing concerns involving the care and use of animals at the institution. Specific protocol elements to be reviewed include minimization of pain and distress, appropriate sedation, analgesia, or anesthesia, appropriate living conditions and veterinary care, adequately trained personnel, and appropriate euthanasia methods. The maximum duration for a protocol following IACUC approval is three years (i.e. a complete de novo review of protocols is required at least every 3 years).

Guide

The *Guide* supports and elaborates on PHS Policy expectations and procedures for IACUC review and oversight of the care and use of animals. Specifically, the *Guide* states that, "*The committee must meet as often as necessary to fulfill its responsibilities and records of committee meeting and results of deliberations should be maintained.*" The following topics should be included in protocols and reviewed by the IACUC:

- rationale and purpose of the proposed use of animals;
- a clear and concise sequential description of the procedures involving the use of animals that is easily understood by all members of the committee;
- availability or appropriateness of the use of less invasive procedures, other species, isolated organ preparation, cell or tissue culture, or computer simulation;
- justification of the species and number of animals proposed; whenever possible, the number of animals and experimental group sizes should be statistically justified;
- unnecessary duplication of experiments;
- nonstandard housing and husbandry requirements;
- impact of the proposed procedures on the animals' well-being;
- appropriate sedation, analgesia, and anesthesia;
- conduct of surgical procedures, including multiple operative procedures;
- postprocedural care and observation (e.g. inclusion of posttreatment or postsurgical animal assessment forms);
- description and rationale for anticipated or selected endpoints;
- criteria and process for timely intervention, removal of animals from a study, or euthanasia if painful or stressful outcomes are anticipated;
- method of euthanasia or disposition of animals, including planning for care of long-lived species after study completion;
- adequacy of training and experience of personnel in the procedures used, and roles and responsibilities of the personnel involved;
- use of hazardous materials and provision of a safe working environment.

The *Guide*, on pages 27–28, describes specific issues which require special consideration by the IACUC: experimental humane endpoints, unexpected outcomes, physical restraint of conscious animals, multiple survival surgical procedures, food and/or water regulation, the use of nonpharmaceutical grade chemicals and other substances, field investigations, and the use of agricultural animals in biomedical research. In addition to the specific topics to be included in animal use protocols, the IACUC is also tasked with evaluating scientific elements of the protocol as they relate to animal welfare (e.g. hypothesis testing, samples size, group numbers, and adequacy of controls). IACUC members named in protocols or who have other conflicts must excuse themselves from decisions concerning these protocols.

The *Guide* describes ongoing IACUC monitoring of approved activities as post approval monitoring (PAM) and states on page 33 that, *"PAM helps ensure the well-being of the animals and may also provide opportunities to refine research procedures."* Various methods of PAM are described: continuing protocol review; laboratory inspections; veterinary or IACUC observations of specific procedures; observations of animals by animal care, veterinary and IACUC staff; and external regulatory inspections. In the US, institutions are required by the AWRs and PHS Policy, to conduct semiannual inspections and program reviews. Depending on the complexity and size of the animal care and use program, the conduct of these required activities may be sufficient to meet the expectations of the *Guide* for PAM. Effective monitoring includes:

- examination of surgical areas, including anesthetic equipment;
- use of appropriate aseptic technique, and handling and use of controlled substances;
- review of protocol-related health and safety issues;
- review of anesthetic and surgical records;
- regular review of adverse or unexpected experimental outcomes affecting the animals;
- observation of laboratory practices and procedures and comparison with approved protocols.

Ag Guide

The *Ag Guide* specifies that an institutional IACUC shall be appointed by the CEO and should consist of at least five members; a scientist who has experience in agricultural animal research or teaching; an animal, dairy, or poultry scientist who has training and experience in the management of agricultural animals; a veterinarian who has training or experience in agricultural animal medicine and who is licensed or eligible to be licensed to practice veterinary medicine; a person whose primary concerns are in an area outside of science; a person who is not affiliated with the institution and who is not a family member of an individual affiliated with the institution; other members as required by institutional needs. One individual may adequately fill more than a single role on the IACUC, but the committee must consist of at least five members. The IACUC is authorized to: review, approve, or disapprove protocols and significant changes in ongoing activities; conduct an inspection of the facilities at least twice each year; study areas and review the overall program of animal care and use providing the IO with a written report of the findings; investigate concerns, complaints, or reports of noncompliance; suspend an activity; make recommendations regarding the development and implementation of policies and procedures to facilitate, support, and monitor humane care and use of animals; and perform other functions required by institutional need and applicable laws, regulations, and policies. The oversight role of the IACUC includes consideration, review, and monitoring of many programmatic aspects of animal care and use. IACUC oversight includes but is not limited to: protocols and written operating procedures; handling and transport; restraint; husbandry, housing, and biosecurity; environmental enrichment and social needs of the animals; personnel qualifications; occupational health and safety; hazardous materials; genetically altered animals; animal procurement, quarantine, acclimation, and stabilization; the program of veterinary care; surgery, anesthesia, analgesia, and postsurgical care; residue avoidance for research or teaching animals that may enter the human food chain; record keeping; drug storage and control; and euthanasia.

REUSE

AWA/AWR's

The animal welfare regulations and standards make no special provisions for reuse of animals other than to prohibit reuse for major operative procedures unless justified for scientific reasons or as part of routine veterinary care.

PHS Policy and OLAW

The PHS Policy and OLAW make no specific recommendations regarding the reuse of animals other than to prohibit reuse for major operative procedures unless justified for scientific reasons or as part of routine veterinary care.

Guide

The *Guide* recommends that procedures which may induce substantial postprocedural pain or impairment should be scientifically justified if performed more than once in a single animal. On page five the *Guide* states, *"Refinement and reduction goals should be balanced on a case-by-case basis. Principal investigators are strongly discouraged from advocating animal reuse as a reduction strategy, and reduction should not be a rationale for reusing an animal or animals that have already undergone experimental procedures especially if the well-being of the animals would be compromised."*

Ag Guide

The *Ag Guide* does not address the reuse of animals except with regard to performing more than one major surgery on a single animal. A distinction is made between multiple surgeries for therapeutic reasons such as for cesarean section or displaced abomasum in cattle, and surgeries necessary to complete the scientific aims of the study. In the latter case, scientific justification is required in order to perform more than one major operative procedure.

SETTING FREE/REHOMING

AWA/AWR's

The animal welfare regulations and standards make no special provisions for setting free/rehoming.

PHS Policy and OLAW

The PHS Policy does not specifically address the use of wildlife and has no setting free/rehoming requirements. OLAW guidance on the use of wildlife states that "Investigators are encouraged to consult relevant professional societies, available guidelines, wildlife biologists, and veterinarians, as applicable, in the design of the field studies" and that "proposed studies are in accord with the *Guide*."[23]

Guide

The *Guide* addresses the issue of rehoming by stating on page 32, that "When species are removed from the wild, the protocol should include plans for either a return to their habitat or their final disposition, as appropriate."

Ag Guide

The *Ag Guide* does not address the topic of setting free/rehoming.

While there is no mandate in the US for setting free/rehoming, "animal adoption" is a practice that institutions may employ when possible. Careful consideration of issues related to transfer of animal ownership, institutional liability, long term health and welfare of the animals, provision of proper husbandry, and the animal's overall needs, such as environment and veterinary care is prudent. Ensuring that animals are conscientiously rehomed with responsible stewardship is essential.

OCCUPATIONAL HEALTH AND SAFETY

AWA/AWR's

The animal welfare regulations and standards make no special provisions for occupational health and safety.

PHS Policy and OLAW

The PHS Policy requires that a health program be in place for personnel who work in laboratory animal facilities or have frequent contact with animals. Further, the PHS Policy requires that programs of animal care and use meet the requirements of the *Guide* which describes the essential elements of an Occupational Health and Safety Program (OHSP). OLAW provides more specific guidance that "An effective occupational health and safety program must encompass all personnel that have contact with animals." And, "minimally, the program should include: pre-placement medical evaluation; identification of hazards to personnel and safeguards appropriate to the risks associated with the hazards; appropriate testing and vaccinations; training of personnel regarding their duties, any hazards, and necessary safeguards; policies and facilities that promote cleanliness; provisions for treating and documenting job-related injuries and illnesses; facilities, equipment, and procedures should be designed, selected, and developed to reduce the possibility of physical injury or health risk to personnel; good personal hygiene practices, prohibiting eating and drinking, use of tobacco products, and application of cosmetics and/or contact lenses in animal rooms and laboratories; and personal protective equipment (PPE)."[24]

Guide

The *Guide* states on page 17, that *"Each institution must establish and maintain an occupational health and safety program as an essential part of the overall program of animal care and use…The nature of the OHSP will depend on the facility, research activities, hazards, and animal species involved. The OHSP must be consistent with federal, state, and local regulations and should focus on maintaining a safe and healthy workplace."* The *Guide* describes many aspects of an occupational health and safety program on pages 18–24, and describes key elements of an OHSP. An OHSP should include programs for control and prevention strategies for the protection of personnel which comprise a hierarchy of importance: (1) appropriate design and operation of facilities as well as the use of safety equipment, (2) the development of standard operating procedures, and (3) the provision of appropriate personnel protective equipment.

Another key feature of an OHSP is hazard identification and risk assessment. The identification of work-related risks and hazards is essential to ensure that risks are reduced to minimal and acceptable levels. The assessment of work-related risks such as biologic agents, radiation, physical hazards, chemical agents, animal bites, scratches and kicks, allergy, and zoonoses is important and should be performed on an ongoing basis. Strategies for minimizing risk should be based on these periodic assessments. The *Guide* states on page 18 that, *"The extent and level of participation of personnel in the OHSP should be based on the hazards posed by the animals and materials used (the severity or seriousness of the hazard); the exposure intensity, duration and frequency (prevalence of the hazard); to some extent the susceptibility (e.g.immune status) of the personnel; and the history of occupational illness and injury in the particular workplace.*"

An OHSP program includes appropriate facilities, equipment, and monitoring. The facilities should preferentially employ engineering controls and equipment to minimize personnel exposure and risk. Aspects of personnel hygiene can also be important to minimize exposure (e.g. changing, washing and showering facilities, and supplies). When biologic agents are used, the *Guide* recommends following the safety guidelines published in the Center for Disease Control and NIH publication, *Biomedical safety in Microbiological and Biomedical Laboratories.*[25]

Personnel training in occupational health and safety is another key aspect to an effective OHSP. The *Guide* states on page 20 that, "*Personnel at risk should be provided with clearly defined procedures, practices and protective equipment to safely conduct their duties, understand the hazards involved and be proficient in implementing the required safeguards. They should be trained regarding zoonoses, chemical, biologic, and physical hazards (*e.g. *radiation and allergies), unusual conditions or agents that might be a part of experimental procedures (*e.g. *the use of human tissue in immune compromised animals), handling of waste materials, personal hygiene, the appropriate use of PPE, and other considerations (*e.g. *precautions to be undertaken during pregnancy, illness, or immunosuppression) as appropriate to the risk imposed by their workplace.*"

Occupational health and safety professionals should be involved in the development and implementation of programs for medical evaluation of personnel and preventive medicine. A preemployment health evaluation is advisable and periodic medical evaluations should occur at appropriate intervals given the nature of risks posed. The *Guide* states, "*Zoonosis surveillance should be part of the OHS. Personnel should be instructed to notify supervisors of potential or known exposures and of suspected health hazards and illness.* Further, the *Guide* states that, "*Clear procedures should be established for reporting all accidents, bites, scratches, and allergic reactions and medical care for such incidents should be readily available.*"

Ag Guide

The *Ag Guide* requires that an occupational health and safety program be established for those who work with agricultural animals. Key elements of an occupational health and safety program include adherence to federal, state, and local regulations; risk assessment by health and safety specialists; consideration of hazards posed by the animals, the materials used, the duration, frequency, and intensity of exposure; the susceptibility of personnel; and the history of occupational injury and illness in the workplace. Health assessments of personnel prior to job placement and periodically thereafter are recommended. Mechanisms for surveillance to ensure protection from health hazards and provision of occupational health care services are also recommended. There should be an educational component of the occupational health and safety program to educate personnel about risks associated with the following: zoonoses; physical hazards; containment of, and protection from hazardous agents; personal hygiene; noise; safety procedures for personnel protection; and special precautions for individuals that may be at higher risk (pregnancy, chemical, and radiation hazards, etc.). Appropriate immunizations should be available and animal caretakers should receive tetanus immunization at least every 10 years. Considerations of allergies and physical injury are important for those working with agricultural animals. Personnel should be knowledgeable about the risks associated with agricultural animals, the specific hazards that may be involved in the study, their overall participation in the occupational health and safety program, and be sufficiently trained so that risks of animal contact are minimized.

Occupational Health and Safety in the Care and Use of Research Animals

In the US there is an additional occupational health and safety reference that is widely used as recommended guidelines for personnel in the animal research environment, "*Occupational Health and Safety in the Care and Use of Research Animals*" (NRC, 1997).[26] These guidelines are important, comprehensive recommendations for the structure and function of occupational health and safety programs involving research animals. The *Guide* and *Ag Guide* reference the use of these guidelines when establishing occupational health and safety programs. Key sections of the guidelines include: program design and management; physical, chemical, and protocol-related hazards; allergens; zoonoses; principle elements of an occupational health and safety program; and occupational health care services. These guidelines are detailed and complete and establish the current standards in the US for occupational health and safety programs involving research animals.

EDUCATION, TRAINING, AND COMPETENCE OF PERSONNEL

AWA/AWR's

It is the responsibility of the research facility to provide training to scientists, animal technicians, and other personnel involved with animal care and treatment. Training should include the use of humane methods of animal maintenance and experimentation; the concept, availability, and use of research or testing methods that limit the use of animals or minimize animal distress; the proper use of anesthetics, analgesics, and tranquilizers for any species of animals used by the facility; and methods whereby deficiencies in animal care and treatment can be reported. In addition, the training should include information on the utilization of library services available to provide information on these subjects.

PHS Policy and OLAW

The PHS Policy requires that institutions describe the "*training or instruction in the humane practice of animal care and use, as well as training or instruction in research or testing methods that minimize the number of animals required to obtain valid results and minimize animal distress, offered to scientists, animal technicians, and other personnel involved in animal care, treatment, or use,*" and that the IACUC ensure that, "*personnel conducting procedures on the species being maintained or studied will be appropriately qualified and trained in those procedures.*" In addition, there is a specific requirement that the veterinarian with "*direct or delegated program authority and responsibility for activities involving animals at the institution*" has "*training or experience in laboratory animal science and medicine.*"

Guide

The *Guide* further elaborates on the tenets of the PHS Policy stating that "*All personnel involved with the care and use of animals must be adequately educated, trained, and/or qualified in basic principles of laboratory animal science to help ensure high-quality science and animal well-being.*" Specific expectations and recommendations for training and qualifications for veterinary, animal care, and research staff, as well as IACUC members are described in Chapter 2. Training expectations for specific activities, such as surgery, euthanasia, anesthesia, or hazards, such as exposure to hazardous agents or workplace risks, are also described in the *Guide*.

Veterinary staff must have experience or training and expertise necessary to evaluate the health and well-being of the animals in the context of the animal use at the institution. Training or experience in other areas may also be important for veterinarians, such as, facility administration and management, facility design and renovation, human resources management, pathology of laboratory animals, comparative genomics, facility and equipment maintenance, diagnostic laboratory operations, and behavioral management. Animal care personnel should be appropriately trained and the institution should provide formal and/ or on the job training. Staff should receive specific training, or have experience, to complete their duties. External training programs for technical staff are available and include the American Association of Laboratory Animal Science (AALAS), and Laboratory Animal Welfare and Training Exchange.

All members of the research team should receive training or have the necessary knowledge, experience, and expertise to conduct the specific procedures with the species used. Training should be aimed at the specific needs of the research groups and specifically include the topics of animal care and use legislation; IACUC function; ethics of animal use and concepts of the 3-R's; methods for reporting concerns about animal use; occupational health and safety issues pertaining to animal use and animal handling; aseptic surgical technique; anesthesia and analgesia; and euthanasia.

The institution also has a responsibility to provide the necessary training for the IACUC members so that they understand their role and responsibilities. Training should include formal orientation to the institutional program of animal care and use; relevant legislation, regulations, guidelines, and policies; animal facilities and laboratories in which animal studies are conducted; and the process of animal protocol review. The overall frequency of training should ensure that all personnel who work with animals are properly trained before they begin working with animals. The IACUC, veterinary, and animal care staffs should be provided ongoing opportunities for training. All personnel training should be documented.

Ag Guide

On page three, the *Ag Guide* states that, "*It is the responsibility of the institution to ensure that scientists, agricultural animal care staff, students, and other individuals who care for or use agricultural animals are qualified to do so through training or experience.*" Training programs should be tailored to user needs and include information about: husbandry; handling; surgical procedures and post-procedural care; methods for minimizing the numbers of animals used; techniques for minimizing pain and distress; proper use of anesthetics, analgesics, tranquilizers, as well as nonpharmacologic methods of pain relief; methods for reporting deficiencies in the animal care program; use of information services; and methods of euthanasia. Records of personnel training should be maintained. Those responsible for animal care should regularly participate in training and education activities relevant to their responsibilities. Both formal and on the job training opportunities should be made available. It is desirable for the animal care staff to undergo professional training and certification. The AALAS and the American Registry of Professional Animal Scientists are two certification organizations that provide such training and certification.

TRANSPORT

AWA/AWR's

The requirements for transportation are covered in the standards sections for the species covered by the AWA. These standards include requirements for consignment to carriers and intermediate handlers, the enclosures used to transport the animals, the cargo space of the primary conveyance, provision of food and water, care in transit, terminal facilities, and handling of the animals to include things such as shelter from sunlight, rain or snow, and cold weather.

PHS Policy and OLAW

The PHS Policy does not specifically address the transport of animals; however, the US Government Principles state that the transportation of animals should be in accordance with the Animal Welfare Act and other applicable federal laws, guidelines, and policies. *"OLAW expects all parties involved in the transportation of animals to apply due diligence in assuring that animals are shipped under appropriate conditions to prevent morbidity or mortality due to temperature extremes or other adverse events. OLAW expects shipping institutions to report adverse events that occur to animals in transit."*[27]

US Government principle number one states, *"The transportation, care and use of animals should be in accordance with the Animal Welfare Act (7 U.S.C. 2131 et. seq.) and other applicable Federal laws, guidelines, and policies."* This reference includes adherence to the *Guide*.

Guide

The *Guide* describes appropriate transport of animals in detail on pages 107–109 stressing the guidance in the AWRs and the need to comply with various governmental agency requirements involving animal transportation. A number of agencies and the requirements for animal transportation are cited including, the National Research Council publication Guidelines for the Humane Transportation of Research Animals (NRC 2006).[11] Careful planning and coordination of transportation procedures should be employed to ensure animal safety and well-being. Personnel responsible for transportation should be well-trained and transit time should be minimized with delivery times occurring during normal business hours. Animals in transit between institutions should be accompanied by appropriate documentation (health certificates, agency permits, sending and receiving institution's addresses, contacts, emergency procedures, and veterinary contact information). The *Guide* indicates that key aspects of transportation include an appropriate level of animal biosecurity while minimizing zoonotic risks, protecting against environmental extremes, avoiding overcrowding, providing for the animals' physical, physiologic, or behavioral needs and comfort, and protecting the animals and personnel from physical trauma. *"Transportation of animals in private vehicles is discouraged because of potential animal biosecurity, safety, health, and liability risks for the animals, personnel, and institution."*

Ag Guide

The *Ag Guide* contains detailed recommendations for the handling and transport of agricultural animals and highlights the goal of calm, respectful, stress-free methods for the benefit of the animals and to promote sound research practices. There are specific guidelines for several species with topics that include: flight zone and behavior principles; aids for moving animals; considerations for animal perception, hearing, vision, and visual distractions; flooring; proper maintenance of equipment; sanitation; and principles to prevent behavioral agitation. Specific recommendations are provided for beef and dairy cattle, horses, sheep, goats, pigs, and poultry. Also, there are specific recommendations for safe animal transport which include: guidelines regarding animals not suitable for transport (near parturition, new born, weak, debilitated, etc.); thermal environment, bedding, and floor surfaces during transport; space provisions during transport (the guidelines published by the National Academy of Sciences are referenced) (ILAR Transportation Guide, 2006)[11]; lairage; adherence to guidelines and federal regulations regarding duration of transportation; and methods for loading and unloading.

HOUSING AND ENRICHMENT

AWA/AWR's

The requirements for housing are contained within the sections of the standards on facilities and operating standards for the species covered by the AWA. Included in this section are specific requirements for facilities in general, including that they be structurally sound and maintained in good repair to protect the animals from injury and to keep them secure. In addition, the site must be maintained in a clean and orderly manner. There are requirements for the surfaces of the facility including the need to maintain and replace surfaces on a regular basis and for cleaning on a regular basis. There are specific requirements for the provision of water and electric power, storage, drainage and waste disposal, and washrooms and sinks for the employees. In addition to the general requirements for facilities, there are specific requirements for indoor, sheltered, outdoor, and mobile facilities which include areas such as heating, cooling, temperature control, ventilation, lighting, and interior surfaces. There are also specific requirements for the design and construction of primary enclosures.

Part three of the AWR's—Standards, describe the requirements for environmental enhancement to promote the psychological well-being of nonhuman primates. Research facilities must develop, document, and follow an environmental enhancement plan. The enhancement plan must include consideration of social grouping; environmental enrichment; special consideration for infants, animals exhibiting psychological distress, animals with IACUC approved restricted activity, animals individually housed, and great apes weighing more than 110lbs; restraint devices; and animals that are exempt from participating in the plan and the reasons for their exemption. The AWRs also require provisions for the compatible social housing for dogs and cats as well as exercise for dogs. Dogs or cats that are housed in the same primary enclosure must be compatible with special consideration given for animals in estrus, young animals, animals with aggressive dispositions, and animals that have or suspected of having a contagious disease. The exercise requirement for dogs includes provisions for dogs housed individually or in groups, and the amount of space that is made available for exercise, as well as the methods and period of providing the opportunity to exercise. Methods shall be determined by the AV and approved by the IACUC. Protocols that involve exercise procedures (swimming, treadmills, etc.) are considered unacceptable for meeting the exercise requirements of the regulations.

PHS Policy and OLAW

While the PHS Policy does not specifically address the topics of housing and enrichment, US Government Principles are endorsed by the policy. Principle number VII states, *"The living conditions of animals should be appropriate for the species and contribute to their health and comfort. Normally, the housing, feeding, and care of all animals used for biomedical purposes must be directed by a veterinarian or other scientist trained and experienced in the proper care, handling, and use of the species being maintained or studied. In any case, veterinary care shall be provided as indicated."* Also, OLAW provides guidance on the use of performance standards to (1) assess and set enclosure space guidelines for rodents and rabbits and (2) assess environmental enrichment issues. *"An institution's environmental enrichment practices must be species-specific and appropriate for the animals."*[28] OLAW categorically states that *"There is universal agreement among oversight agencies that nonhuman primates should be socially housed."*[29] Furthermore, it provides guidance on associated issues such as staff training on nonhuman primate socialization and IACUC and veterinary review of exemptions to the social housing expectation. OLAW does not specifically address aquatics other than to state, when discussing management of Heating, Ventilation, and Air Conditioning System (HVAC) failures, that *"Institutions are also responsible to ensure the welfare of fishes, amphibians, and other vertebrates whose environment is aquatic, with the emphasis on water temperature and quality, including oxygenation, circulation, and filtration."*[30]

Guide

The *Guide* provides extensive and specific requirements and recommendations in regard to housing, environmental enrichment, and behavioral management of all vertebrate species, including aquatics. Guidelines and recommendations are described for the provision of adequate temperature and humidity; ventilation and air quality; illumination; noise and vibration; terrestrial housing; microenvironment of the primary enclosure; environmental enrichment and social housing; sheltered or outdoor housing; space requirements; behavioral and social management; husbandry; population management; and water quality and life support systems for aquatic species.

Chapter three of the *Guide* provides information regarding a number of key topics related to housing and enrichment. Specifically, the *Guide* offers recommendations regarding the microenvironment or primary enclosures. Animals should be housed under conditions that provide sufficient space and supplementary structures and resources to meet their physical, physiologic, and behavioral needs. The social needs of animals are prominently featured.

The primary enclosure should provide a secure environment and be constructed of materials that are safe, durable, and sanitizable. Flooring should be solid, perforated or slatted, and slip resistant. Adequate bedding or substrate should be provided as well as structures for resting or sleeping.

Opportunities for species typical behaviors such as burrowing and nest building should be considered and accommodated whenever possible. Considerations for animal biosecurity and minimization of airborne particles that may transfer between cages may be important and the potential need for specialized caging systems may be necessary. Appropriate housing strategies should be developed and implemented by animal care and use management in consultation with investigators and the veterinarian, and be reviewed by the IACUC. Recommendations for sheltered, outdoor, or naturalistic housing strategies are described.

A series of tables are provided as recommended minimum cage space guidelines for a variety of species including: rodents; rabbits; dogs; cats; poultry; nonhuman primates; sheep; goats; swine; cattle; and horses. Of equal or greater importance are the performance criteria describing adequacy of cage space. The *Guide* acknowledges that determining the appropriate size for primary enclosures includes the consideration of many aspects and not simply body weight/size alone. The performance criteria described in the *Guide* provide recommendations for functional cage space given a variety of factors that should be considered, recognizing that cage dimensions alone represent only one of the factors necessary when determining adequacy of cage space. Some of the performance criteria for adequacy of cage space include space for the animals to rest away from areas soiled by urine and feces, ability to make normal postural adjustments, consideration of ancillary equipment/structures that may otherwise limit floors space, sufficient cage height for arboreal nonhuman primates, opportunities to escape aggression, and increased space for breeding animals.

The *Guide* also provides a detailed discussion of the need for environmental enrichment of animals. The *Guide* states, "*The primary aim of environmental enrichment is to enhance animal well-being by providing animals with sensory and motor stimulation, through structure and resources that facilitate the expression of species-typical behaviors and promote psychological well-being through physical exercise, manipulative activities, and cognitive challenges according to species-specific characteristics.*" Examples of enrichment are provided. Well-conceived enrichment provides animals with choices and some degree of control over their environment which affords the ability to better cope with environmental stressors. Recommendations for the cautious application of enrichment items are provided so as to avoid the unintentional introduction of stressors as some animals may be averse to novel items. Similarly, enrichment structures may act as fomites and increase the potential for disease transmission. The *Guide* specifies that the enrichment program should be reviewed by the IACUC, researchers, and veterinarian on a regular basis to ensure it functions as intended, keeping in mind the benefit to the animals and the scientific goals. Enrichment can be a source of environmental variation if not judiciously applied and the *Guide* recommends the careful implementation of a uniformly applied program that avoids an undesirable impact on scientific outcomes while improving animal well-being.

The need to consider social housing is emphasized in the *Guide*. Appropriate social interactions among conspecifics are considered essential for normal development and well-being. Consideration of behavior such as animals being naturally territorial or communal are important when determining whether they are best housed singly, in pairs, or in groups. The *Guides* states that, "*Single housing of social species should be the exception and justified based on experimental requirements, or veterinary-related concerns about animal well-being.*" Not all members of a social species are compatible, and in such cases single-housing may also be justified. The *Guide* further indicates that animals should be housed under conditions that meet their behavioral and social needs and that social animals should be housed in stable pairs or groups of compatible individuals. When animals are socially housed, often structural accommodations within the enclosure are necessary (perches, visual barriers, refuges, and multiple opportunities for obtaining food and water). Socially housed animals require careful and frequent monitoring to identify and manage antagonistic interactions among conspecifics, should they occur. Whenever animals are single-housed, it should be limited to the minimum period necessary and where possible, visual, auditory, olfactory, and tactile contact with compatible conspecifics should be provided. The housing needs of animals, particularly those in single housing, should be reviewed on a regular basis by the IACUC and the veterinarian.

Ag Guide

The *Ag Guide* provides extensive recommendations regarding housing conditions for agricultural animals. With regard to facilities and environment, there are guidelines regarding the widespread options for housing (pasture to confinement) and the need to understand the impact of environmental stressors on the animal and on the research activity. Criteria for well-being are detailed. Topics include: macro- and microenvironment; thermal indices/comfort; temperature, vapor pressure, and ventilation; air quality; pasture, feedlot, and confinement housing; social needs; protection from the elements for animals housed outdoors; husbandry; feed and water; bedding; ventilation; lighting; provision for space; standard agricultural practices; waste management; and special considerations. There are a wide range of recommendations for the enrichment of agricultural animals and enrichment is often considered part of the refinement effort for animals used in research and teaching. Five categories of enrichment are described: social; occupational (enrichment that provides both psychological benefits and physical challenges to encourage exercise); physical enrichment (altering the size and complexity of the enclosure or the addition of accessories); sensory (stimuli that are visual, auditory, tactile, olfactory, or taste); and nutritional enrichment (providing varied or novel food types or the methods of delivery). Methods of enrichment for each species which employ these five basic strategies are provided. The specific methods recommended vary with the species of animal, but strategies for social enrichment are emphasized, except with boars and the careful application of social housing for males of other agricultural species is also described. The importance of calm and gentle human interaction with animals is emphasized. The benefits of paddock and pasture environments in promoting natural grazing behaviors and allowing for appropriate social interactions, as well as the need to recognize and minimize abnormal behaviors is provided.

HUSBANDRY AND ENVIRONMENT

AWA/AWRs

The standards contain sections for the covered species on animal health and husbandry. The environmental requirements are contained in the section on facilities and operating standards. The husbandry standards include requirements for feeding, watering, cleaning, sanitization, housekeeping, and pest control. In addition, the standards require that licensed and registered facilities have enough employees to carry out the husbandry and care required by the standards and be supervised by someone with the appropriate knowledge, background, and experience with the species being maintained. The employer must also ascertain that the staff can perform to the prescribed standards. The standards for some species also address issues such as compatibility, exercise, classification and separation, and environmental enrichment.

PHS Policy and OLAW

The PHS Policy does not specifically address daily care activities or environmental conditions other than to state in US Government Principle VII that, *"The living conditions of animals will be appropriate for their species and contribute to their health and comfort. The housing, feeding, and nonmedical care of the animals will be directed by a veterinarian or other scientist trained and experienced in the proper care, handling, and use of the species being maintained or studied."* Similarly, OLAW does not go into specifics of daily care activities or environmental conditions other than to reference the *Guide* and specify the need for planning for HVAC failures.

Guide

The *Guide* provides extensive and specific requirements and recommendations in regard to the provision of daily animal care. Key sections include temperature and humidity; ventilation and air quality; illumination; noise and vibration; space provisions; food, water, bedding, and nesting materials; sanitation procedures and frequency of bedding changes; cleaning and disinfection activities; assessing efficacy of sanitation; waste disposal; pest control; weekend, emergency and holiday care; identification methods; record keeping; breeding; genetics; and nomenclature. There are detailed descriptions of environmental enrichment and social housing as well as appropriate overall environmental conditions for all vertebrate species including aquatics.

For terrestrial animals, there are species-specific recommended ranges for establishing set points for temperature and humidity, with the expectation that day to day variation around the set point be minimal so that unnecessary demands are not placed on the animals which require that they adapt to changes in environmental temperature and humidity. Similarly, there is a need to minimize temperature and humidity variation because of the potential impact such variation may have on scientific outcomes. Ventilation and air quality are also considered important aspects of the animal's environment. Recommendations for air exchange previously emphasized engineering criteria, and now the *Guide* stresses the importance of performance-based criteria to determine and assess the adequacy of ventilation rates. Factors such as species housed, animal density, management and husbandry practices, and type of housing may impact the requirements for adequate ventilation of animal areas. Likewise, the scientific aims and specific animal models employed may effect ventilation and supply and exhaust air requirements. Animals in barrier and containment facilities may require a high degree of air filtration as well as areas for surgery, and necropsy. The *Guide* cautions the use of recycled air as supply for animal areas because of risks of biosecurity and the potential for the spread of pathogens. The *Guide* provides specific recommendations for illumination, light intensity, consideration of photoperiod, and monitoring the function of lighting systems to ensure they are appropriate for the animals and adequately support scientific needs. The inadvertent exposure of light (during the dark phase) is of particular concern as many physiologic functions can be affected. There are recommendations regarding noise and vibration in the animal environment with specific details provided about the need to identify and dampen sources of vibration and minimize sources of noise (at frequencies important for the species of animal housed).

Recommendations for the overall husbandry program include detailed considerations and requirements for the source and provision of food and water. There are descriptions of dietary formulations and their potential uses for conventional, microbiologically defined, and immune-compromised animals. Recommendations for water quality, monitoring, and methods of drinking water delivery systems are provided. The requirements for bedding and nesting materials are provided as are the considerations of types of bedding and their potential impact on animal well-being and scientific outcomes. Nesting materials can be beneficial for a number of animal species and the *Guide* recommends careful consideration for the provision of nesting materials as a means of providing environmental enrichment (to promote species-typical behaviors), as well as a means to improve the animal's behavioral repertoire for thermoregulation and rearing young.

The *Guide* provides detailed information regarding goals, methods, and monitoring of sanitation practices. This section applies to sanitation of the animal facility, secondary enclosures, as well as primary enclosures. Specific performance outcomes are defined while providing latitude for a particular institution's best approach to achieve them. Specific procedures of cleaning, sanitation, and disinfection are provided. There are guidelines regarding the methods and intervals of sanitation. A regular, microbiologically-based method to verify the effectiveness of sanitation is recommended. The *Guide* also contains recommendations for waste disposal and the need to effectively manage waste while ensuring safe and efficient means of personnel and public protection. Practices must be in accordance with all applicable federal, state, and local regulations. Hazardous waste must be handled in a manner that ensures personnel safety. The *Guide* states that, *"Hazardous wastes must be rendered safe by sterilization, containment, or other appropriate means before their removal from the facility."* Adherence to the regulatory requirements of local and federal authorities is of particular importance for hazardous wastes.

The *Guide* also contains recommendations for programs of pest control, emergency, weekend and holiday care, population management, and record keeping. Pest control programs are essential in the animal environment and the *Guide* indicates that a *"regularly scheduled and documented program of control and monitoring should be implemented."* The use of pesticides can be an effective means to minimize or eliminate arthropod pests, but the *Guide* recommends they be used with caution as the effects of these chemicals may impact animal health and scientific outcomes. These chemicals should be used in the facility only when necessary and researchers should be consulted prior to their use to ensure that there will be no impact on the studies. The *Guide* requires that animals be cared for by qualified personnel every day, including weekends and holidays. Emergency veterinary care must be available at all times. A disaster plan that includes the needs of both animals and personnel must be in place. Population management of research animals includes a proper means of identification and record keeping. A discussion of the various methods is provided and ranges from procedures to identify individual animals to cage card information and radio frequency identification. Record keeping is an important aspect of the animal facility and program management. The *Guide* states that, *"animal records are essential for genetic management and historical assessments of colonies. Records of rearing and housing histories, mating histories, and behavioral profiles are useful for the management of many species, especially nonhuman primates."* Medical records for individual animals of several species are important and should contain pertinent information regarding clinical and diagnostic procedures, inoculations, history of surgical procedures, and postoperative care. Information regarding research procedures and necropsy findings may also be important.

The *Guide* has recommendations regarding the breeding, genetics, and nomenclature for research animals and underscores the importance of these aspects of the animal care and use program. There are recommendations regarding the use of outbred stocks and genetic management of such colonies. With regard to the use of inbred animals, the *Guide* emphasizes the need for careful monitoring of F1 progeny to ensure appropriate genetic selection. Similarly, the careful monitoring of newly created genetically modified animals is an important aspect of day to day animal care and use which ensures that unexpected phenotypes are appropriately identified and managed so as to minimize potential pain and distress that may result. Careful, detailed records of pedigree and genetic monitoring are essential.

The most recent update of the *Guide* provides recommendations for the care and use of aquatic species. Similar to the terrestrial animal section, the sections for aquatic animals include guidelines for: microenvironment and macroenvironment; water quality; life support systems; temperature, humidity, and ventilation; illumination; and noise and vibration. There are details provided for: environmental enrichment of aquatic species; sheltered, outdoor, and naturalistic housing; space requirements; behavioral and social management; husbandry; sanitation; pest control; emergency, weekend, and holiday care; population management; and record keeping. Many of the fundamental tenets of aquatic animal care and use are similar to those for terrestrial animals even though the species requirements for housing and management are quite different and diverse. As in all other sections of the *Guide*, the performance outcomes are defined with the understanding that the means to achieve the outcomes can vary based on the specific needs of a particular species and the scientific aims of the research.

Ag Guide

As previously described, there are extensive recommendations in the *Ag Guide* for each species with regard to facilities and husbandry practices, feed and water, common agricultural practices, fencing, predator control, and management techniques to promote health and safety of the animals.

VETERINARY CARE

AWA/AWRs

The regulations require that each research facility employ an attending veterinarian with appropriate authority to ensure the provision of adequate veterinary care and to oversee other aspects of animal care use. Each research facility must establish and maintain a program of adequate veterinary care that includes: appropriate facilities, personnel, and equipment; methods to control, diagnose, and treat diseases; daily observation and provision of care; guidance to personnel on the use of anesthetics, analgesics, and euthanasia procedures and pre and postprocedural care. The AV must be a graduate of an accredited school of veterinary medicine or have equivalent formal education, received training and/or experience in the care and management of the animals under his/her care, and have direct or delegated authority for activities involving animals at the facility. The attending veterinarian shall be the voting member of the IACUC.

PHS Policy and OLAW

Neither the PHS Policy nor OLAW specifically address veterinary care activities other than to state in Section IV.C.e. of the policy that "Medical care for animals will be available and provided as necessary by a qualified veterinarian."

Guide

The *Guide* provides extensive and specific requirements and recommendations regarding the provision of veterinary care, preventive medicine, animal procurement, health monitoring, and the conduct of surgery. An entire chapter is devoted to the topic of veterinary care. Programs of veterinary care are based on the requirement stated on page 112 that, *"All animals should be observed for signs of illness, injury, or abnormal behavior by a person trained to recognize such signs. As a rule, such observation should occur at least daily."* The *Guide* further states that *"An adequate veterinary care program consists of assessment of animal well-being and effective management of:*

- *animal procurement and transportation;*
- *preventive medicine (including quarantine, animal biosecurity, and surveillance);*
- *clinical disease, disability, or related health issues;*
- *protocol-associated disease, disability, and other sequelae;*
- *surgery and perioperative care;*
- *pain and distress;*
- *anesthesia and analgesia."*

The *Guide* places a significant amount of authority and responsibility on the AV. Chapters 2 and 4, provides an overall description of the role of the AV that encompasses many topics and prominently features the AV as a leader, in collaboration with the IO and IACUC, of the animal care and use program. The *Guide* indicates that the animal care and use program is the collaborative responsibility of the IO, the AV, and the IACUC. The AV is responsible for the well-being and clinical care of the animals throughout all phases of the animal's life. The program of veterinary care must uphold the highest standards of care and ethics and there must be sufficient authority of the AV to provide treatment and relieve pain and distress, including euthanasia. Although the AV is imbued with a significant level of institutional authority, the *Guide* emphasizes the need for a collaborative approach between the AV and investigators to achieve the best possible outcome for the animals and the needs of the science. While the AV's areas of responsibility essentially encompass all aspects of animal care, the *Guide* acknowledges that other individuals in the institution may assume responsibility for some of the duties related to veterinary care, but this must be done under the authority and direction of the AV. Regarding the qualifications, the *Guide* stipulates that veterinarians must have experience, training, and expertise necessary to appropriately evaluate the health and well-being of the species used. Veterinarians should remain knowledgeable about the latest practices in order to ensure the highest quality of veterinary medical care. Further, the veterinarian must have access to all animals and the provision of veterinary medical care must be available at all times. The veterinarian must have oversight of programmatic aspects such as preventive medicine, health surveillance, medical treatment, establishment of sedation, anesthetic and analgesic, and handling guidelines and should oversee other aspects of the program such as husbandry and housing. The AV is also tasked with providing guidance for surgery and perioperative care. The *Guide* indicates that the AV or designee has responsibilities related to the IACUC. There should be ongoing, regular communication among the AV, the IACUC, and the IO. The veterinarian has definite responsibilities during protocol review and particularly for the development of study removal criteria and pain and distress management plans. The IACUC and AV together are responsible for assessing the qualifications for those performing surgery and anesthesia, as well as the classification of major versus minor surgical procedures.

The American College of Laboratory Animal Medicine (ACLAM) has developed guidelines which describe expectations for adequate veterinary care.[31] The guidelines were developed in 1996 and provide a detailed description of the key aspects related to the role of veterinarians and programs of veterinary care and these guidelines are cited as a reference document in the *Guide*. Important features of these guidelines include the requirement that institutional veterinarians be qualified through postgraduate training or experience in laboratory animal science and medicine; institutional authority of the veterinarian must be provided to allow the veterinarian to fulfill their duties; a veterinarian must be a full member of the IACUC and be actively involved in the review of protocols, projects, and institutional programs involving animals, and inspection of facilities. Further, the ACLAM guidelines specify that the provision of adequate veterinary care involves: disease detection and surveillance, prevention, diagnosis, treatment, and resolution; handling and restraint; anesthetics, analgesics and tranquilizer drugs; methods of euthanasia; surgical and postsurgical care; animal well-being; and the appropriate use of animals in research and testing. The guidelines also describe other aspects related to veterinary care which include: participating in the development and administration for staff training; assisting institutional health officials in establishing and maintaining an occupational health program; monitoring for zoonotic disease; advising and monitoring of standards of hygiene; and advising and monitoring of biohazard control policies and procedures.

Ag Guide

The program of veterinary care is the responsibility of the AV and includes: having direct or delegated authority to develop and oversee the program of veterinary care and other aspects related to husbandry, nutrition, sanitation, zoonoses, and hazard containment; have access to all animals; establish guidelines for anesthetic and analgesics; involvement in protocol review and study removal criteria; involvement in training for animal users and responsible conduct of research activities; and available to provide guidance and advice to research personnel. The program of veterinary care should consist of elements such as: preventive medicine; recognition of sick or injured animals so that timely veterinary medical care can be provided; maintenance of health records; considerations for multiple, major survival surgeries; oversight of surgery personnel; surgical facilities and aseptic technique; recognizing pain and distress; anesthesia and analgesia; postprocedural care; zoonoses; biosecurity and vermin control; and euthanasia.

CONDUCT OF EXPERIMENTAL PROCEDURES

AWA/AWRs

The AWA prohibits the promulgation of rules, regulations, or orders related to the design, outlines, or guidelines of actual research or experimentation. That said, the research facility is required to follow professionally acceptable standards for the care, treatment, and use of animals during actual research or experimentation. These standards specifically apply to procedures that minimize pain and distress including adequate veterinary care to include the use of tranquillizers, analgesics, anesthetics, the provision of pre- and postsurgical care consistent with established veterinary medical and nursing procedures, and the prohibition of the use of paralytics without anesthesia. If withholding of tranquillizers, analgesics, anesthetics, or euthanasia is scientifically necessary, it shall be done only for the necessary period of time. On an annual basis, each registered research facility must submit a report that assures that the facility is adhering to the standards and regulations governing the care and use of animals including the appropriate use of anesthetics, analgesics, and tranquilizers, and that investigators considered alternatives to painful procedures. They must also report the number of animals that were held and not used, used for procedures that caused no more than momentary pain and distress, involved pain and distress that were relieved by the use of anesthetics, analgesics, or tranquilizers, and/or involved pain and distress that could not be relieved by the use of anesthetics, analgesics, or tranquilizers because such use would have adversely affected the data being generated.

PHS Policy, OLAW

The Health Research Extension Act which provides the statutory mandate for the PHS Policy describes some fundamental aspects regarding the conduct of animal studies. The proper use of tranquilizers, analgesics, anesthetics, paralytics, and euthanasia agents/procedures is required. The requirements for appropriate pre- and postsurgical care as well as veterinary medical and nursing care are also described. There is also a requirement that scientists, animal technicians, and other personnel involved with animal care, treatment, and use have available to them instruction or training in the humane practice of animal maintenance and experimentation, as well as the testing methods that employ concepts that limit use of animals or animal distress. Also described in the PHS Policy are the US Government Principles, all of which relate to the conduct of animal studies. The PHS Policy lists the fundamental requirements in the conduct of animal studies: that procedures avoid or minimize animal discomfort, distress, and pain; procedures that cause more than momentary or slight pain or distress will employ appropriate sedation, analgesia, or anesthesia unless withholding of such agents is scientifically justified; description of procedures designed to assure that discomfort and injury to animals will be limited to that which is unavoidable in the conduct of scientifically valuable research; animals that would otherwise experience severe or chronic pain or distress will be humanely euthanatized with a description of the methods used; the living conditions of the animals will be appropriate and contribute to their health and comfort; rationale for involving animals, and for the appropriateness of the species and numbers used; and a complete description of the proposed use of the animals;

Further, the IACUC is tasked with conducting continuing reviews of each previously approved ongoing activity at appropriate intervals. Finally, the PHS Policy requires that conduct of PHS-funded animal studies must be in accordance with the institution's assurance statement, the AWA, and the *Guide*. The PHS Policy and OLAW do not specify classification of animal activities based on severity, either prospective or retrospective. Refinement was discussed above in the subchapter on "Principles" in association with the 3-Rs.

Guide

The *Guide* extensively discusses appropriate conduct of experimental procedures and endorses the following principles (excerpted from a larger list): consideration of alternatives (in vitro systems, computer simulations, and/or mathematical models) to reduce or replace the use of animals; design and performance of procedures on the basis of relevance to human or animal health, advancement of knowledge, or the good of society; avoidance or minimization of discomfort, distress, and pain; and establishment of humane endpoints. Essentially, all of Chapter 2 of the *Guide* relates to the proper conduct of animal studies. The *Guide* places particular emphasis on the need to develop detailed, well-written, comprehensive animal study protocols which describe all animal procedures and their anticipated impact on the animals. A thorough review of the proposed protocols must be provided by the IACUC. Once protocols are approved, the IACUC must provide ongoing PAM to ensure that the studies conducted are in accordance with the approved protocol. Should a need arise to revise animal procedures once a protocol has been approved, the proposed changes must be described in writing by the PI and approved by the IACUC before the revised procedures are conducted. The IACUC has the authority to investigate concerns related to animals use and suspend animal activities.

With regard to pain and distress during the conduct of experiments, the *Guide* states that "*the proper use of anesthetics and analgesics in research animals is an ethical and scientific imperative.*" based on the concept that, unless the contrary is known or established, it should be considered that procedures that cause pain in humans may also cause pain in other animals.

Ag Guide

A central feature of the recommendations in the *Ag Guide* is the IACUC and its breadth and scope of responsibility. In essence, all research and teaching procedures conducted on animals must be approved by the IACUC. Procedures must be described, in detail, in either the protocol or in written operating procedures. The IACUC must approve the procedures (protocols or written operating procedures) before the initiation of the research or teaching activity. The IACUC should perform a complete review of ongoing studies at least once every three years, or more often at an interval deemed necessary by the IACUC. The IACUC is authorized to investigate concerns, complaints, or reports of noncompliance (e.g. the conduct of procedures that are not described in an approved protocol). The IACUC is further authorized to suspend an activity when it is not in compliance with the approved protocol or written operating procedures.

EUTHANASIA

AWA/AWRs

Methods of euthanasia must be consistent with the definition contained in the regulations. That definition requires that the method used produce rapid unconsciousness and subsequent death without evidence of pain or distress or a method that utilizes anesthesia produced by an agent that causes painless loss of consciousness and subsequent death. The USDA's *Animal Care Resources Guide*, which is intended to clarify the regulations and standards, indicates that the methods used should be consistent with the current guidelines on euthanasia of the American Veterinary Medical Association[32] (http://www.aphis.usda.gov/animal_welfare/downloads/policy).

AVMA Guidelines on Euthanasia

The American Veterinary Medical Association published its most recent guidelines in 2013 titled, *AVMA Guidelines for the Euthanasia of Animals: 2013 Edition*.[7] The guidelines are a comprehensive set of recommendations that are widely used by veterinarians and IACUCs as the standard for euthanasia procedures and techniques. There are detailed recommendations for species used in research settings with guidance on those methods that are considered "acceptable," "acceptable with conditions," "adjunctive," and "unacceptable." With regard to the agents used, there are recommendations for inhaled and non-inhaled as well as physical methods. There are considerations for dangerous or fractious animals, emergencies, age or life stage, and animals in free-range conditions. Traditional laboratory animal species are included as are ruminants, equids, fish, amphibians, reptiles, birds, embryos, fetuses, and neonates as well as some wild species. Apart from the extensive recommendation for animals in research, there are sections detailing consideration for companion animals, free-ranging wildlife, captive, and free-ranging marine mammals. The guidelines also contain recommendations regarding euthanasia as a humane technique and the professional judgment and decision making process when considering the euthanasia of animals. Veterinary medical ethics, animal and human behavior, stress and distress, as well as pain and its perception are all topics that warrant appropriate consideration.

Species-specific sections have been added and expanded to include terrestrial and aquatic species in a variety of environments. The three main parts to these guidelines include Part I—Introduction and General Comments, which details: the historical context of euthanasia; a precise definition of euthanasia; veterinary medical ethics; and discussion regarding many aspects of the methods of euthanasia such as consciousness and unconsciousness, pain and distress, and human behavior among other topics. Part II—Methods of Euthanasia details aspects related to inhaled agents, noninhaled agents, and physical methods, providing an array of options for consideration depending on species, environment, safety, efficacy, etc. Part III—Methods of Euthanasia by Species and Environment, includes a thorough discussion related to companion animals, laboratory animals, animals farmed for food and fiber (including finfish and free-ranging nondomestic animals), marine mammals and embryos, fetuses, and neonates. Each section provides detailed information regarding acceptable methods, methods that are acceptable with conditions, adjunctive methods, and unacceptable methods. In many cases, there are also descriptions of specific technical aspects of proper technique for various methods, anatomy and physiology of a variety of species, restraint, and final disposition of animal carcasses. Extensive references are provided.

Overall, the guidelines stress the importance of proper, respectful animal handling for calm, stress-free methods. Key criteria for euthanasia methods include: (1) *"ability to induce loss of consciousness and death with a minimum of pain and distress"*; (2) *"time required to induce loss of consciousness"*; (3) *"reliability"*; (4) *"safety of personnel"*; (5) *"irreversibility"*; (6) *"compatibility with intended animal use and purpose"*; (7) *"documented emotional effect on observers or operators"*; (8) *"compatibility with subsequent evaluation, examination, or use of tissue"*; (9) *"drug availability and human abuse potential"*; (10) *"compatibility with species, age, and health status"*; (11) *"ability to maintain equipment in proper working order"*; (12) *"safety for predators or scavengers should the animal's remains be consumed"*; (13) *"legal requirements"*; and

(14) *"environmental impacts of the method or disposition of the animal's remains"*. Most IACUC's in the US require that euthanasia be conducted in a manner that is consistent with recommendations of these important guidelines.

PHS Policy and OLAW

In Section IV.C.1.g the PHS Policy states that *"Methods of euthanasia used will be consistent with the recommendations of the American Veterinary Medical Association Panel on Euthanasia [https://www.avma.org/KB/Policies/Documents/euthanasia.pdf], unless a deviation is justified for scientific reasons in writing by the investigator."* OLAW provides further guidance on the procedural methods for use of carbon dioxide as a euthanasia agent[33] (http://grants.nih.gov/grants/olaw/faqs.htm#useandmgmt_1 and http://grants.nih.gov/grants/guide/notice-files/NOT-OD-02-062.html).

Guide

The *Guide* discusses euthanasia criteria and methods in detail, but does not recommend specific agents other than to state that *"The selection of specific agents and methods for euthanasia will depend on the species involved, the animal's age, and the objectives of the protocol. Generally, chemical agents (e.g. barbiturates, non-explosive inhalant anesthetics) are preferable to physical methods (e.g. cervical dislocation, decapitation, use of a penetrating captive bolt)."* The *Guide* also endorses the AVMA Guidelines on Euthanasia. Methods used should induce rapid unconsciousness and death without pain, distress, or anxiety. Methods should also be reliable and irreversible and appropriate for the species, compatible with research objectives, and safe for personnel. The *Guide* also acknowledges the importance of considering the emotional effects on humans. When euthanasia is a planned procedure at the end of a protocol, criteria for euthanasia should be in accordance with protocol-specific endpoints that have been carefully developed and approved by the IACUC. Humane endpoints described in protocols must include four key criteria: (1) a precise description/definition of the endpoint; (2) frequency of animal observations; (3) the training and expertise of personnel to ensure recognition of the endpoint; and (4) the appropriate action to be taken once the endpoint is reached. Special considerations are required for fetuses and neonates and the specific selection of agents and methods will be determined in part, by the age of the animal. The *Guide* acknowledges the controversy of carbon dioxide as an agent for euthanasia of rodents and recommends careful consideration of its use with the general requirement that specific methods be compatible with the *AVMA Guidelines for the Euthanasia of Animals*.

Ag Guide

The *Ag Guide* specifies that protocols for euthanasia should follow the recommendations of the AVMA Guidelines on Euthanasia. There is emphasis on personnel training, experience, skill, and familiarity with the normal behavior of agricultural animals and how handling and restraint affect their behavior. Further, the AV or qualified scientist should ensure that all personnel performing euthanasia have demonstrated proficiency. Acceptable methods are those that initially depress the central nervous system to ensure insensitivity to pain and techniques should minimize stress and anxiety of the animal and result in rapid unconsciousness followed by cardiac or respiratory arrest and the ultimate loss of brain function. Equipment and devices must be properly maintained and personnel adequately trained in their use. Agents that result in tissue residues cannot be used in animals intended for food unless approved by the Federal Food and Drug Administration. The slaughter of animals entering the human food chain must be accomplished in compliance with regulations of the federal Humane Methods of Slaughter Act (9 CFR. 313.1-90; CFR, 1987),[8] which provide regulations for the humane treatment of livestock before and during slaughter.

EQUIPMENT AND FACILITIES

AWA/AWRs

The requirements for facilities are contained within the sections of the standards on facilities and operating standards for the species covered by the AWA. General requirements for housing facilities, as well as specific requirements for indoor, sheltered, outdoor, and mobile housing facilities are included in this section. The specific standards address the construction of the structure, surfaces, water and electric power, storage, and drainage and waste removal. For indoor and sheltered housing facilities, there are requirements for heating, cooling, ventilation, temperature controls, lighting, and interior surfaces. There are requirements for shelter from the elements for sheltered and outdoor facilities. There are specific space requirements for the primary enclosures of the species covered by the regulations. There are no specific specifications for equipment other than the space requirements for primary enclosures.

PHS Policy and OLAW

The PHS Policy defines an animal facility as "*Any and all buildings, rooms, areas, enclosures, or vehicles, including satellite facilities, used for animal confinement, transport, maintenance, breeding, or experiments inclusive of surgical manipulation. In Section III.B of the Policy, a satellite facility is any containment outside of a core facility or centrally designated or managed area in which animals are housed for more than 24 h.*" The PHS Policy and OLAW do not specifically address equipment or facility characteristics.

Guide

Chapter five of the *Guide* discusses equipment and facilities in detail providing guidance and expectations on construction guidelines for animal housing (corridors, floors, drainage, walls and ceilings, heating, ventilation, air conditioning, power and lighting, storage areas, noise control, vibration control, facilities for sanitizing materials, and environmental monitoring) and support space such as surgical facilities; security and access control; hazardous agent containment; facilities for imaging and whole body irradiation, barrier housing, behavioral studies, and aquatic species housing.

The *Guide* defines the major functional areas of an animal facility as: animal housing and care; receipt, quarantine, separation, and/or rederivation of animals; separation of species or isolation of individual projects when necessary; storage; and specialized laboratories/areas for surgery, intensive care, necropsy, irradiation, diet preparation, experimental procedures, behavioral testing, imaging, clinical treatment and diagnostic laboratory procedures. Other areas often required include containment facilities and equipment; barrier facilities; receiving and storage of equipment and supplies; areas for washing and sterilizing equipment; waste storage; cold storage; space for personnel areas such as offices, training areas, showers, lockers, toilets, and break rooms; and areas for maintenance and repair of specialized equipment. Personnel spaces should be separate from animal areas.

Corridors should be sufficiently wide to facilitate the movement of personnel and equipment and should be designed to be durable and withstand regular cleaning and disinfection. Floor wall junctions should also be designed to facilitate cleaning. Corridors should provide sound attenuation by the location of double door vestibules which isolate the noisy areas of the facility. Animal room doors should be of appropriate size to allow easy passage of equipment and doors should fit tightly within their frames and be self-closing. Doors should open into animal rooms. Door security is important at points of ingress to the animal facility and holding areas in which access should be limited. Windows in animal holding areas may provide benefit, but concerns related to control of photoperiod, temperature variation, and security should be considered. Floors in the animal facility should be "*moisture resistant, nonabsorbent, impact resistant, relatively smooth, although textured surfaces may be required in some high moisture areas and for some species (e.g. farm animals).*" Floors should be durable and able to withstand regular cleaning with chemical agents, resist cracking, gouging, or pitting to promote proper cleaning.

A key feature of the animal facility is the HVAC. The *Guide* states that, "*A properly designed and functioning HVAC system is essential to provide environmental and space pressurization control.*" Proper temperature and humidity control are critical to maintain a suitable environment for animals and minimizing environmental variability for the conduct of quality science. Adequate ventilation is essential to remove waste gases and heat associated with animal housing areas. Guidelines for ventilation rates and temperature and humidity settings are provided for several species. Directional airflow is critical in several areas of the animal facility for the protection of both animals and personnel (e.g. containment areas, surgery, necropsy, barriers, isolation, washrooms, and quarantine).

Specific design features of modern HVAC system are described with recommendations for variable air volume systems, safeguards such as reheat coils that fail in the closed position, and automated alarm systems to notify appropriate personnel which malfunctions or environmental parameters are out of normal ranges. The HVAC system should be designed with sufficient redundancy or back-up systems so that minimal function is maintained should an overall malfunction occur. An automated system to monitor temperature and humidity at the room level, and which notifies personnel of abnormalities/malfunction, should be in place. Similar systems for monitoring the lighting and photoperiod are also recommended. A source of emergency power should be available to provide function to critical areas of the facility during a power failure.

Overall, facilities should be designed and operated to minimize vibration and noise. Facilities for sanitation of cages and equipment should be properly designed and operated with consideration of the following criteria:

- location with respect to animal rooms and waste disposal and storage areas;
- ease of access, including doors of sufficient width to facilitate movement of equipment;
- sufficient space for staging and maneuvering of equipment;
- soiled waste disposal and prewashing activities;
- ease of cleaning and disinfection of the area;
- traffic flow that separates animals and equipment moving between clean and soiled areas;
- air pressurization between partitioned spaces to reduce the potential of cross-contamination between soiled and clean equipment;
- insulation of walls and ceilings where necessary;
- sound attenuation;
- utilities, such as hot and cold water, steam, floor drains, and electric power;
- ventilation, including installation of vents or canopies and provisions for dissipation of steam and fumes from sanitizing processes;
- vibration, especially if animals are housed directly above, below, or adjacent to the washing facility;
- personnel safety, by ensuring that safety showers, eyewash stations, and other equipment are provided as required by code; exposed hot water and steam lines are properly insulated; procedures with a propensity to generate aerosols are appropriately contained; and equipment, such as cage/rack washers, and bulk sterilizers, which personnel enter, are equipped with functioning safety devices that prevent staff from becoming trapped inside.

The *Guide* also provides specific recommendations for specialized areas such as surgery facilities. Surgical facilities should contain at least five separate areas: surgical support; animal preparation; surgeon scrub; the operating room; and postoperative recovery. Also specific recommendations are provided for the design and construction of: barrier facilities; imagining areas; whole body irradiation; hazardous agent containment; behavioral studies; and aquatic species housing. Recommendations for animal facility security and access control are also provided.

Ag Guide

Guidelines regarding equipment and facilities are described in the previous section, "Housing and Enrichment".

MISCELLANEOUS

AWA/AWRs

The AWR contain a section entitled miscellaneous which includes requirements for furnishing business information, access to and inspection of records and property, publication of the names of licensed and registered facilities, inspection for missing animals, confiscation and destruction of animals, handling, identification of dogs and cats, health certificates, holding of animals, and holding period and compliance with standards and prohibitions. In addition, there are subparts of the regulations on the scope and applicability of the rules of practices including summary actions and stipulations.

REFERENCES

1. AWA. Animal Welfare Act of 1966 (Pub. L. 89-544) and subsequent amendments 1966. U.S. Code. Vol. 7, Secs. 2131-2157 et seq.
2. PHS Policy. *Public health service policy on humane care and use of laboratory animals*. Office of Laboratory Animal Welfare, National Institutes of Health, Public Health Service; 2002.
3. Health Research Extension Act of 1985 (Pub. L. 99-158), November 20, 1985 "Animals in Research."
4. U.S. Government Principles. OER Home Page – Grants Web, Office of Laboratory Animal Welfare: PHS Policy on Humane Care and Use of Laboratory Animals. Available at: http://grants.nih.gov/grants/olaw/references/phspol.htm#USGovPrinciples.
5. National Academy of Sciences (NAS). *Guide for the care and use of laboratory animals*. Washington, D.C: National Academy Press; 2011.
6. FASS. *Guide for the care and use of agricultural animals in research and teaching*. 3rd ed. Federation of Animal Science Societies (FASS); January 2010.
7. AVMA 2013. American Veterinary Medical Association guidelines for the euthanasia of animals. 2013 Edition. Available at: https://www.avma.org/KB/Policies/Documents/euthanasia.pdf.
8. Humane Methods of Livestock Slaughter Act of 1958 (Pub. L. 85-765) and subsequent amendments 1978. 7 U.S. C. 1901 et seq.
9. Horse Protection Act of 1970 (Pub. L. 91-540) and subsequent amendments 1976. (15 U.S.C. 1821 et seq.)
10. Twenty-Eight Hour Law. 1873 (49 USC, Section 80502) and later amendments 1994.
11. National Academy of Sciences (NAS). *Guidelines for the humane transportation of research animals*. Washington D.C: National Academy Press; 2006.
12. Sikes RS, Gannon WL, the Animal Care and Use Committee of the American Society of Mammalogists. Guidelines of the American Society of Mammalogists for the use of wild mammals in research. *J Mammol* 2011;**92**(1):235–53.

13. Gaunt AS, Oring LW. In: Fair JM, Paul E, Jones J, editors. *Guidelines to the use of wild birds in research*. 3rd ed. The Ornithological Council; 2010.

14. National Academy of Sciences (NAS). *The psychological well-being of nonhuman primates*. Washington, DC: National Academy Press; 1998.

15. Workman P, Aboagye EO, Balkwill F, et al. Guidelines for the welfare and use of animals in cancer research. *Br J Cancer* 2010;**102**:1555–77.

16. National Academy of Sciences (NAS). *Guidelines for the care and use of mammals in neuroscience and behavioral research*. Washington DC: National Academy Press; 2003.

17. American Society of Ichthyologists and Herpetologists 1987. *Guidelines for the use of live amphibians and reptiles in field research*. American Society of Ichthyologists and Herpetologists. Available at: http://www.asih.org/files/hacc-final.pdf.

18. AWIC 2006. *Environmental enrichment for nonhuman primate resource guide*. Series No. 32 United States Department of Agriculture, National Agricultural Library, Animal Welfare Information Center. Available at: http://www.nal.usda.gov/awic/pubs/Primates2009/primates.shtml#ack.

19. United States Department of Agriculture, Animal Plant Health Inspection Service. Animal Welfare Regulations. Code of Federal Regulations, Title 9, vol. 1. 2008. Available at: http://www.aphis.usda.gov/animal_welfare/awr.shtml.

20. Council of Europe 2006. Appendix a of the European convention for the protection of vertebrate animals used for experimental and other scientific purposes (ETS No. 123). Guidelines for accommodation and care of animals (Article 5 of the convention). Approved by the Multilateral consultation. Cons 123 (2006) 3.

21. NRC 2002. *Animal biotechnology: science-based concerns*. Washington, DC: National Academies Press. Available at: http://www.nap.edu/openbook.php?isbn=0309084393&page=R1.

22. FDA. Guidance for Industry 187 Regulation of Genetically Engineered Animals Containing Heritable Recombinant DNA Constructs 2009. Federal Register Volume 74, Number 11.

23. OLAW 2006–2013. National Institutes of Health, Office of Extramural Research. Frequently asked Questions. A6 – Does the PHS Policy apply to animal research that is conducted in the field? Available at: http://grants.nih.gov/grants/olaw/faqs.htm#App_6.

24. OLAW 2006–2013. National Institutes of Health, Office of Extramural Research. Frequently asked Questions. G2 – What is required for an occupational health and safety program? Available at: http://grants.nih.gov/grants/olaw/faqs.htm#instresp_2.

25. DHHS 2009. In: Chosewood LC, Wilson DE, editors. *Biosafety in microbiological and biomedical laboratories*. 5th ed. Department of Health and Human Services, Public Health Service Center for Disease Control and Prevention, National Institutes of Health; CDC 21-1112. Available at: http://www.cdc.gov/biosafety/publications/bmbl5/BMBL.pdf.

26. NRC. *Occupational health and safety in the care and use of research animals*. National Academies Press; 1997.

27. OLAW 2006–2013. National Institutes of Health, Office of Extramural Research. Frequently asked Questions. F2 – What are the institution's responsibilities in ensuring that animals are shipped safely and in reporting adverse events that occur in shipment of animals to or from the institution? Available at: http://grants.nih.gov/grants/olaw/faqs.htm#useandmgmt_12.

28. OLAW 2006–2013. National Institutes of Health, Office of Extramural Research. Frequently asked Questions. F17-May performance standards determine environmental enrichment issues? Available at: http://grants.nih.gov/grants/olaw/faqs.htm#useandmgmt_12.

29. OLAW 2006–2013. National Institutes of Health, Office of Extramural Research. Frequently asked Questions. F14 – Is social housing required for nonhuman primates when housed in a research setting? Available at: http://grants.nih.gov/grants/olaw/faqs.htm#useandmgmt_12.

30. OLAW 2006–2013. National Institutes of Health, Office of Extramural Research. Frequently asked Questions. F6 – How can institutions assure animal welfare when HVAC systems malfunction or fail? Available at: http://grants.nih.gov/grants/olaw/faqs.htm#useandmgmt_12.

31. ACLAM 2013. *American College of Laboratory Animal Medicine position statement on adequate veterinary care*. Available at: http://www.aclam.org/Content/files/files/Public/Active/position_adeqvetcare.pdf.

32. USDA 2000. *Information resources for institutional animal care and use committees 1985–1999*. AWIC Resources Series No.7. United States Department of Agriculture, Agricultural Research Service, National Agricultural Library animal Welfare Information Center. Available at: http://www.nal.usda.gov/awic/pubs/IACUC/iacuc.htm.

33. OLAW 2006–2013. National Institutes of Health, Office of Extramural Research. Frequently asked Questions. F1 – Is the use of carbon dioxide an acceptable euthanasia agent? Available at: http://grants.nih.gov/grants/olaw/faqs.htm#useandmgmt_12.

Chapter 3

Canada's Oversight of Animal Use in Science

Gilly Griffin and Michael Baar

Canadian Council on Animal Care, Ottawa, ON, Canada

Chapter Outline

GENERAL FRAMEWORK

The oversight of animals used in science has evolved in a somewhat different manner in Canada than in other countries. In 1961, a committee of the Canadian Federation of Biological Societies published the first Canadian standards as *"Guiding Principles on the Care of Experimental Animals."*[1] A few years later, in 1966, the then Medical Research Council (MRC) and the National Research Council (NRC) undertook a study to determine how the implementation of national standards should be overseen in Canada. The subsequent report recommended the formation of a Canadian Council on Animal Care (CCAC), to provide standards and a quality assurance program for all aspects of the care and use of experimental animals.

Laboratory Animals. http://dx.doi.org/10.1016/B978-0-12-397856-1.00003-9

The Canadian Council on Animal Care (CCAC) was officially founded in 1968 as a standing committee of the Association of Universities and Colleges of Canada, the national organization representing Canadian universities. In line with recommendations made by the MRC/NRC in the report, the Canadian program developed as a peer review system, selected to draw on the strengths of many individuals and organizations to reach its goals, and continues to operate through the involvement of over 2000 volunteers and 22 member organizations on Council drawn from federal granting agencies; federal government departments/agencies using animals or supporting animal-based research; national charitable organizations funding animal-based research; institutionally-based national academic associations; national scientific and academic associations; national organizations representing pharmaceutical companies; and national organizations representing animal welfare and animal care. The Canadian Federation of Humane Societies (CFHS) was included from the outset as a representative of the animal welfare movement in Canada.

Since 1968, the CCAC Program has brought about high standards for experimental animal care and use through the development of relevant standards and a code of ethics, and through education and the certification of institutional programs. The *CCAC policy statement on: the ethics of animal investigation*,[2] is the fundamental policy statement for the CCAC. In addition, other policy statements establish the ground rules and basic requirements for each institutional program within the Canadian system of oversight of animal care and use. Prior to 1997, CCAC guidance for the care and use of experimental animals consisted of two volumes of the CCAC Guide to the Care and Use of Experimental Animals.[3,4]

Subsequently, and in particular with the advent of electronic publication, general and specific guidelines documents have been written, revised, and expanded in response to changes in scientific and societal attitudes to the use of animals in research, teaching, and testing (http://www.ccac.ca). This approach permits guidelines to be updated on a more regular basis where needed, recognizing that some areas of animal care and use change more rapidly than others. The full list of policy statements and guideline documents which support the implementation of the Canadian system of oversight can be found on the Standards and Guidance section of the CCAC website (http://www.ccac.ca/en_/standards).

Each institution wishing to participate in the CCAC Program must have at least one active and effective animal care committee (ACC), responsible for ensuring ethical animal use and optimal levels of animal care. The CCAC oversees the work of over 220 local ACCs in 184 institutions across Canada.

ACCs, composed of researchers, teachers, veterinarians, animal care technicians, community representatives, students, animal facility managers, and support staff, must ensure that every reasonable safeguard is in place to minimize pain and distress. (As described in the *CCAC policy statement on: terms of reference for animal care committees*[5]).

ACCs typically report to and derive their authority from the senior administrator responsible for animal care and use at the institution (i.e. Vice-President of Research for most academic institutions) and have a number of important responsibilities including animal use protocol and standard operating procedure (SOP) review, policy development, postapproval monitoring, and facility inspection. Most ACCs spend the majority of their time on animal use protocol review and approval, with the largest institutions in the program reviewing several hundred protocols a year.

Protocol review and approval is based on the principles of the Three Rs: Replacement, Reduction, and Refinement. Before a protocol is reviewed for its ethical merit, it has to undergo review for scientific, pedagogical, or regulatory merit. ACCs base their decisions on CCAC standards, including guidelines and policies, as well as on professional judgment and common sense. Perspectives from all ACC members are taken into account, including community representatives. Decisions are made on a consensus basis to avoid marginalizing non-animal users.

Since 1998, the CCAC has been awarding its Certificate of GAP-Good Animal Practice® to institutions that meet its national animal care and use standards. The certificate is a prerequisite to receiving research funding from Canadian and American granting agencies, from charitable organizations that fund animal-based research, and is a requirement for institutions wishing to import certain species of research animals into Canada.

The CCAC is managed by a Council whose members represent 22 national organizations with an interest in the use of animals in science. Council member representatives include researchers, ethicists, veterinarians, teachers, animal care technicians, and animal welfarists. Council representatives also sit on one of five standing committees: Guidelines, the Three Rs, Education, Training and Communication, Assessment and Certification, or Planning and Finance. Each standing committee functions according to Standing Rules of the CCAC.

The Assessment and Certification Committee (AACC) oversees the work of the Assessment and Certification sector (full time staff that includes Assessment Directors and support staff) and is the peer group that awards certificates to deserving institutions. The AACC includes at least two representatives from the Canadian Federation of Humane Societies. The Guidelines Committee is responsible for overseeing the work of the Guidelines Program (including Secretariat staff) in developing and revising all CCAC guidelines documents.

A legal opinion commissioned by the CCAC in 1998, Legislative Jurisdiction over Animals Used in Research, Teaching, and Testing[6] and an independent study commissioned by Health Canada in 2000, The Protection of Animals Used for the Purpose of Xenotransplantation in Canada,[7] both reach the conclusion that under Canada's Constitution Act 1867,[8] the federal government does not have jurisdiction to legislate with respect to experiments involving animals, as this is a provincial jurisdiction. However, there are three areas in which the federal government has taken action in this regard. Firstly, Section 446 and 447 of the Criminal Code of Canada[9] protects animals (in general) from cruelty, abuse, and neglect. This section of the Criminal Code has been under review for several years. Secondly, the Health of Animals Act (1990)[10] and its regulations primarily aim to protect Canadian livestock from a variety of infectious diseases that would threaten both the health of the animals and people, and Canadian trade in livestock with other countries. This act is used both to deal with named disease outbreaks in Canada, and to prevent the entry of unacceptable diseases that do not exist in Canada. Lastly, the other mechanism through which the federal government has lent its support to the humane treatment of animals while not strictly speaking legislative in nature, is one of the most powerful instruments available to the federal government for setting national standards. The federal government makes the awarding of research grants through the two main granting agencies contingent on an institution holding a CCAC Certificate of GAP-Good Animal Practice®. Where the government itself awards a contract on an academic or nonacademic institution, clause A9015C of Public Works Standard Acquisition Clauses and Conditions Manual[11] imposes conditions related to the care and use of experimental animals in public works and government services.

All Canadian provinces have legislated in the area of animal welfare in some form or another, however, only certain provinces have specifically occupied the field of animals acquired and used for research, teaching, and testing purposes. These are Alberta, Manitoba, Saskatchewan, Ontario, New Brunswick, NewFoundland and Labrador, Nova Scotia, Québec, and Prince Edward Island. In general, these provincial Acts require compliance with CCAC standards, and list them in the regulations. Explanation of the details of each Act can be found on the CCAC website: http://www.ccac.ca/en_/education/niaut/stream/cs-guidelines#alberta. New Brunswick has a slightly different approach than the other provinces in that no one may be found guilty of offenses such as withholding food or water, or failing to provide adequate shelter and care, as long as he or she has complied with the CCAC *Guide to the Care and Use of Experimental Animals*. Ontario is the only province which has its own *Animals for Research Act,*[12] and is unique in Canada in that it creates a system of control based on the registration of research facilities and the issuance of licenses for supply facilities. Specifically, it applies to premises where animals are used in research and includes premises used for the collecting, assemblying or maintaining of animals in connection with a research facility, it also covers supply facilities. Among the provisions of the *Animals for Research Act*, one should note the duty to establish an ACC, the responsibilities and powers of which are similar to those required under the CCAC system, and the requirement for any operator of a research facility to submit to the person designated by the Ontario Minister of Agriculture, Food, and Rural Affairs (OMAFRA) a report respecting the animals used in the research facility for research. The *Regulation 24—Research Facilities and Supply Facilities* also provides minimum standards for the housing and care of animals and *Regulation 25—Transportation*, prescribes the conditions for transporting animals used or intended for use by a research facility.

One of the general concerns about the seemingly "voluntary" participation by institutions in the national CCAC program has been that some private and provincial government units have not subscribed to the CCAC program, and thus may not have external assessment of the care and use of animals. Many private and some provincial government units have embraced the CCAC program, recognizing the scientific and public relations benefits that a monitored animal care and use program brings. In an attempt to provide a more universal oversight of animals in research, eight of the nine provinces listed above amended the regulations to their respective legislation to make reference to CCAC standards. Although we believe it to be unlikely, there is the possibility that a private company, or a provincial government laboratory operating within a province where there is no requirement for adherence to CCAC standards could carry out research or testing without any oversight; i.e. British Columbia, Nunavut, or the Northwest Territories.

THE PRINCIPLES

The overarching policy on the scientific use of animals in Canada is the *CCAC policy statement on: the ethics of animal investigation,* which sets the principles for Canadian animal-based science. It states, *"The use of animals in research, teaching, and testing is acceptable ONLY if it promises to contribute to understanding of fundamental biological principles, or to the development of knowledge that can reasonably be expected to benefit humans or animals."*[12] Like most national systems of oversight of animal experimentation, the CCAC incorporates the principles of the Three Rs[13] in its overarching policy which also states, *"Animals should be used only if the researcher's best efforts to find an alternative have failed. A continuing sharing of knowledge, review of the literature, and adherence to the Russell-Burch "3R" tenet of "Replacement, Reduction, and Refinement" are also requisites. Those using animals should employ the most humane methods on the smallest number of appropriate animals required to obtain valid information."*[2] The policy also sets requirements for meeting animals' physical and psychological requirements, and points to restrictions for various types of procedures (burns, freezing, fractures, trauma, staged predator/prey encounters). It does not permit certain types of procedures, for example, the use of neuromuscular blockers without anesthetics, studies where death is the endpoint, and painful experiments or serial invasive procedures conducted solely for the purpose of instruction of students.

In 2009, the CCAC also established a Three Rs Program.[14] The program has two main functions: (1) to promote the principles of the Three Rs; and (2) to support the implementation of the Three Rs. The first function is largely addressed through the CCAC Three Rs Microsite: http://3rs.ccac.ca/. The Microsite enables the CCAC to post information which goes beyond requirements outlined in the CCAC guidelines. The CCAC considers this to be a means of clearly distinguishing between requirements, which may form the basis of recommendations in CCAC assessment reports, and best practices which support improvements in animal welfare, or which lead to efficiencies in animal use, or perhaps a different research strategy including the use of nonanimal methods. The program itself carries out research, and builds tools to support the second function of implementation of the Three Rs. To date, these include a Three Rs Search Guide, papers on the extent of analgesia withholding,[15] the importance of synthesis of evidence in animal-based studies,[16] and an alternatives methods table, all of which can be found on the Three Rs Microsite.

SCOPE/APPLICABILITY

The CCAC program extends to any institutions using animals for scientific purposes. This includes academic institutions where institutions sign an *Agreement on the Administration of Agency Grants and Awards by Research Institutions*, which includes the requirement for an institution to hold a CCAC Certificate of GAP-Good Animal Practice® in order to be eligible for Canadian granting agency funds to carry out animal-based research (http://www.nserc-crsng.gc.ca/NSERC-CRSNG/Policies-Politiques/MOURoles-ProtocolRoles/3-ResearchwithAnimals-RechercheavecAnimaux_eng.asp). It also includes government laboratories: both federal and provincial, and private companies. For government laboratories, the CCAC has signed a Memoranda of Understanding with the science-based departments and agencies, to ensure they meet CCAC standards and hold a valid CCAC Certificate of Good Animal Practice®. In the case of private companies, there is no legal or policy requirement for participation in the CCAC program, and hence these institutions could in practice operate without any oversight, except for inspection by provincial Societies for Prevention of Cruelty to Animals (SPCAs), if there is thought to be grounds to suspect cruel treatment. However, there is recognition of the importance of public accountability for the use of animals in science. The vast majority of these organizations choose to participate in the CCAC program and use the CCAC Certificate of GAP-Good Animal Practice® as a tangible demonstration of their commitment to high standards for the ethical use of animals. In this respect, institutions holding a CCAC Certificate of GAP-Good Animal Practice® may choose to have their names listed on the CCAC website http://www.ccac.ca/en_/assessment/certification/holders_list.

The CCAC program does not cover the use of animals in primary or secondary school science classes, although Canada's Youth Science Foundation does reference CCAC guidelines (http://www.youthscience.ca).

The CCAC program extends to any use of animals (vertebrates and cephalopods) in science by participating institutions. All vertebrates and cephalopods used for research, teaching or testing, or for display purposes or eventual use in research, teaching or testing must be the subject of a written animal use protocol to be approved by the institutional ACC, and therefore fall under the purview of the CCAC.[17]

Animal use is categorized into six main purposes, described in Appendix D of the CCAC Interpretation Bulletin—Purpose of Animal Use (PAU)[17] and includes:

PAU 0—Animals held in breeding colonies or on holding protocols (e.g. fish, rodents, farm animals) that have not been assigned to a particular research, teaching, or testing protocol;

PAU 1—Studies of a fundamental nature in sciences relating to essential structure or function (e.g. biology, psychology, biochemistry, pharmacology, physiology, etc.), including basic biomedical, biological, or agricultural research;

PAU 2—Studies for medical purposes, including veterinary medicine, that relate to human or animal diseases or disorders, including applied research to develop therapies;

PAU 3—Studies for regulatory testing of products for the protection of humans, animals, or the environment, including vaccine efficacy trials and testing of new therapeutic compounds for new drug submissions;

PAU 4—Studies for the development of products or appliances for human or veterinary medicine; and

PAU 5—Education and training of individuals in postsecondary institutions or facilities.

As protocol review is the responsibility of the institutional animal care committees, the CCAC provides guidance on the level of invasiveness of protocols, with a view to indicate which activities require a review by an animal care committee, as well as activities that should be reported to the CCAC as part of the annual submission of animal use data.[18] Categories of invasiveness are assigned prospectively, taking a precautionary approach; ie, considering the potential greatest degree of pain and distress likely to be experienced by the animals. There are five categories of invasiveness A–E, with A being the least severe and level E the most severe:

A. most invertebrates or live isolates
B. little or no discomfort or stress
C. minor stress or pain of short duration
D. moderate to severe distress or discomfort
E. severe pain near, at, or above the pain tolerance threshold of unanesthetized conscious animals.

Category A protocols involve experiments on most invertebrates or on live isolates or observation of wildlife. Possible examples include: the use of tissue culture and tissues obtained at necropsy or from the slaughterhouse; the use of eggs, protozoa, or other single-celled organisms; experiments involving containment, incision, or other invasive procedures on metazoan; or observation of migratory routes. The number of animals involved in these types of protocols do not have to be recorded, or reported to the CCAC; however, they do permit ACCs to reflect on relative replacement—i.e. whether investigators are moving toward replacing animals of greater sentiency such as vertebrates, with animals that current scientific evidence indicates have a significantly lower potential for pain perception, such as some invertebrates.[19] Conversely, permitting Category E protocols to be considered ensures that ACCs discuss, reject, or approve protocol procedures which are deemed to be at the limits of ethical acceptability, based on scientific necessity. Some institutions will refuse to permit any E level protocols to be carried out.

In Ontario, the Animals for Research Act[12] extends to all nonhuman vertebrate animals.

AUTHORIZATION OF USER-BREEDING INSTITUTIONS/ INSPECTION/PENALTIES BY COMPETENT AUTHORITIES

The CCAC works at the institutional level, as it is the institution itself which receives the CCAC Certificate of GAP-Good Animal Practice®. Institutions that wish to obtain a CCAC Certificate must have the following elements in place:

- an active ACC, whose composition, authority, responsibilities, and functioning are defined in written terms of reference based on the *CCAC policy statement on: terms of reference for animal care committees*[5];
- an animal use protocol form (or forms), according to the guidance contained in the *CCAC policy statement on: terms of reference for animal care committees;*
- complete protocols submitted by animal users for all existing and planned animal-based work, with the ACC having reviewed and made decisions on all ongoing protocols and any protocols planned for the near future. The protocol reviews should be based on relevant guidance and documented in the minutes of one or more ACC meetings;
- an effective postapproval monitoring program;
- a formal agreement or agreements for veterinary services, based on the main elements of the *Canadian Association for Laboratory Animal Medicine (CALAM) Standards of Veterinary Care;*[20]
- trained, qualified personnel in sufficient numbers to care for all groups of animals seven days a week;
- programs based on CCAC guidance for:
  - the training of animal users
  - occupational health and safety, to cover all animal project related risks
  - crisis management;
- where animal facilities are needed, they must either meet relevant CCAC guidance or a detailed plan with timelines must be in place describing how they will be improved to meet the guidance; and

- the ACC must have visited all facilities for animal care or use, whether currently in use or designated for use, and must have approved of their use in one or more written site visit report(s).

CCAC assessment panels visit institutions at least once every three years to ensure that:

- the overall structure of the animal care and use program is sound, effective, and well supported, and that it meets the specific needs of the institution with regard to the scope and complexity of the program;
- their ACC remains active and functional, meeting at least twice every year, visiting all animal facilities at least once every year and fulfilling all of the responsibilities described in the *CCAC policy statement on: terms of reference for animal care committees*,[5] including post-approval monitoring of animal care and use;
- their veterinary and animal care services continue to meet institutional needs and CCAC standards;
- their training, occupational health, and safety and crisis management programs are relevant, complete, up-to-date and in line with CCAC guidance; and
- their facilities (if required) meet institutional needs and CCAC standards.

In Ontario, under the Animals for Research Act,[12] all research premises, and supply premises are required to be licensed. Those responsible for a registered research facility or facilities are also required to establish an ACC, one of the members of which shall be a veterinarian.

Participating institutions in the CCAC Program are required to submit animal use data to the CCAC on an annual basis. Numbers of animals used are categorized according to species, purpose of use (6 categories), and level of invasiveness (5 levels, A–E), as described above. This permits the CCAC to publish reasonably comprehensive pooled data on an annual basis. Annual statistics, along with notes providing some analysis of the data, are made available on the CCAC website within one calendar year of collection http://www.ccac.ca/en_/publications/audf. Complete data has been published since 1996, with a more limited data set published in the previous years beginning in 1976.

Research facilities in Ontario are required to submit a separate annual report to the OMAFRA which includes (1) the total number of every species of animal used for research in the research facility in the year; (2) the total number of dogs and the total number of cats purchased or otherwise acquired from, (a) other research facilities, (b) pounds, (c) supply facilities, and (d) other sources; and (3) the total number of dogs and the total number of cats that in any experiment or surgical procedure did not recover from anesthesia. R.R.O. 1990, Reg. 22, s. 4 (1).[21]

NONHUMAN PRIMATES: SPECIAL CONSIDERATIONS AND RESTRICTIONS

Nonhuman primates continue to be used for research and testing purposes in Canada. Because of their high level of sentiency and the difficulties in meeting the psychological and behavioral needs of these complex animals, ACCs are expected to take particular care in providing an ethics review of any protocols involving their use. No great apes are currently used in Canada, and while there is no official moratorium on their use, there would have to be a considerable scientifically justified need for them to be used. As for any other animals used for scientific purposes in Canada, the CCAC guidelines for their care and maintenance focus on meeting the physical and psychological needs of the animals. Currently, the published CCAC guidance for nonhuman primates is quite out of date, as Volume 2 of the Guide to the Care and use of Experimental Animals is under review.[3] The nonhuman primate subcommittee of the CCAC Guidelines Program is currently revising Chapter XX which should be published in 2014.

In Ontario, there are special mention of nonhuman primate requirements within the Animals for Research Act.[12] These relate to ensuring that the animals are tested for tuberculosis (TB) upon arrival in the research facility and either isolated or euthanized if found to be carrying TB, unless it is a necessary element of the research study. In addition, no person who is known to have active TB may be employed in the care of nonhuman primates. There are also requirements for isolation upon arrival in order to permit the animal to become accustomed to the normal research environment. Where nonhuman primates are housed in a communal cage or pen, no more than 25 nonhuman primates are permitted to be housed in the cage or pen. The regulations to the Act also make restrictions on the use of restraint chairs, in that they may only be used to the extent necessitated by the nature of the experiment. R.R.O. 1990, Reg. 24, s. 23 (6).[22]

GENETICALLY ALTERED ANIMALS: SPECIAL CONSIDERATIONS

In 1997, the CCAC published *CCAC guidelines on: transgenic animals,*[23] recognizing that there was a rapid increase in the creation and use of these animals, and that the public was beginning to express concern about genetic modification of animals.[23] While the CCAC does not currently enumerate genetically-engineered animals separately from other animals, it is clear from the increase in Category D level of invasiveness protocols (the category required for any genetically altered animal creation protocols), and the increase in the numbers of mice and fish over the past 15 years, that the creation and use of genetically altered animals continues to expand. This was further substantiated by a bibliometric study of animal studies in the scientific literature.[24]

The 1997 guidelines recognized that "the creation and use of genetically modified animals is a rapidly evolving field of research, therefore, these guidelines will be subject to regular review."[23] The CCAC Guidelines Program's subcommittee on biotechnology is working on a revision of this document which will be published as *CCAC guidelines on: genetically engineered animals used in science* within the next few months.

INSTITUTIONAL AND DESIGNATED PERSONNEL RESPONSIBILITIES

Within each institution, the senior administrator holds the ultimate responsibility for the operation of the institution's animal care and use program. The responsibilities of the senior administrator are laid out in the *CCAC policy statement for: senior administrators responsible for animal care and use programs.*[25]

The CCAC oversees a wide variety of institutions which fall into three main categories: academic, government, and private. Despite the variety of type and size of institutions, each is required to have the following main elements.

1. At least one functional ACC, which is composed of scientists/teachers experienced in animal use; at least one veterinarian with experience/ training with regard to the species being used and the types of work; an institutional member who does not use animals in his/her work; a minimum of one community representative with no links to the institution or to animal use in science (most institutions' workloads require more than one community representative); the director/manager of the animal facilities (where there are several directors or managers, their representation on the ACC should be worked out in a manner consistent with the program structure); technical staff representation; student representation in the case of academic institutions; and the ACC coordinator, or person paid by the institution who assists the ACC in its work.
2. Competent veterinary and animal care service providers whose numbers and expertise match the nature and scope of the institutional program. This usually includes the Director of the animal facilities, a clinical veterinarian, and an appropriate number of animal health/veterinary technicians and animal care technicians.

Investigators are expected to cooperatively work with the ACC at their institutions to prepare animal use protocols according to the requirements of their institutions, following the CCAC guidelines and policies. As a prerequisite to completion of the protocol form, it is expected that they will carry out a search for potential Three Rs alternatives. The CCAC has developed a Three Rs Search Guide to assist with this requirement: http://3rs.ccac. ca/en/searches-and-animal-index/guide/. Investigators are also expected to collaborate with the institutional laboratory animal veterinarian both when designing their experiments and when carrying them out, to ensure that the animals receive appropriate housing and husbandry, and that any potential pain and distress is minimized. Investigators are required to complete both theoretical and practical training, according to the *CCAC guidelines on: institutional animal user training,*[26] prior to carrying out any animal-based studies. Responsibilities of investigators are covered in general in Chapter 1 of the *CCAC Guide to the Care and Use of Experimental Animals,* Volume 1,[4] and more specifically in many of the CCAC guidelines documents.

Under the Ontario Animals for Research Act, operators of research facilities are required to register the facility with the Ministry. The operator is also responsible for ensuring that there are an appropriate number of individuals competent in the care of animals to care properly for every animal. The operator of every research facility is also required to maintain a record of all animals in the research facility, preserving the record within the research facility for at least two years in the case of dogs and cats and one year from the date that the animals entered the research facility for other animals.

OVERSIGHT AND ETHICAL REVIEW PROCESS

The CCAC provides oversight of animal care and use programs through its Assessment and Certification program, which carries out regular visits to institutions, typically on a three-year basis. Assessment visits are usually conducted by CCAC assessment panels. When conducting a visit, a CCAC assessment panel assesses the structure and resources of the animal care and use program, the composition, functioning, and effectiveness of the ACC, and the appropriateness of animal care and use practices, procedures, and facilities. Assessments are based on CCAC guidelines, policy statements, and associated documents. Each assessment panel is composed of at least one scientist and one veterinarian, selected for their experience in animal experimentation and care relevant to the institution to be visited. Each panel also includes a community representative, selected by an assessment director from a shared CCAC—CFHS registry of potential community representatives, usually drawn from the geographical area of the institution. A CCAC assessment director is present at every assessment visit as an ex officio member of the assessment panel.

Assessment visits are carried out in a spirit of collaboration and education, with a focus on encouraging institutions to improve their animal care and use programs to ensure that Canadian societal expectations for the ethical use of animals and high standards of animal welfare are met.

Following an assessment visit to an institution, the CCAC produces an assessment report which generally includes recommendations that must be answered satisfactorily in order for the institution to be certified. The assessment report contains general remarks on the quality of the animal care and use program and details any concerns, as evaluated by the CCAC assessment panel through the following:

- reviews of the completed CCAC *Animal Care and Use Program Review Form,* submitted to the CCAC prior to the assessment visit and other relevant documents;
- meetings with members of the ACC, members of the administration, and other personnel involved in the animal care and use program; and
- a site visit of the animal facilities.

Formal recommendations are presented based on CCAC policy statements and guidelines and other CCAC-recognized standards (see the *CCAC policy statement on: definitions of recommendations made in CCAC reports*).[27] Excellent conditions, practices, or personnel are recognized in formal commendations. The assessment report is reviewed by the assessment panel and the Assessment and Certification Committee prior to being forwarded to the senior administrator responsible for the institution's animal care and use program, usually within 10 weeks of the assessment visit (15 weeks for the largest institutions).

Certification of an institution's animal care and use program is based on the institutional response to CCAC recommendations, which must be detailed in an implementation report sent to the CCAC by the senior administrator within the time limits specified in the assessment report. Institutions are typically given three months to respond to a serious recommendation and six months to respond to a regular recommendation. Major recommendations require a prompt response from the institution, but the allowable time limits can vary.

The implementation report(s) are reviewed by the assessment panel(s) that participated in the assessment and by the Assessment and Certification Committee. If the recommendations are not completely addressed, the CCAC will request that the institution provide additional or updated information and a special follow-up visit may be conducted if necessary. Unsatisfactory institutional responses or the absence of a response can lead to a probationary certification, and eventually to certificate removal. For academic institutions this could have the effect of rendering them ineligible for granting agency funding. Details of the certification process can be found in the *CCAC policy statement on: the certification of animal care and use programs.*[28]

The Ontario for Research Act[12] ensures that institutions comply with the legislation through unannounced inspections of research and supply facilities. Currently, there is only one inspector to cover the entire province of Ontario.

The CCAC operates a devolved system of oversight, so the responsibility for the oversight of animal-based studies is carried out at the institutional level. All animal-based work must be included on an animal use protocol and submitted to an institutional ACC for ethics review, prior to commencement of the work.

Before an ACC can review a protocol, there must be evidence of scientific, pedagogical, or regulatory merit. In the case of research studies *"Information provided within the protocol review form should provide the ACC with a clear sense of the need for the experimental project, and of the relationship between the proposed experiment and the overall objective."*[29] This is generally provided by the peer review process of the main granting agencies; however, where evidence of good peer review is absent, the institutional ACC must put in place a peer review mechanism with appropriate independence and expertise, the *CCAC policy statement for: Senior Administrators responsible for animal care and use programs*[25] requires *"that the institution constitute a pool of reviewers with expertise in the fields in which the members of the institution work. The various departments and researchers may be called upon to assist in identifying reviewers, including as many external scientists as possible, especially in the case of smaller institutions. Two independent reviewers at least one of whom must be external to the ACC can then be selected for each project. The reviews should be provided (without identifying the reviewers) to the author of the protocol and any questions or concerns addressed before the relevant information is forwarded to the ACC."* The new *CCAC policy statement on: scientific merit and ethical review of animal based research* provides further guidance.[30] For the use of animals in teaching, the pedagogical merit for the use of animals should be established, and for testing projects, an indication that the testing has been planned according to the most current regulatory requirements, using guidelines acceptable to the Canadian regulatory agency(ies) and which meet the requirements of the *CCAC policy statement on: ethics of animal investigation;*[2] that the planned animal use not exceed the requirements of the regulatory authorities—if it does, justification for the additional animal use must be provided. The elements that must be included in an animal use protocol are listed in the *CCAC guidelines on: animal use protocol review*[29] and the *CCAC policy statement on: terms of reference for animal care committees.*[5] Animal use protocols should include: the project title and descriptive procedures; name(s) of the principal investigator and all personnel who will be handling the animals, along with their training qualifications; department affiliation; proposed start and end dates; funding sources; an indication of scientific or pedagogical merit, or requirement for regulatory purposes; lay summary; indication of the use of any hazardous materials, with institutional approvals; categories of invasiveness (as defined by CCAC); evidence of addressing the Three Rs—Replacement, description of why sentient animals must be used, including search for alternatives, Reduction, justification for the numbers and species of animals selected, Refinement, description of anesthesia, analgesia, medical treatments, housing husbandry, improvements to procedures, improvements to the duration of the experiment and holding times; description and time course of procedures (flow diagram where possible); description of endpoint; description of capture, etc. for field studies; method of euthanasia; ultimate fate of the animals; and any other relevant information, such as relevant information from studies previously carried out. Decisions concerning the acceptability of the protocol are generally taken by consensus of the ACC members.

Once a protocol has been approved by the ACC, the investigator is permitted to proceed with his/her research, teaching, or testing studies. He/she is expected to report back to the ACC on an annual basis with a progress report and requesting any amendments to the protocol. ACCs are also required to implement an effective post-approval monitoring program, designed to meet the specific needs of the institution, to include the following elements:

1. day to day collegial work of the veterinarian(s) and animal care staff to help animal users remain compliant with approved protocols and institutional and CCAC standards, focusing more specifically on ensuring that: (a) individual animal users are comfortable handling animals and carrying out procedures successfully, and that they are able to do so in appropriate conditions; (b) endpoints are applied as approved by the ACC to avoid unnecessary distress to the animals;
2. availability of the ACC coordinator or other ACC members to assist animal users with their work, and to facilitate the process of amending a protocol when it cannot be successfully continued in practice as originally approved for technical or logistical reasons;
3. site visits and discussion of protocols with animal use teams by ACC members or other colleagues, to address concerns through good communication; and
4. careful assistance and follow-up for new procedures and for procedures more likely to result in animal pain or distress.

The Ontario Animals for Research Act also requires that every person or body of persons having control of a registered research facility or facilities establish an ACC, one of the members of which must be a veterinarian. R.S.O. 1990, c. A.22, s. 17 (1).[12] The ACC responsibilities under this Act include the coordination and review of (a) the activities and procedures relating to the care of animals; (b) the standards of care and facilities for animals; (c) the training and qualifications of personnel that are engaged in the care of animals; and (d) procedures for the prevention of unnecessary pain including the use of anesthetics and analgesics.

REUSE

Reuse of animals is permitted within the Canadian system. Specifically, guidelines for reuse are laid out in the *CCAC guidelines on: procurement of animals used in science.*[31] When considering the procurement of animals from another scientific institution or from another animal user within the same institution, a request should be made for documentation detailing the original source of the animal and the history of the animal while in captivity (e.g. conditioning, housing, nutrition, previous use in research, etc.). For transfers between institutions, a health certificate for the animals should be provided, and the animal care services at the receiving institution should be notified of the transfer and be given a copy of the health certificate. These animals should only be procured if they are suited to their intended use and to the conditions under which they will be housed, and they have not been subjected to procedures that would preclude their use. In general, animals are not permitted to be reused if they have been subjected to invasive procedures. Since categories of invasiveness (CI) are precautionary and describe potential pain or distress, CIs are not the only criteria considered when approving animal reuse.

Reuse, in general, occurs more frequently in the case of larger animals, and so specific guidance is included within guidelines documents relevant to these species. For example, from the *CCAC guidelines on: the care and use of farm animals in research, teaching and testing*[32] Guidelines 16 states: "*Animals subjected to invasive surgery must not be used in additional studies, without explicit approval of the animal care committee.*" The text goes on to explain that "*Occasionally, animals that have been used for a study and have not been subjected to invasive procedures may be used for a further scientific study. As well, a second major surgery may be performed on an animal if it is a nonsurvival procedure. Minor procedures such as biopsies may be performed more than once, but only if they can be done with effective anesthesia and analgesia and do not significantly impact the well-being of the animal. Complete recovery between procedures is recommended if possible.*"

Larger species of animals are also often reused for teaching purposes,[32] so guidelines referring to these animals also make recommendations for restricting the number of times that they should be reused. For example, Section 8.2 of the *CCAC guidelines on: the care and use of farm animals,*[32] Guideline 33: Frequency-of-Use, states that "*When planning student exercises with animals, the instructor must carefully weigh the pedagogical merit of the procedure against the invasiveness of the procedure and how often it will be carried out on each animal,*" and the supporting text explains that: "*Even relatively innocuous procedures, when done repeatedly, can be harmful to animals. Special precautions need to be taken when potentially painful or distressing procedures (e.g. rectal palpation, tail or jugular bleeding, etc.) are taught. Endpoints need to be set in advance by the course instructor and approved by the ACC. Records that document animal use under approved conditions should be kept to prevent an excess of manipulations, especially of those animals that, by their natures, may be more disturbed by such repeated manipulations. As far as possible, student practical sessions should be timed to coincide with routine husbandry (e.g. worming, metabolic profiles, etc.). The local ACC must be presented with sound reasons for any given student-animal ratio. Student-animal ratios and instructor-student ratios must be such that there is adequate supervision and monitoring of student performance and of animal use and discomfort levels.*"… "*Animals should not be maintained indefinitely for teaching purposes. There should be an established length of time and/or number of training procedures that an animal is involved in before disposition of the animal and replacement. This will include consideration of a number of factors, including the level of invasiveness of the procedure and whether the animal is used for other purposes."*

SETTING FREE/REHOMING

Investigators are required to indicate the intention for the disposition of the animals at the end of the study. In general, rehoming of laboratory animals is not encouraged; however, some institutions have put in place programs which aim to rehome animals that have been used. Normally, these are companion animals (dogs and cats), which have been socialized and are able to make the transition to a home environment.

Rehoming is likely to be considered more specifically as the Chapters in Volume 2 of the CCAC Guide[3] are revised. For example, the *CCAC guidelines on: the care and use of fish in research, teaching and testing*[33] details expectations if fish are to be rehomed: "*Some institutions release healthy research fishes (not GM fishes) that are commonly accepted pet or companion species to individuals with the knowledge and ability to provide adequate care. No GM fish may be removed from research facilities to private premises. If fishes are to be released to the care of an individual as companion animals, the institution should develop an appropriate policy describing the conditions that need to be fulfilled before their release.*"

OCCUPATIONAL HEALTH AND SAFETY

The CCAC takes occupational health and safety as seriously, as those working with experimental animals risk exposure to physical hazards (e.g. heat, noise, radiation), chemical hazards (e.g. disinfectants, cleaning solutions), as well as intestinal parasites, enteric bacteria, pathogenic organisms, and animal bites.[34] Zoonotic diseases are a particular concern depending on the species. Chapter VIII of Volume 1 of the CCAC Guide[4] "Health and Safety in the Workplace" provides information about regulatory requirements, dealing with biological hazards, zoonotic diseases, working with nonhuman primates safely, allergies, physical injuries, chemical hazards and radiation, and ultraviolet light. In terms of regulatory requirements, for Canada, The Workplace Hazardous Materials Information System (WHMIS) is a national information system designed to protect Canadian workers by providing safety and health information about hazardous workplace materials. The key elements of WHMIS are cautionary labeling of containers of hazardous materials, the provision of material safety data sheets and worker education programs. The system aims to balance the worker's right to know with industry's right to protect confidential business information.

Guidelines for working with biohazards (e.g. bacteria, viruses, parasites, fungi, and other infectious agents) are provided in the Canadian Biosafety

Standards and Guidelines.[35] These guidelines include such items as biohazard containment, laboratory design, personal hygiene, and safety facilities, and can be used to provide training for employees as mandated by WHMIS, and support the implementation of the Human Pathogens and Toxins Act.[36] In addition, the Canadian Food Inspection Agency (CFIA) *Containment Standards for Facilities Handling Aquatic Animal Pathogens*[37] applies to aquatic facilities involved in research and testing.

Additional guidance related to occupational health and safety will be included in each of the CCAC species-specific guidelines, as they are revised. For example, see the *CCAC guidelines on: the care and use of farm animals in research teaching and testing.*[32]

EDUCATION, TRAINING, AND COMPETENCE OF PERSONNEL

In Canada, all personnel from institutions participating in the CCAC program and involved with the ethical use of animals in science must be competent and adequately trained in the principles of ethical use and care of animals. The CCAC ensures that appropriate institutional training programs are implemented as an essential component of any institutional animal care and use program.

Animal users (including investigators, graduate students, postdoctoral fellows and research staff, study directors), animal health professionals, ACC members, and institutional officials (senior administrators) directly responsible for animal care and use programs have different educational and training needs. Therefore, the CCAC develops different educational and reference materials to support the training needs of all these audiences, or makes materials available through external links. These resources and events include a training syllabus, training modules, and other educational materials, workshops, and web-based seminars, and are described below.

For animal users, the CCAC *guidelines on: institutional animal user training*,[26] makes it a requirement for all animal users to acquire: an adequate knowledge of the principles of experimental animal science relevant to their area of work (laboratory, field, or agriculture); the technical skills required for any procedure they must carry out; and an appreciation of the ethical issues of using animals for scientific purposes in Canada. While the CCAC provides access to training modules to support the CCAC recommended syllabus for an institutional animal user training program, institutions are expected to provide practical hands-on training as appropriate and to ensure that animal users are competent to carry out any animal-based procedures required for their research.

Animal health professionals are experts in the field of appropriate care of animals, but institutions are expected to provide continuing education opportunities as required in the *CCAC policy statement for: senior administrators responsible for animal care and use programs,*[25] the *CALAM Standards of*

Veterinary Care,[20] and the *CCAC policy statement on: terms of reference for animal care committees.*[5]

TRANSPORT

Under the *Health of Animals Act*[10] and its regulations, the CFIA is responsible for the humane transportation of animals in Canada, and details requirements for such elements as the provision of food, water, rest, protection from adverse weather, use of proper containers and transport vehicles, and segregation of incompatible animals. This Act and its regulations also specify that livestock, poultry, animal embryos, and animal semen exported from Canada must be accompanied by a health certificate issued or endorsed by a CFIA veterinary inspector, and that CFIA is responsible for testing, inspection, permit issuing, and quarantine activities for live animals imported to Canada. Guidelines specific to the transportation of animals for research purposes are included in the *CCAC guidelines on: procurement of animals used in science.*[31] These are based on recommendations made by the US National Research Council *Guidelines for the Humane Transportation of Research Animals*[38] and the UK Laboratory Animal Science Association *Guidance on the Transport of Laboratory Animals.*[39] The CCAC guidelines recommend that these two documents be consulted for more detailed information on best practices for the transportation of animals. The guidelines refer not only to transportation of animals between institutions, but also within an institution, recognizing that transportation can be an acutely stressful event for laboratory animals. Further recommendations regarding transportation are also made in the *CCAC guidelines on the care and use of farm animals in research, teaching and testing*[32] and the *CCAC guidelines on: the care and use of fish in research, teaching and testing,*[33] recognizing that the transportation of these species poses unique challenges for these species groups.

The OMAFRA has made specific regulations concerning transportation under the Ontario Animals for Research Act. These regulations include requirements for the vehicles, the shipping containers, feed, water, and duration of the transportation, as well as for personnel, both to accompany the animals and to be present to receive the animals at their destination.

HOUSING AND ENRICHMENT

The CCAC guidance for housing many of the species of animals used in science dates back to 1984.[3] For this reason, the chapters in Volume 2 of the CCAC Guide are under revision as individual *guidelines on: the care and maintenance of* (marine mammals, rats, mice, and nonhuman primates–with other species to follow). Enclosure dimensions in Volume 2 are recognized to be out of date, and so typically these are only used as bare minimum when making recommendations to institutions. There is no legal minimum cage size requirement: rather, right from the first Canadian guidelines document, requirements have always focused on giving animals sufficient space to "exercise."[1]

The CCAC has never had a strong focus on cage sizes, preferring instead to focus on appropriate housing that meets the physical and psychological needs of the animals. To support this approach, best practice information is provided on the CCAC Three Rs Microsite: http://3rs.ccac.ca/en/searches-and-animal-index/ai-animal-index/, which will be used as a starting point for the development of the species-specific care and maintenance guidelines. For species where this has already been accomplished, namely farm animals and fish,[32,33] guidance is provided on enclosure dimensions, but more specifically on the needs for the particular animals in question. So for example, where dairy cattle are to be kept in tie stalls the *CCAC guidelines on: the care and use of farm animals*[32] state in Guideline 59 "*Dairy cows kept in tie-stalls should be allowed a period of exercise every day,*" and the supporting text states that "*Cows kept in tie-stalls should be allowed a period of exercise every day unless experimental procedures or inclement weather preclude it. This also allows assessment of mobility and other health problems. Moving cows in and out of tie-stalls should be done with care and slowly so that they do not fall and injure themselves.*"

The CCAC has had a strong focus on meeting the social and behavioral needs of animals. Prior to the revision of the CCAC Guide to the Care and Use of Experimental Animals in 1993,[4] the CCAC implemented a *policy statement on: social and behavioural requirements of experimental animals,*[40] which was subsequently revised and included in Chapter VI of the CCAC Guide Volume 1.[4] Social housing is expected to be the norm for animals housed within Canadian institutions. Protocols which involve single housing must describe proposed measures for meeting the social requirements of the isolated animal (e.g. where appropriate, increased positive human contact).

During the revision of the *CCAC guidelines on: the care and use of farm animals in research, teaching and testing*[32] consideration was given to the term "environmental enrichment." Although this has become a term routinely used, the subcommittee of experts developing the guidelines was concerned that enrichment really implies that an animal's environment is already meeting its physiological and psychological needs. Historically, this has rarely been the case, so the subcommittee of experts responsible for the development of the *CCAC guidelines on: the care and use of farm animals*[32] decided that, at least within the CCAC *guidelines on: farm animals* the term *environmental improvement* should be applied to manipulations or additions to an animal's environment that address areas where the animal may otherwise experience some degree of suffering. For example, young calves are highly motivated to suck and the use of artificial teats during or just after feeding provides an appropriate outlet for this motivation. It has also been shown that providing artificial teats increases relaxation in the calf after feeding. However, where artificial teats are not provided, calves engage in abnormal behavior, such as cross-sucking among group-housed calves, which can have a negative impact on their welfare. They consider that the term *environmental enrichment* should be reserved for those improvements that provide additional benefit to the animal, but whose absence will not result in suffering. An example of environmental enrichment for cattle would be the provision of grooming devices, also known as "scratchers".

These concepts have been further explored in a recent article.[41] Moving forward, as the guidelines on the care and maintenance of the various species groups are revised (Chapters from Volume 2 of the CCAC Guide to the care and use of experimental animals[3]), emphasis will be placed on environmental improvement as the norm. Environmental enrichment, in the sense of providing positive experiences for the animals that improve their quality of life is a laudable goal; however, it is important to ensure that the enrichment strategies that are used do not interfere with scientific goals (see for example Refs 42–44).

The CCAC *guidelines on: the care and use of fish in research, teaching and testing*[33] does not include any recommendations concerning environmental enrichment for fish; however supportive information is available on the CCAC Three Rs Microsite aimed at assisting institutions to provide appropriate environments for the various fish species housed in their facilities: http://3rs.ccac.ca/en/searches-and-animal-index/ai-animal-index/fish.html.

The Ontario Animals for Research Act includes basic requirements for housing. No cage sizes etc. are given, rather the regulations to the Act make generalized statements regarding the behaviors that the animals should be able to perform "that, (a) except in the case of fish and snakes, every animal in the cage, tank or pen may comfortably, (i) extend its legs to their full extent, (ii) stand, (iii) sit, and (iv) lie down, and in the case of animals other than livestock, turn around and lie down in a fully extended position; (b) in the case of fish and snakes, every animal in the cage, tank or pen shall have adequate room for its health, welfare and comfort; (c) it is not likely to harm any animal therein; (d) every animal therein may be readily observed unless the natural habits of the animal otherwise require; (e) any animal therein cannot readily escape therefrom; (f) it minimizes as nearly as practicable the transfer of pathogenic agents; and (g) it may be readily sanitized. R.R.O. 1990, Reg. 24, s. 13."[22]

HUSBANDRY AND ENVIRONMENT

General guidelines in relation to daily care of animals in a laboratory animal facility are included in the *CCAC Guide to the Care and Use of Experimental Animals* Volume 1,[4] Chapter V—Laboratory Animal Care. More specific information on a species by species basis is provided in Volume 2,[3] this includes detailed requirements for the nutritional requirements of each species. More up-to-date information is provided through the CCAC website by direct links to the relevant sections of the CCAC Three Rs Microsite (see for example http://www.ccac.ca/en_/standards/guidelines/additional/vol2_mice). As the CCAC moves through its guidelines revision process, relevant information for the various species groups will be included within the relevant species guidelines. This has already been completed for farm animal species[32] and for fish.[33] However, in particular in the case of fish the guidelines acknowledge upfront that the greatest challenge in providing *guidelines on: the care and use of fish* is the wide variety of fish used in Canada, and the diversity of their habitats, behavior, life history, and environmental and husbandry requirements. It is also acknowledged that the scientific information required to define the preferred conditions for fish well-being is limited.

In general, the CCAC approach to developing guidelines is to provide the framework for the implementation of good animal practices, rather than stating the details of those practices, in order to allow for the evolution of refinements to animal-based procedures, etc. This approach also allows for coverage of a large range of species, even in situations where scientific information on preferred conditions (such as for fish well-being) is limited. Within the guidelines, where scientific evidence is currently lacking to support the implementation of good animal practices, efforts are made to define the most appropriate conditions, based on expert opinion, and approaches to identify those conditions are described. The CCAC provides links to additional information concerning improved practices as it becomes available, and encourages institutions to submit best practice information for peer review and publication on the CCAC website.

For general environmental parameters such as room temperature, humidity, and ventilation, basic recommendations are included in the *CCAC guidelines on: laboratory animal facilities—characteristics, design and development.*[45] More specific information relevant to individual animal species are included in the species-specific guidelines, which can be accessed on the CCAC website: http://www.ccac.ca/en_/standards/guidelines.

Similarly general requirements are included in the Regulations to the Ontario Animals for Research Act.[12]

VETERINARY CARE

In general, the CCAC refers to the *CALAM Standards of Veterinary Care*[20] as the national standard in this area. The CALAM Standards include the CALAM Position on Standards of Veterinary Care which states that a *"veterinarian with authority and responsibility for supporting an institutional animal care and use program must be involved in all issues and activities that relate in any way to animal care and use."* The types of institutions and facilities widely vary, and are recognized in the position statement which also states that *"The extent of the program of veterinary care will depend on several factors, such as the number and species of animals used and the nature of the activities that involve animal use. The structure of the veterinary program, including the number of licensed veterinarians and their background and training must be appropriate to fulfill the programs' requirements and to ensure that the CALAM Standards of Veterinary Care are met. This will vary by institution."*

The importance of having veterinary expertise available at all times is also recognized, *"In all cases, formal arrangements must be made by the senior administration of the institution to ensure that veterinary services are readily available at all times to meet both routine and emergency needs. Animal caregivers and users must be able to report an animal health or welfare concern (e.g. injury, ill health, or death) at any time, and a veterinarian must be available to respond to the concern."*

The CCAC uses the CALAM standards during its assessment visits as the basis for recommendations made in CCAC Assessment Reports to institutions. In addition, throughout CCAC guidelines documents there are recommendations relating to areas where the services of a veterinarian should be used. For example, the *CCAC guidelines on: the procurement of animals used in science*[31] indicates in Guideline 5 that *"the institutional veterinarian should have ultimate responsibility for ensuring procurement of healthy animals."* The text goes on to define these responsibilities to included, but not be limited to: *"identifying potential sources and suppliers of animals; developing or assisting in developing in-house quarantine and conditioning programs and other relevant Standard Operating Procedures (SOPs), e.g. health and safety policies; educating members of the ACC and researchers/animal users on the pros and cons of utilizing various sources, including public perception and related concerns; providing assistance to researchers in selection of animal models; and overseeing routine inspection of animal suppliers (including record keeping) when warranted (e.g. pound source dogs, private breeding colonies, etc.), and developing a good working relationship with suppliers to negotiate appropriate holding times, delivery and transportation methods, and animal selection routines."*

While the responsibilities relating to veterinary care for "traditional" laboratory animals is well-defined in the *CALAM Standards for Veterinary Care*, for some of the other species groups, the CCAC guidelines make particular reference to the responsibilities of the veterinarian. In addition, the guidelines reinforce the importance for veterinarians to receive adequate training in the relevant species. For example from the CCAC *guidelines on: the care and use of farm animals in research, teaching and testing,*[32] *"Veterinarians attending farm animals should have special training in farm animal health management in research, teaching or testing environments"* and the CCAC *guidelines on: the care and use of fish in research, teaching and testing*[33] *"Veterinarians working at institutions with large populations of fishes are encouraged to have special training in fish health management in research, teaching or testing environments."* The care and use of wildlife represents another special case where oftentimes, the researchers are the most knowledgeable about the species in question. The *CCAC guidelines on: the care and use of wildlife*[46] states in Guideline 17 *"Consultation and/or participation of veterinarians having experience with wildlife should be sought in projects involving potential animal health concerns, such as translocation of animals and medical or surgi-*

cal procedures. Consultation with veterinarians having experience with wildlife or experienced wildlife professionals should also be sought for immobilization activities." The supporting texts also emphasizes that, *"In general, veterinarians remain liable for the use of pharmaceuticals dispensed and for veterinary care. This means that a veterinarian should be an integral part of research where they have prescribed pharmaceuticals for the use of investigators and/or when medical or surgical procedures are involved."*

The Ontario Animals for Research Act states that "The Lieutenant Governor in Council may make regulations...classifying research facilities, requiring the operators of any class of research facility to provide for the services of a veterinarian in connection with the care of animals in the research facility and prescribing the terms and conditions on which such services shall be provided in respect of any such class;"[12] however, regulations in respect of veterinary services have yet to be made.

CONDUCT OF EXPERIMENTAL PROCEDURES

As described above, prior to the conduct of any animal-based studies in Canada, an animal use protocol must be approved by the institutional animal care committee, as described in the *CCAC guidelines on: animal use protocol review*.[29] The expectation is that investigators will carry out a Three Rs search prior to drafting the protocol in order to ensure that the possibility for using replacement, reduction, or refinement alternatives has been considered.[5] To assist in this regard, the CCAC has developed a Three Rs Search Guide which investigators can use to help develop a search strategy and collect the relevant information, prior to completion of an animal use protocol: http://3rs.ccac.ca/en/searches-and-animal-index/guide/. The CCAC *policy statement on: the ethics of animal investigation*[2] requires investigators to minimize pain and distress for the animals, and to ensure that the appropriate numbers of animals will be used.[2] At the time of their approval, animal use protocols are also assigned a category of invasiveness based on the potential pain and distress that could be experienced by the animals.[18] Categories range from A to E, with level A being the least and E the most invasive. The classification of pain and distress is mostly prospective. The *CCAC guidelines on: transgenic animals*[23] require that protocols involving generation of genetically-engineered animals be assigned to category of invasiveness level D, in recognition of the potential for pain and distress both from the procedures and from unanticipated welfare impacts. This category should be revisited once the actual impact of the procedures on the animals has been evaluated, and reduced where warranted.[47] This appears to have been a challenge for institutional ACCs, with the result that numbers of animals reported associated with level D of invasiveness in the annual CCAC census has risen since 1997, giving a false impression of the numbers of animals actually subject to pain and distress.

EUTHANASIA

The *CCAC guidelines on: euthanasia of animals used in science*[48] is based on recommendations made by the International Council for Laboratory Animal Science (ICLAS) Working Group on Harmonization[49] and the two international reference documents on euthanasia recommended by ICLAS: the American Veterinary Medical Association (AVMA) *Guidelines on Euthanasia*[50] and *Recommendations for Euthanasia of Experimental Animals,* Part 1[50] and Part 2;[51] with some modification to fit the Canadian context. An overview of acceptable methods of euthanasia for common species used for research, teaching, and testing based primarily on the two major reference documents is provided and information is included on other methods of euthanasia that are not considered best practice but that may be acceptable for specific purposes providing they comply with the general guiding principles and receive the approval of the ACC reviewing the application.

Table 3.1 is a summary chart of the acceptable methods of euthanasia, and Table 3.2 is a summary chart of the conditionally acceptable methods, as published in the CCAC *guidelines: on euthanasia.*

Further details for each of the methods are provided within the text of the guidelines. In addition, the CCAC is in the process of preparing additional information on the potential impact of each euthanasia method on research data. This will be available on the CCAC website in association with the guidelines document.

The Ontario Animals for Research Act makes specific statements regarding euthanasia in regulation 24 Research facilities and Supply Facilities.[22]

"Where euthanasia is carried out with respect to any animal in a research facility or supply facility, it shall be carried out, (a) by a person or persons properly trained in the euthanasia procedure to be used; (b) in such manner that the death of the animal occurs without unnecessary pain, delay or discomfort; and (c) in a manner that does not endanger or disturb other animals in the research facility or supply facility." R.R.O. 1990, Reg. 24, s. 28(1).[22]

"No person shall use a euthanasia procedure with respect to any animal in a research facility or supply facility unless it is a procedure that is permitted under section 29, 30, 31 or 32." R.R.O. 1990, Reg. 24, s. 28(2).[22]

The following euthanasia procedures are permitted:

For amphibia and reptiles, the insertion of a sharp instrument between the skull and atlas and into the cranial cavity. In the case of fish, the striking of a strong blow to the head behind the eyes. In the case of all cold-blooded animals, decapitation. In the case of all cold-blooded animals, cervical dislocation. R.R.O. 1990, Reg. 24, s. 29 (2).[22]

Euthansia of cold-blooded animals by the use of chemicals is permitted as follows:

For amphibia or reptiles, (a) injection of barbiturates; (b) injection of procaine hydrochloride; (c) oral administration of tribromoethanol; (d) the

TABLE 3.1 Summary Chart of Acceptable Euthanasia Methods for Experimental Animals, *according to CCAC guidelines on: euthanasia of animals used in science*[48]

Classification and Common Name	Acceptable Methods	Details and Cautions
Class Amphibia (Amphibians)		
Frog, toad	Immersion or injection of buffered tricaine methanesulfonate (TMS; also known as MS222, tricaine)	Section 4.1.1 & addendum
	Immersion or injection of benzocaine	Section 4.1.1 & addendum
	SC injection of barbiturates into lymph sac	Section 4.1.1 & addendum
	Overdose of inhalant anesthetics (for species that do not breath hold), followed by another method to ensure death	Section 4.1.1 & addendum
Class Reptilia (Reptiles)		
Turtle, snake, lizard	IV or IP injection of barbiturates	Section 4.1.1 & addendum
	Penetrating captive bolt (for larger species)	Section 4.1.2.1 & addendum
Class Osteichthyes (Bony Fishes) **Class Chondrichthyes (Cartilaginous Fishes)**		
Fish	See also *CCAC guidelines on: the care and use of fish in research, teaching and testing*	
	Immersion or injection of buffered tricaine methanesulfonate (TMS; also known as MS222, tricaine)	Section 4.1.1.3 & addendum
	Benzocaine*	Section 4.1.1.3 & addendum
	Etomidate* Metomidate (also known as Marinil™)	Section 4.1.1.3 & addendum
	Clove oil*	Section 4.1.1.3 & addendum
	Maceration (for fish less than 2 cm in length)	Section 4.1.2.2 & addendum

Class Aves (Birds)		
Chicken, pigeon, etc.	IV or IP injection of barbiturates with local anesthetic	Section 4.1.1.2 & addendum
	Inert gases (Ar, N₂) for poultry	Section 4.1.1.1 & addendum
	Overdose of inhalant anesthetics (for species that do not breath hold), followed by another method(s) to ensure death	Section 4.1.1.1 & addendum
	Captive bolt concussion stun/killing for poultry only	
Class Mammalia (Mammals)		
Order rodentia mouse, rat, hamster, gerbil, guinea pig	IP injection of buffered and diluted barbiturates with local anesthetic	Section 4.1.1.2 & addendum
	Overdose of inhalant anesthetics (for species that do not breath hold), followed by another method(s) to ensure death	Section 4.1.1.1 & addendum
Order lagomorpha rabbit	IV injection of barbiturates	Section 4.1.1.2 & addendum
	Captive bolt	Section 4.1.2.1 & addendum
	Overdose of inhalant anesthetics, followed by another method(s) to ensure death	Section 4.1.1.1 & addendum
Order carnivora (felidae) cat	IV injection of barbiturates	Section 4.1.1.2 & addendum
	Overdose of inhalant anesthetics, followed by another method(s) to ensure death	Section 4.1.1.1 & addendum
Order carnivora (canidae) dog	IV injection of barbiturates	Section 4.1.1.2 & addendum
	Overdose of inhalant anesthetics, followed by another method(s) to ensure death	Section 4.1.1.1 & addendum

Continued

TABLE 3.1 Summary Chart of Acceptable Euthanasia Methods for Experimental Animals, *according to CCAC guidelines on: euthanasia of animals used in science*[48]—cont'd

Order carnivora (mustelids) ferret, skunk	IP injection of barbiturates with local anesthetic	Section 4.1.1.2 & addendum
	Overdose of inhalant anesthetics (for species that do not breath hold), followed by another method(s) to ensure death	Section 4.1.1.1 & addendum
Order artiodactyla (hoofed animals) ruminants sheep, cattle, goats	IV injection of barbiturates	Section 4.1.1.2 & addendum
	Penetrating captive bolt or free bullet, followed by exsanguination or destruction of the brain	Section 4.1.2.1 & addendum
Order artiodactyla (hoofed animals) swine	IV injection of barbiturates (or IP injection with local anesthetic)	Section 4.1.1.2 addendum
	Penetrating captive bolt, followed by exsanguination or pithing	Section 4.1.2.1 & addendum
	Electrical stunning, followed by exsanguination or pithing	Addendum
	Overdose of inhalant anesthetics, followed by another method(s) to ensure death	Section 4.1.1.1 & addendum
	Argon (under tightly controlled conditions)	Section 4.1.1.1 & addendum
Order perissodactyla (hoofed animals) horse, donkey	IV injection of barbiturates	Section 4.1.1.2 & addendum
	Penetrating captive bolt, followed by exsanguination or pithing	Section 4.1.2.1 & addendum
Order primates (nonhuman primates) monkeys	IV injection of barbiturates	Section 4.1.1.2 & addendum
	Overdose of inhalant anesthetics, followed by another method(s) to ensure death	Section 4.1.1.1 & addendum
Wildlife free-ranging mammals free-ranging birds	See CCAC guidelines on: the care and use of wildlife	
Marine mammals	Barbiturates	Section 4.1.1.2 & addendum

SC—subcutaneous; IV—intravenous; IP—intraperitoneal.

*Currently only TMS and metomidate are registered for veterinary use in Canada for fish that may be consumed by humans; investigators are individually responsible for the use of other anesthetic agents that have not been approved for such use.

TABLE 3.2 Summary Chart of Conditionally Acceptable Methods of Euthanasia for Experimental Animals, according to *CCAC guidelines on: euthanasia of animal used in science*[48]

Species	Method	Details and Cautions
Fish	Concussion (emergency killing for other species)	Section 5.5 & addendum
Birds	CO_2	Section 5.1 & addendum
	Cervical dislocation	Section 5.6 & addendum
	Decapitation	Section 5.7 & addendum
	Maceration (chicks within 2 days of hatching only)	Addendum
Rodents	CO_2	Section 5.1 & addendum
	Argon/Nitrogen	Section 5.2 & addendum
	Cervical dislocation	Section 5.6 & addendum
	Decapitation	Section 5.7 & addendum
Rabbits	Cervical dislocation	Section 5.6 & addendum
	T-61	Section 5.3 & addendum
Cats, dogs	T-61	Section 5.3 & addendum
Cattle, sheep, goats, horses, donkeys	Gun shot	Addendum
Pigs	CO_2	Section 5.1 & addendum

administration of chloroform by inhalation; (e) the administration of ether by inhalation; and (f) injection of chlorobutanol saturated solution.

For fish, (a) the suspension in water of tricaine methanesulfonate; (b) the suspension in water of 2-methylquinoline; and (c) prolonged bubbling into the tank of a high concentration of carbon dioxide.

For amphibia, the suspension in water of tricaine methanesulfonate. R.R.O. 1990, Reg. 24, s. 30.[22]

The following euthanasia procedures are permitted for warm-blooded animals: exsanguination, but only where the animal is completely anaesthetized prior to and during the procedure.

In the case of birds and rodents, cervical dislocation. In the case of livestock and dogs, electrocution, but only where the electrocution equipment is approved by the Director (of the Animals for Research Act inspectorate). In the case of rodents, decapitation, but only with equipment that is approved by the Director.

In addition, the following euthanasia of warm-blooded animals by the use of chemicals is permitted. R.R.O. 1990, Reg. 24, s. 32 (1).[22]

Administration of barbiturates intravenously, intracardially, intrathoracically, or intraperitoneally; Administration of tribromoethanol rectally or orally other than in the case of dogs; Slow intravenous administration of Hoechst Pharmaceutical product T-61; Administration of chloral hydrate intraperitoneally, intravenously, or orally; Administration of ether by inhalation; Administration of carbon dioxide by inhalation; Administration of chloroform by inhalation. R.R.O. 1990, Reg. 24, s. 32 (2).[22]

Some of the methods of euthanasia are no longer considered to be acceptable by the CCAC. For those institutions in Ontario which are participants in the CCAC program, more humane practices, as defined in the CCAC *guidelines on: euthanasia of animals used in science*[48] are expected to be used. In general, the Ontario Ministry inspector is well versed in the requirements of the CCAC guidelines, so these tend to be used as a basis for inspections, given that they are regularly updated.

EQUIPMENT AND FACILITIES

The CCAC requires that facilities for the care and use of all animals in research, teaching, and testing must be conducive to the well-being and safety of the animals, provide an appropriately-appointed and safe workplace for personnel, and establish a stable research environment. The *CCAC guidelines on: laboratory animal facilities—characteristics, design and development*,[45] is intended to assist the users and designers of laboratory animal facilities to achieve these objectives. The goal is to promote optimal levels of animal care and facilitate good research without curtailing new and innovative ideas for facility design. Therefore, these guidelines are intended to be viewed as a tool for achieving acceptable standards and not as mandatory instructions.

However, the renovation or new construction of laboratory animal facilities must meet certain basic functional criteria to be in compliance with the spirit and intent of the CCAC. These basic criteria are outlined in the text of this guidelines document.

The CCAC *guidelines on: laboratory animal facilities—characteristics, design and development*[45] applies to animals such as rats, mice, rabbits, dogs, and cats held in controlled environments, but not to those used in field settings. Facilities for farm animals, fish and short-term holding of captive wildlife are described in other CCAC guidelines (CCAC *guidelines on: the care and use of farm animals in research, teaching and testing;*[32] CCAC *guidelines on: the care and use of fish in research, teaching and testing*[33] and CCAC *guidelines on: the care and use of wildlife*).[46] However, many of the general principles described within the guidelines on: laboratory animal facilities are applicable to most species maintained in captive environments for the purposes of research, teaching, and testing.

These guidelines do not attempt to address building codes or safety codes and standards. It is the responsibility of consultant architects and engineers to address these issues in concert with the responsible institutional officials.

If institutions are planning and designing biosafety containment facilities, the CCAC *guidelines on: laboratory animal facilities*[45] should be used in conjunction with the relevant biosafety guidelines, such as the Canadian Biosafety Standards and Guidelines[35] or the CFIA Containment Standards for Facilities Handling Aquatic Animal Pathogens.[37] The above biosafety guidelines must be implemented whenever facilities will be used to house animals that are experimentally infected with human and/or animal pathogens.

The CCAC *guidelines on: laboratory animal facilities—characteristics, design and development*[45] refers to barrier systems for reducing or minimizing cross-contamination since these are important concepts in all animal facilities. Barriers are commonly used to separate animals of different or unknown disease statuses, such as dogs, cats, mice, specific pathogen-free animals, and genetically modified animals.

Bare minimum requirements for research facilities are also included in regulation 24 to the Ontario Animals for Research Act "Research Facilities and Supply Facilities."[24]

REFERENCES

1. Canadian Federation of Biological Societies—CFBS. *Guiding principles on the care of experimental animals*. Ottawa, ON: CFBS; 1961. p.1.
2. Canadian Council on Animal Care—CCAC. *CCAC policy statement on: ethics of animal investigation*. Ottawa, ON: CCAC; 1989. p.2.
3. Canadian Council on Animal Care—CCAC. *Guide to the care and use of experimental animals*, vol. 2. Ottawa, ON: CCAC; 1984. p. 372.
4. Canadian Council on Animal Care—CCAC. *Guide to the care and use of experimental animals*, vol. 1. Ottawa, ON: CCAC; 1993. p. 212.
5. Canadian Council on Animal Care—CCAC. *CCAC policy statement on: terms of reference for animal care committees*. Ottawa, ON: CCAC; 2006.
6. Wilson P. *Legislative jurisdiction over animals used in research, testing and testing*. Canadian Council on Animal Care Commissioned Legal Opinion. Ottawa, ON: Osler, Hoskin & Harcourt; 1998.
7. Létourneau L. The protection of animals used for the purpose of xenotransplantation in Canada, 2000 [unpublished, archived at Health Canada].
8. Canada. Constitution Act/Lois Constitutionnelle. 30 & 31 Victoria, c 3 (UK). Ottawa, ON: Department of Justice Canada; 1867.
9. Canada. Criminal code/code criminal RSC, 1985, c C-46. Ottawa, ON: Department of Justice Canada; 1985.
10. Canada. Health of Animals Act/Loi sur la santé des animaux. SC 1990, c 21. Ottawa, ON: Department of Justice Canada; 1990.
11. Canada. *Standard acquisition clauses and conditions manual (SACC)*. Ottawa, ON: Public Works and Government Services Canada; 2010.
12. Ontario, Canada. Animals for Research Act/Loi sur les animaux destinés à la recherche RSO 1990, Chapter A22. Ottawa, ON: The Ontario Ministry of Agriculture Food and Rural Affairs; 1990.

13. Russell WMS, Burch RL. *The principles of humane experimental technique*. London UK: Special edition published by Universities Federation for Animal Welfare—UFAW; 1959/1992.
14. Griffin G. Establishing a three Rs programme at the Canadian Council on Animal Care. *Altern Lab Anim* 2009;**37**(S2):63–7.
15. Fenwick N, Tellier C, Griffin G. *The characteristics of analgesia-withholding in animal-based scientific protocols in Canada*. Ottawa, ON: Canadian Council on Animal Care—CCAC; June 2010. p.15.
16. Leenaars M, Ritskes-Hoitinga M, Griffin G, Ormandy E, 2011. Background to the Montréal Declaration on the Synthesis of Evidence to Advance the 3Rs Principles in Science, as Adopted by the 8th World Congress on Alternatives and Animal Use in the Life Sciences, Montréal, Canada, on August 25, 2011. In: ALTEX Proceedings 1/12: p. 35–38.
17. Canadian Council on Animal Care—CCAC. *Interpretation bulletin no. 1-1 animal use data form*. Ottawa, ON: CCAC; 2006. p. 20.
18. Canadian Council on Animal Care—CCAC. *CCAC policy statement on: categories of invasiveness in animal experiments*. Ottawa, ON: CCAC; 1991. p.2.
19. Fenwick N, Ormandy E, Gauthier C, Griffin G. Classifying the severity of scientific animal use: a review of international systems. *Anim Welfare* 2011;**20**:281–301.
20. Canadian Association for Laboratory Animal Medicine—CALAM. *CALAM/ACMAL standards for veterinary care*. CALAM; 2007. p.15.
21. Ontario, Canada. General/Dispositions Générales R.R.O. 1990 Regulation 22 of the Animals for Research Act. Ottawa, ON: Department of Justice Canada; 1990.
22. Ontario, Canada. Research Facilities and Supply Facilities/Services de Recherche et Animaleries R.R.O. 1990, Regulation 24 of The Animals for Research Act. Ottawa, ON: Department of Justice Canada; 1990.
23. Canadian Council on Animal Care—CCAC. *CCAC guidelines on: transgenic animals*. Ottawa, ON: CCAC; 1997. p. 9.
24. Ormandy EH, Schuppli CA, Weary DM. Worldwide trends in the use of animals in research: the contribution of genetically-modified animal models. *Altern Lab Anim* 2009;**37**(1):63–8.
25. Canadian Council on Animal Care—CCAC. *CCAC policy statement for: senior administrators responsible for animal care and use programs*. Ottawa, ON: CCAC; 2008. p. 30.
26. Canadian Council on Animal Care—CCAC. *CCAC guidelines on: institutional animal user training*. Ottawa, ON: CCAC; 1999. p. 10.
27. Canadian Council on Animal Care—CCAC. *CCAC policy statement on: recommendations made in CCAC assessment reports*. Ottawa, ON: CCAC; 2012. p.2.
28. Canadian Council on Animal Care—CCAC. *CCAC policy statement on: certification of animal care and use programs*. Ottawa, ON: CCAC; 2010.
29. Canadian Council on Animal Care—CCAC. *CCAC guidelines on: animal use protocol review*. Ottawa, ON: CCAC; 1997. p. 12.
30. Canadian Council on Animal Care—CCAC. *CCAC policy statement on: scientific merit and ethical review of animal-based research*. (in press).
31. Canadian Council on Animal Care—CCAC. *CCAC guidelines on: the procurement of animals used in science*. Ottawa, ON: CCAC; 2007. p. 27.
32. Canadian Council on Animal Care—CCAC. *CCAC guidelines on: the care and use of farm animals in research, teaching and testing*. Ottawa, ON: CCAC; 2009. p. 135.
33. Canadian Council on Animal Care—CCAC. *CCAC guidelines on: the care and use of fish in research, teaching and testing*. Ottawa, ON: CCAC; 2005. p. 87.
34. Soave O, Brand CD. Employer responsibility for employee health in the animal environment. *Lab Anim* 1991;**20**(2):41–6.

35. Public Health Agency of Canada and Canadian Food Inspection Agency–PHAC/CFIA. *Canadian Biosafety Standards and Guidelines.* Ottawa ON: Government of Canada; 2013. p. 343.

36. Canada. Human Pathogens and Toxins Act/Loi sur les agents pathogènes humains et les toxines. Ottawa, ON: Department of Justice Canada; 2009.

37. Biohazard Containment and Safety Science Branch. *Containment standards for facilities handling aquatic animal pathogens.* 3rd ed. Canadian Food Inspection Agency—CFIA; 2010. p. 65.

38. Institute for Laboratory Animal Research—ILAR. *Guidelines for the humane transportation of research animals.* Washington DC: National Academies Press; 2006. p. 164.

39. Swallow J, Anderson D, Buckwell AC, Harris T, Hawkins P, Kirkwood J, et al. Guidance on the transport of laboratory animals. *Lab Anim* 2005;**39**(1):1–39.

40. Canadian Council on Animal Care—CCAC. *CCAC policy statement on: social and behavioural requirements of experimental animals.* Ottawa, ON: CCAC; 1990. p.1.

41. Griffin G. Evaluating environmental enrichment is essential. *Enrich Rec* 2012:29–33.

42. Cao L, Liu X, Lin E-JD, Wang C, Choi EY, Riban V, et al. Environmental and genetic activation of a brain-adipocyte BDNF/leptin axis causes cancer remission and inhibition. *Cell* 2010;**142**(1):52–64.

43. Hamm RJ, Temple MD, O'Dell DM, Pike BR, Lyeth BG. Exposure to environmental complexity promotes recovery of cognitive function after traumatic brain injury. *J Neurotrauma* 1996;**13**(1):41–7.

44. Passineau MJ, Green EJ, Dietrich WD. Therapeutic effects of environmental enrichment on cognitive function and tissue integrity following severe traumatic brain injury in rats. *Exp Neurol* 2001;**168**:373–84.

45. Canadian Council on Animal Care—CCAC. *CCAC guidelines on: laboratory animal facilities—characteristics, design, and development.* Ottawa, ON: CCAC; 2003. p. 108.

46. Canadian Council on Animal Care—CCAC. *CCAC guidelines on: the care and use of wildlife.* Ottawa, ON: CCAC; 2003. p. 66.

47. Griffin G, Dansereau M, Gauthier C. Categories of invasiveness—a precautionary approach. *Altern Anim Test Exp* 2007;**14**(special issue):715–20.

48. Canadian Council on Animal Care—CCAC. *CCAC guidelines on: euthanasia of animals used in science.* Ottawa, ON: CCAC; 2010. p. 32.

49. Demers G, Griffin G, De Vroey G, Haywood JR, Zurlo J, Bedard M. Harmonization of animal care and use guidance. *Science* 2006;**312**(5774):700–1.

50. American Veterinary Medical Association—AVMA. *AVMA guidelines on euthanasia.* Schaumburg IL: AVMA; 2007. p. 39.

51. Close B, Banister K, Baumans V, Bernoth EM, Bromage N, Bunyan J, et al. Recommendations for euthanasia of experimental animals: Part 1. *Lab Anim* 1996;**30**(4):293–316.

52. Close B, Banister K, Baumans V, Bernoth EM, Bromage N, Bunyan J, et al. Recommendations for euthanasia of experimental animals: Part 2. *Lab Anim* 1997;**31**(1):1–32.

Laboratory Animal Science Legislation in Latin America

Ekaterina Akimovna Botovchenco Rivera[1], Cecilia Carbone[2], Rafael Hernandez Gonzalez[3], and Juan Manuel Baamonde[4]

[1]*Federal University of Goias, Brazil,* [2]*Facultad de Ciencias Veterinarias, Universidad Nacional de La Plata, La Plata, Buenos Aires, Argentina,* [3]*National Institute for Medical Science and Nutrition-SZ, Animal Resources Department, Mexico City, Mexico,* [4]*Centro de Estudios Científicos (CECs), Valdivia, Chile*

GENERAL FRAMEWORK

Latin America is a multicultural region due to its people's different ethnic features, colonial background, religion, language, and level of educational and financial development. Therefore, there are many different ways of thinking. Almost every country has Spanish as their common language, with exception to Brazil, but these countries greatly differ in the development of their technological, educational, health, and economic systems. Of course, this reflects the

Laboratory Animals. http://dx.doi.org/10.1016/B978-0-12-397856-1.00004-0

development of science too, and we may state that we have centers of excellence in some places and in other places science is just starting to bloom.

This discrepancy may be seen in some countries like Argentina, which as early as 1891, was concerned to regulate the protection of animals, and a national law against animal cruelty called "Law Sarmiento" n° 2786[1] was approved by the government. Also in Brazil, in 1934, a Federal Decree 24.645 was issued to assure the protection of experimental animals (http://funed.mg.gov.br/wpcontent/uploads/2010/05/Decreto-lei-24645-34-maus-tratos-animais.pdf[2]) but this Decree did not come into force.

On the other hand, these same countries have fairly new cities (for example Brazil's capital, Brasilia is only 60 years old) where universities and technical institutions, are also very young, and as a consequence, science is too.

Thus, it is not difficult to imagine that regulations on the scientific use of experimental animals is a new subject in Latin America, and in a universe of 23 countries, only three have a national law specific to the use and care of experimental animals. However, substantial effort has been placed in developing regulations, knowing that this is a very important aspect in developing a successful laboratory animal science, but also for it to be internationally harmonized.

In order to better understand the evolution of laboratory animal science in Latin American countries and the current scenario, a little bit of history might be useful.

Brazil had no regulatory or other normative issues related to the use of animals in research, teaching, or testing until 1985.

The first attempt to call the attention of the scientific community on how research should be conducted, i.e. based on Ethical Principles was in 1983, when a veterinarian called Dr Fernando Sogorb invited some scientists, mainly veterinarians who worked with laboratory animals, to create an association called the National College on Animal Experiments (COBEA; www.cobea.org.br/). One of the first activities of this association was to issue the COBEA Ethical Principles, based on the CIOMS (Council for International Organizations of Medical Science) principles: http://cioms.ch/publications/guidelines/1985_texts_of_guidelines.htm.[3]

In 2008, members of Brazilian College of Animal Experiments (COBEA) decided to change its name to Brazilian Laboratory Animal Science Association (SBCAL), because it was in fact an association and not a college, and also because people thought COBEA performed animal experiments giving rise to many protests from activists.

Brazilian scientists knew about laws in other countries and felt uneasy, as they were aware of the importance of having some type of regulations if animals were to be used in their scientific projects. Since there were no laws, COBEA/SBCAL and scientists from different universities thought something had to be done and organized a task force to implement ethics committees in their institutions, using the Canadian model as a reference (http://www.ccac.ca). These were easier to implement than having a law passed in Congress. No efforts were spared to achieve these goals, and thus, with lectures, seminars, and congresses, the word was spread. And the first Ethics Revision Committee in the Use of experimental animals was in place at the Federal University of Goiás, in 1992 (http://www.hc.ufg.br/pages/24568).

We can say, without any doubt, that this was a very important first step toward better science and better animal care in Brazil. It completely changed the approach scientists had toward laboratory animals. As a consequence of such drastic changes, the Brazilian Scientific Council (CNPq; www.cnpq.br), the most important scientific body in the country, decreed that no grant was to be given to researchers if their project had not passed by an Ethics Revision Committee. The main scientific publications also started to accept only articles in which the research had been approved by an Ethics Committee.

Although Brazil had no specific law on the Use of Laboratory Animals, there were some tentative federal law projects against the cruelty to animals in experiments. These projects were: the already mentioned Federal Decree n° 24.645 which assured protection to experimental animals (not enacted); and the Law n° 6638/79[4] (http://www.lexml.gov.br/urn/urn:lex:br:federal:lei:1979-05-08;6638) on the protection of experimental animals. However, it was not enacted.

In 1995, a physician called Dr Sergio Arouca, also a federal representative, coordinated a group of scientists who issued the first draft of a law project concerning the scientific use of animals. Although one might be tempted to shake one's head in disbelief, the law was approved 13 years later, with lots of amendments, and without the supervision of Dr Arouca, who died in 2003.

As Brazilian law does not cover all the important topics related to the use and care of experimental animals, the National Council on the Control of Animal Experiments (CONCEA) and its group of ad hocs are working on preparing the *Brazilian Guide on the Care and use of Laboratory Animals* that will cover the following items: ethics and animal welfare, environment, education, training and competence of personnel, transport, housing, enrichment, husbandry and environment, veterinary care, conduct of experimental procedures, equipment, and facilities.

Meanwhile, CONCEA accepts international guides as references, such as:

- Australian Guide for the Care and Use of Laboratory Animals.[5]
- National Institutes of Health Guide for the Care and Use of Laboratory Animals.[6]
- CCAC Guide for the Care and Use of Experimental Animals.[7,7a]

Mexico also has a specific law on the care and use of experimental animals. This country differs greatly from other countries in the area. The Mexican economy is one of the ten largest world economies and is a member of the Organization for Economic Development, while other countries of the region have a totally emerging economy. For example, Cuba has a socialist economic model and Puerto Rico has a political and economic system under the jurisdiction of the United States of America (USA). Since Puerto Rico applies USA's legislation, its case will not be considered herein.

Additionally, Mexico has the largest North American university, the greatest technological development among this group of Hispanic countries, and has the most demanding population of health and education services.

Mexico, Cuba, and Costa Rica produce the majority of PhD graduates in different knowledge areas and have health coverage for almost 100% of their people. The three countries have developed a scientific research system recognized in the region and have centers of reference for laboratory animals.

Mexican official norm specific for laboratory animals is NOM 062/2008.[8] An important aspect of this law is that it derived from the initiative of veterinarians who worked with laboratory animals, scientists interested in regulating the use of these animals, and the Ministry of Agriculture, Livestock, Rural Development, Fishery, and Food (SAGARPA)'s officials, who were all interested in improving the use of laboratory animals in Mexico. The norm has its roots in three documents internationally accepted as reference:

- Guide for the Care and Use of Laboratory Animals. National Academy of Sciences of the United States of America.
- Euthanasia Guidelines. American Veterinary Medical Association.[9]
- Guide for the Care of Laboratory Animals. Canadian Council on Animal Care. Canada.

Uruguay, like Brazil and Mexico, also has a law specific to experimental animals, Law n° 18.611/2009.[10] Despite the small number of scientists, Uruguay has an important background in laboratory animal matters. In 2001, a group of professionals identified the need for a framework at a national level that would regulate the care and use of experimental animals. This working group was named The Honorary Committee for Experimental Animals, that worked until 2008 under the umbrella of the Secretary of Science and Technology from the Universidad de la República.

It can be said that other Latin American countries are trying to develop their own specific laboratory animal legislations such as **Argentina** and Colombia.

In 2011 the Argentinean Association for Laboratory Animal Science and Technology (AACyTAL; www.aacytal.com.ar) presented a new project for a national law taking as examples the European Directive 2010/63/EU[11] and the Guide for the Care and Use of Laboratory Animals, Eighth edition. It was discussed and agreed by the Argentine biomedical community in 2012 and now it is being revised by a national committee.

All private or national institutions that use animals for scientific purposes in the national territory will be covered by this law.

The 3Rs (Replacement, Reduction, and Refinement) are addressed as general principles that support the document. Occupational health and safety, housing, enrichment, husbandry and environment; education and training of personnel and veterinary care are included in the law project.

There were two Law Project drafts, in **Colombia**, from 2009 to 2011 to modify the general animal protection law: Law n° 84, 27 December, 1989 on the Protection of Animals (http://www.worldcat.org/title/lei-84-codigo-civil-nacional-espedido-por-el-congreso-de-los-estados-unidos-de-colombia-en-sus-sesiones-de-1873/oclc/427927866[12]). But, there is no specific law on the use of laboratory animals in research and teaching, not even a draft of a project.

Other countries in Latin America do not have a specific law on the care and use of laboratory animals, but have laws of general scope which include articles related to them.

In **Peru** there are two main official documents which address the care and use of laboratory animals: one is the national Law n° 27265/2000 for the "Protection of Domestic and Wild Animals Maintained in Captivity"[13]; the other official document is an Institutional one, the Ethics Code for the Care and Use of Experimental Animals[14] approved by the National Institute of Health in Lima in April 2012. It covers all the studies, diagnostic assays and experiments with animals that are developed in this institution. It is The Public Health Service Policy on Humane Care and Use of Laboratory Animals and describes the Institutional Animal Care and Use Committee (IACUC) functions.

Similar to the previously mentioned countries, **Venezuela** has no specific law related to experimental animals. But the Law for the Protection of the Fauna: Domestic, Wild, and in Captivity Animals, n° 39.338, 4 January, 2010—Asamblea Nacional de la República Bolivariana de Venezuela[15] has some specific articles covering these animals. In Venezuela, this law is applied when needed.

There is no specific law on the care and use of laboratory animals in **Colombia**. However, it has the Federal Law n° 84, 27 December, 1989 on the Protection of Animals[12] which states that "all animals in the national territory must be protected from pain and suffering caused by man." *Animal* meaning any domesticated or wild animal, no matter its habitat.

In Chapter II, Art. 6, item "s" states that "vivisection is banned unless it is scientifically sounded and performed in authorized places."

In **Chile**, the Law n° 20.380/2009 on the Protection of Animals[16] is a generic Law with a wide range of application, starting from Article 1, which states that "the Act is intended to establish rules to know, protect and respect animals as living beings and part of nature, with the purpose to give proper treatment and to avoid unnecessary suffering." There is no clear definition if this applies to vertebrates, invertebrates, nonhuman primates and if it includes fetal forms in their final stages, among other omissions.

In 1994, the Act of Animal Welfare was published in **Costa Rica**, as a law of general applicability which, in its Chapter III, Articles 10-11-12-13, refers to the use of experimental animals and covers ethical considerations for their use.[17]

We have seen that Brazil and Mexico have specific laws on the use of laboratory animals, but it is also worth mentioning other nonspecific laws where articles on laboratory animals can be found.

In **Brazil** there is another law called Law of Environmental Crimes, also known as The Law of Nature (IBAMA-Brazilian Institute of the Environment) n° 9605/98 (http://www.planalto.gov.br/ccivil_03/leis/L9605.htm) regulated by the Decree n° 3179/99[18] (http://www.planalto.gov.br/ccivil_03/decreto/D3179. htm) which is not specific for laboratory animals, but it has one article (Article n° 32) which states that "anyone who treats animals in experiments with cruelty is subjected to the penalties of this law." It is very vague and difficult to interpret. However, it was the first Law to address the care and use of experimental animals.

In **Mexico** there are also laws of general and unspecified scope which cover, among others, the use of experimental animals:

1. Act of General Health and Its Regulation.[19]
2. Act of Animal Welfare.[20]
3. NOM-042-SSA-2006 or Mexican Official Norm for the Disease Prevention and Control. Health Specifications for Canine Control Centers.

The NRC Guide for the Care and Use of Laboratory Animals[6] is used as a reference in almost all Latin American countries, mainly in the countries where there is no regulation on the use of experimental animals. Ecuador also follows the OIE (World Organisation for Animal Health) Principles from the Terrestrial Code.[21]

There is no information available on regulations/laws/guidelines on the care and use of experimental animals from the following countries: *Paraguay, Bolivia, Guiana, Suriname, French Guiana, Panamá, Honduras*, and *El Salvador.*

Guatemala has no specific legislation; only the University of San Carlos has an Internal Regulation of the Main Laboratory Animal Facility, the "Reglamento Interno del Bioterio Central de la Universidad de San Carlos." It describes general procedures for the proper management of laboratory animals issued by some scientists working in research.

COUNTRIES WITH A SPECIFIC LAW IN THE CARE AND USE OF EXPERIMENTAL ANIMALS

Brazil, **Mexico,** and **Uruguay** are the three Latin American countries with specific laws on the care and use of experimental animals and they all have the same core principles with some slight differences due to regional demands and different legal systems.

In **Mexico**, in addition to some domestic regulations which are subdivided in other laws of general applicability which relate, among many other aspects, to the use of laboratory animals, the specific document addressing the laboratory animals is the NOM-062-ZOO-1999 or Mexican Official Norm for The Production, Care And Use Of Laboratory Animals.[8] This document will be discussed in this chapter.

The **Brazilian** Law concerning the scientific use of animals is Law n° 11.794 from October 8, 2008, called Arouca Law[22] (http://www.planalto.gov. br/ccivil_03/_ato2007-2010/2008/lei/l11794.htm). The law is administered by the Ministry of Science, Technology, and Innovation.

This law creates, in its Article 4, the CONCEA (http://www.mct.gov.br/index.php/content/view/310553.html). This is an advisory, deliberative, and appellate body.

CONCEA has various competences, among others:

- to issue regulations and guidelines on the ethical and humane use of animals in research, teaching, and testing;
- to see if these regulations are being followed;
- to approve and register institutions which breed or use animals for experimental purposes;
- to monitor and evaluate the introduction of alternatives to replace the use of animals;
- to keep an updated data bank of the institutes, scientists, and researchers; and
- to assess the Executive Power, when asked, about the activities treated in this law.

CONCEA is composed of members and substitutes from 12 Institutions, including representatives, among others, of the Ministry of Education, Ministry of Health, and the Brazilian Academy of Sciences, and who are nominated by the Ministry of Science, Technology, and Innovation.

The members must be Brazilian citizens with a PhD degree or recognized knowledge in research and teaching.

In **Uruguay** the National Law is Law n° 18.611, called "Use of Animals in Experiments, Teaching and Research Activities,"[23] published in October, 2009.

Like in Brazil, the Uruguayan Law created (Articles 4,5) a statutory national body responsible for the control of the activities in which animals are used named the National Commission for Animal Experimentation (CNEA).

The CNEA is responsible for the revision of guidelines on the accommodations and care of animals used for experimental and other scientific purposes. Animal breeders, facilities, and research institutions where animals are used for scientific purposes or teaching should be registered in a National Register (Article 8).

THE PRINCIPLES

The 3Rs Principles are the core of the laws regulating the use of animals in all the countries having a specific law on experimental animals and also in those general laws on the protection of animals in other countries.

The overarching policy on the scientific use of animals in Mexico is reflected in the following NOM 062 policy statement: "The use of animals in research, teaching, and testing is acceptable only if it promises to contribute to development of knowledge that can be expected to benefit humans or animals." It incorporates the principles of the 3Rs when stating, "Animals should be used only if the researcher cannot find an alternative method." A continuing sharing of knowledge, review of the literature, and adherence to the 3Rs principle of "Replacement, Reduction, and Refinement" are also requisites. Those using animals should employ "the best existing methods according to animal welfare principles on the smallest number of appropriate animals required to obtain valid information."

In the Brazilian Law (2010/2008/lei/l11794.htm), the 3Rs are clearly mentioned as noted in:

- Chapter II, Article 5, Item III: CONCEA has to evaluate the introduction of alternatives that will replace the use of animals in teaching and research (Replacement).
- Chapter IV, Article 14, paragraph 3: use of alternatives instead of animals in teaching procedures (Replacement).
- Chapter IV, Article 14, paragraph 4: use the minimal possible number of animals to achieve the expected results (Reduction).
- Chapter IV, Article 14, paragraph 5: If an experiment may cause pain and distress sedatives, analgesia or anesthesia must be used (Refinement).

The 3Rs principles are not included in detail in the Uruguayan Law. Only the principle of reducing the number of animals used in research and testing is mentioned in Article 17. Alternative methods and their validation are not included.

In 1994, the Act of Animal Welfare was published in Costa Rica as a law of general applicability and in its Chapter III, Articles 10-11-12-13, it refers to the use of experimental animals and covers ethical considerations for their use. The 3Rs are the basis of this law when referring to the use of laboratory animals. Article 10 paragraph *b* refers to the Reduction on the number of animals used, paragraphs *c*, and *d,* and *e* of the same Article refer to Refinement and Article 11 treats specifically on alternatives.

In Peru the national Law n° 27265/2000 for the "Protection of Domestic and Wild Animals Maintained in Captivity" in its Chapter IV contains regulations on the use of animals for scientific purposes mainly emphasizing the use of alternative methods (Article 10).

In Venezuela, the law for the Protection of the Fauna: Domestic, Wild, and in Captivity Animals, n° 39.338, 4 January, 2010—Asamblea Nacional de la República Bolivariana de Venezuela states: "Laboratory Practices. The use of domestic animals is permitted only in technical schools, universities, for the development of laboratory practices included in the curricula approved by the competent authorities when no other methods or techniques are available to obtain the same results."

Some places are allowed to breed laboratory animals if there is no alternative.

In the draft project from Argentina the 3Rs are addressed as a general principle that supports the document.

SCOPE/APPLICABILITY

In the Brazilian Law,[22] Chapter 1, Article 1, paragraph 1 indicates that the use of animals for teaching is restricted to academic institutions and technical schools of the biomedical area. And paragraph 2 of the same article considers scientific research all the activities related to basic science, applied science, technological development, drugs production and control, food, immune procedures, instruments, or any other procedure using animals.

Paragraph 3 emphasizes that all zootechnical practices are not considered as research. So, they do not need approval of an ethics committee.

Article 2 of this same Chapter specifies the animal species covered by the law: the law regulates the use of animals (Phylum Chordata, Subphylum Vertebrata) in research, teaching, and testing.

In a similar manner, in Uruguay the Law[23] regulates the use of animals in research, teaching, and testing and the animals included in the scope of its law in its Article 3 are animals of the Phylum Chordata, Subphylum Vertebrata.

The Mexican NOM 062[8] only covers species directly referred to in the document: laboratory rodents (mice, rat, gerbil, hamster, and guinea pig), rabbits, dogs, cats, pigs, and nonhuman primates. All these animals used for research, teaching, or testing, or for display purposes or eventual use in research, teaching or testing must be the subject of a written animal use protocol to be approved by IACUC.

AUTHORIZATION OF USER/BREEDING INSTITUTIONS

In Mexico the NOM 062[8] applies to all institutions using animals for scientific purposes. This includes academic institutions, government laboratories, and private companies. The Mexican Agriculture Department has a list of institutions with a commitment to practice NOM 062 standards for the ethical use of animals.

Both the Animal Welfare Act and the NOM 062 do not allow the use of animals in primary or secondary school science classes.

Only institutions accredited by CONCEA can breed or use experimental animals in Brazil and to be accredited, they must have an active Ethics Committee, called the Commission on the Ethical Use of Animals (CEUA) (Chapter III, Article 8).

In Uruguay, universities, research institutes, and technical schools can use animals under the conditions of Law n° 18.611/2009 (Article 2).[10]

In Venezuela the Law for the Protection of the Fauna: Domestic, Wild, and in Captivity Animals, n° 39.338, 4 January, 2010—Asamblea Nacional de la República Bolivariana de Venezuela[15] in its Chapter III—Article 52 states that: "It is permitted to use domesticated animals in research only in licensed places, legally authorized by the competent authorities, if these animals are necessary for the advancement of science in areas like diagnosis, prevention and treatment of diseases which affect man and other living beings."

In Colombia the Federal Law n° 84, 27 December, 1989 on the Protection of Animals[12] refers to authorized places in its Chapter II, Article 6, item "s": "vivisection is banned unless it is scientifically sounded and performed in authorized places."

Law n° 20.380 on the Protection of Animals in Chile[16] also states that experiments should be performed in suitable facilities according to the species and status of the animal.

The Law on the Protection of Animals from 1994 in Costa Rica[17] in its Article 12 indicates that all animal experiments must be registered in the Ministry of Science and Technology, except in the case of wild animals which have to follow Law n° 7317, 30 October, 1992. The experiments not registered will be subjected to the penalties covered by this law.

INSPECTIONS

The Mexican Agriculture Department provides oversight of animal care and use programs through its Verification–Certification program, which carries out regular visits to institutions, typically on a 3 year basis. Assessment visits are usually conducted by a representative of the Mexican Agriculture Department. When conducting a visit, the representative assesses the structure and resources of the animal care and use program, the composition, functioning and effectiveness of the IACUC, and the appropriateness of animal care and use practices, procedures, and facilities. Assessments are based on NOM 062 guidelines and associated documents.[8]

In Uruguay the statutory national body responsible for the control of the activities in which animals are used is the CNEA (Articles 4,5).[10] Animal breeders, facilities, and research institutions where animals are used for scientific purposes or teaching should be registered in a National Register (Article 8).

In Brazil Chapter V, Article 21 states: "The inspection of all the activities regulated by this law is under the responsibility of the Ministry of Agriculture, Ministry of Education, Ministry of the Environment, and Ministry of Science."[22]

In Chile there are no regulatory agencies to audit the establishments using animals for research and teaching.

PENALTIES

Following chapter V, Article 17 of the Brazilian Law,[22] all Brazilian institutions that develop activities regulated by this law are subject, if they transgress its requirements, to several potential penalties. They include, in increasing range of importance: advertence, fines, temporary suspension of activities, and also of governmental financial support, and at last, definitive suspension of activities or closing of the facilities.

The observance of the norm's implementation in Mexico relies on the following bodies:

- the laboratory's Institutional Animal Care and Use Committee (IACUC).
- Ministry of Agriculture, Livestock, Rural Development, Fishery and Food (SAGARPA).
- State and Mexico City, Federal District's Governments.

However, it does not set forth any penalty for the noncompliance of the provisions declared under NOM-062.[8]

In Uruguay, Articles 18–22 of the Law[10] correspond to the penalties and punitive measures that the *law* imposes for the performance of an act that is proscribed.

In Chile, infringements related to this subject will be sanctioned by the corresponding Ministerial Regional Secretary of Education.

NONHUMAN PRIMATES

Specification for the care and use of nonhuman primates are described in Mexico in the NOM 062.[8] Animal health monitoring must ensure that the animals are tested for tuberculosis upon arrival in the research facility and either isolated or euthanized if found to be carrying tuberculosis, unless it is a necessary element of the research study. In addition, no person who is known to have active tuberculosis may be employed in the care of nonhuman primates. There are also requirements for isolation upon arrival in order to permit the animal to become used to the research environment.

IACUCs are expected to take particular care in providing an ethical review of any protocol involving their use. No Great apes are currently used in Mexico.

Nonhuman primates are not mentioned in the Law in Uruguay.[10]

The Brazilian Law[22] has no special considerations or restrictions on the use of nonhuman primates. However, there are a large number of primates being used in research in this country and ethical considerations are of concern in the use and care of these animals. So, the role of regulating their use stays with the Ethics Committees (CEUAs) of every institution.

GENETICALLY ALTERED ANIMALS

In 2005, the Mexican government published the Biosafety Act for genetically modified organisms,[24] recognizing that there was a rapid increase in the creation and use of these animals, and that the public was beginning to express concern about the genetic modification of animals. In 2008, the Mexican government published a policy within the Biosafety Act for Genetically Modified Organisms and in 2009, a new edition was published with some amendments. The Act does not separately enumerate genetically-engineered bacteria, virus, plants, or animals.

The 2009 Biosafety Act creates the National Register of Genetically Modified Organisms in Mexico.

In Brazil, the Law n° 11794/08[22] does not cover genetically altered animals because there was already a specific law in existence. This is referred to as Law n° 8.794, 5 January, 1995 and it created the National Technical Commission on Biosafety (CTNBIO; www.ctnbio.gov.br).[25] CTNBIO is the organization that proposes regulations, guidelines, the Ethics Code on the use of genetically altered animals, and all other procedures related to these animals.

Genetically modified animals are not mentioned in the Uruguayan Law.[10]

INSTITUTIONAL AND DESIGNATED PERSONNEL RESPONSIBILITIES

In Mexico and Brazil, they are almost the same, with minor changes but the same core. Within each institution, the senior director holds the ultimate responsibility for the operation of the IACUC or the CEUA. The responsibilities of the senior director are laid out in the NOM 062[8] and in Law 11 794/08,[22] respectively.

Despite the variety of type and size of animal facilities, all institutions are required to have the following main elements:

1. At least one functional institutional animal care and use committee (IACUC–CEUA), composed of scientists/teachers experienced in animal use; at least one veterinarian with experience/training with regard to the species being used and the types of work; an institutional member who does not use animals in his/her work; the director/manager of the animal facility and the IACUC–CEUA coordinator. Also, one person paid by the institution who assists the IACUC–CEUA in its work.
2. Competent veterinary and animal care service providers whose numbers and expertise match the nature and scope of the institutional program and an appropriate number of laboratory animal care technicians.

The NOM 062 demands the registration of the facility with the Mexican Agriculture Department and in the case of Brazil with CIUCA, Ministry of Science, Technology, and Innovation.

The operator of every research facility is also required to maintain a record of all animals in the research facility preserving the record within the research facility for at least one to five years from the date that the animals entered the research facility.

The quality and professional competence of the personnel supervising procedures in Uruguay, as well as of those performing procedures or supervising those taking care of the animals on a daily basis is not described in the Law.[10]

ETHICAL REVIEW PROCESS

In Mexico, Brazil, and Uruguay the Ethical Review Process is mandatory by law as mentioned before and they all have the same requests:

Each research institution using laboratory animals must have at least one active and effective Institutional Animal Care and Use Committee (IACUC, CEUA or CICUAL respectively), responsible for ensuring ethical animal use and optimal levels of animal care. These committees are composed of researchers, veterinarians, animal care technicians, and animal facility managers. Each committee typically reports to and derives their authority from the senior administrator of the institution, and has a number of important responsibilities including reviewing animal use protocols, defining and reviewing standard operating procedures, policy development, as well as postapproval monitoring and facility inspection.

Protocol review and approval are based on the principles of the 3Rs: Replacement, Reduction, and Refinement. Protocols are reviewed for its ethical merit. The committees base their decisions on the existing law and also guidelines, professional judgment, and common sense.

In addition, each committee has to submit an annual report to the Mexican Agriculture Department in Mexico, to CONCEA in Brazil, and to CNEA in Uruguay, respectively.

In the Brazilian Law,[22] the composition, authority, responsibilities, and functioning of the Ethical Committees (CEUAs) are defined in written terms of reference in Chapter III, Articles 9 and 10 and also reinforced in Chapter IV Article 13, paragraphs 1 and 2.

The CEUAs are to be composed of at least one veterinarian, biologists, researchers and teachers of specific areas, and one representative of an animal protection society.

The CEUAs have the responsibility of assuring the compliance of this law when animals are used in teaching and research. They have to make an ethical review of all the research protocols of the institution, keep an updated register of all the research conducted and that of researchers, and send a copy to CONCEA. Also, if needed, the CEUAs are to issue certificates to financial organizations or scientific journals and to immediately notify CONCEA and sanitary authorities if any accident occurs with research animals and what actions are to be taken.

If there is a problem with a CEUA decision, the question can be sent to CONCEA.

In Mexico, as well as in Brazil, there must be an IACUC in every institution using animals in research, to which must be submitted, for ethical review, all the protocols prior to the commencement of the work.

Before the IACUC–CEUA can review a protocol, there must be evidence of scientific, pedagogical, or regulatory merit. In the case of research studies, information provided within the protocol review form should provide the IACUC–CEUA with a clear sense of the need for the experimental project, and of the relationship between the proposed experiment and the overall scientific objective. For the use of animals in teaching, the pedagogical merit for the use of animals should be established, and for testing projects, an indication that the testing has been planned according to the most current regulatory requirements. Animal use protocols should include: the project title and descriptive procedures; name(s) of the primary investigator and all personnel who will be handling the animals; department affiliation; proposed start and end dates; funding sources; an indication of scientific or pedagogical merit, or requirement for regulatory purposes; summary and indication on the use of any hazardous materials; evidence of addressing the 3Rs; a description of why sentient animals must be used; justification for the numbers and species of animals selected; anesthesia and analgesia; medical treatments; housing and husbandry; improvements to procedures; improvements to the duration of the experiment and holding times; a description and time course of procedures (flow diagram where possible); a description of humane endpoint; a description of capture, etc., for field studies; method of euthanasia; ultimate fate of the animals; and any other relevant information, such as relevant information from studies previously carried out. Decisions concerning the acceptability of the protocol are generally taken by consensus of the IACUC–CEUA members.

Once a protocol has been approved by the IACUC–CEUA, the investigator is allowed to initiate the research, teaching, or testing studies. Researchers should report back to the IACUC–CEUA on an annual basis with a progress report and requesting any amendments to the protocol.

IACUCs–CEUAs are also required to implement an effective postapproval monitoring program, designed to meet the specific needs of the institution, including actions such as:

- seeing if individual animal users are comfortable handling animals and successfully carrying out procedures, and that they are able to do so in appropriate conditions;
- assessing if endpoints are applied as approved by the IACUC–CEUA to avoid unnecessary distress to the animals;
- conducting site visits and discussing the protocols with animal users by IACUC–CEUA members; and
- addressing concerns through good communication.

The Uruguayan Law[10] states that an IACUC should be created in all the institutions in Uruguay where animals are used for experiments or teaching (Article 9). The *Institutional Animal Care and Use Committees* are of central importance to the application of the law to animal research in Uruguay. They review all projects involving animals to ensure that they are justified by their benefits and minimize any animal pain or suffering that might occur. This includes research, teaching, and testing. The IACUC regularly inspects all projects using animals and all projects housing animals. The IACUC is comprised of at least one veterinarian, one scientist or teacher, users of laboratory animals, and one representative of the community (Articles 10, 11).

In more recent years, several institutions in countries that do not have specific laws on the use and care of experimental animals have decided to create Ethics Committees, also called CICUAL or IACUC to oversee and evaluate all aspects of the institution's animal care and use program. This is a way to have some kind of control on the use of laboratory animals.

In Bolivia and Ecuador, few institutions working with laboratory animals have an Ethics Committee. In Ecuador, they are called IACUCs and follow the norms of the IACUCs as in the USA.

In Chile, the IACUCs came in to use to compensate for the gaps in the existing animal protection law, and to comply with international recommendations and the pressure exercised by national agencies such as the National Fund for Scientific and Technological Development. This is a public program administered by the National Commission for Scientific and Technological Research. These IACUCs often have problems in their compositions and in some institutions they only serve as a committee to review scientific projects that involve the use of laboratory animals, and not to implement all activities of an IACUC as in the USA.

An important fact in Argentina is the voluntary establishment of IACUCs, which are of central importance to the application of general principles to animal research. Many institutions that use animals for scientific purposes have formed an IACUC. As a consequence, an IACUC network was created on May 2011 and it has coordinated the member activities ever since.

In Venezuela, since 2008, there have been two active ethics committees which analyze animal protocols: one in Caracas (Hospital Vargas) and the other one in the Universidad de los Andes (Mérida).

In Colombia, there is an ethics committee in almost every university, but there are no special regulations on the composition and functioning of these committees. The CICUALES is the Colombian net of ethics committees and is trying to define criteria on how to establish ethics committees at the institutional level, thus promoting the correct use of experimental animals.

Since 1992, the "Code of Ethics for Science Workers" has existed in Cuba and is mandatory setting forth the workers' responsibility to prevent any unnecessary injury to the animals.

The Vadi Resolution 4/00, of 2000 establishes the incorporation of ethics committees in the Cuban health institutions working on basic, clinical, and social investigation. Chapter V and VIII specifically mention the use of laboratory animals. Regulation 64/12, of 2004 requires and defines the creation of the Laboratory's Institutional Animal Care and Use Committee (IACUC or CICUAL for its acronym in Spanish), which revises the animal protocols used in every institution.

Peruvian Law n° 27265/2000 for the "Protection of Domestic and Wild Animals Maintained in Captivity"[13] in Article 10 contemplates the functioning and composition of the ethical committees, called IACUC's.

In Chile the Law n° 20.380/2009 on the Protection of Animals[16] says that within 30 days after the publication of the Law, the Council of Deans from the Chilean Universities will have the obligation to designate the members of an Animal Bioethics Committee comprising the following members:

a. Two academics appointed by the Council of Deans from the Chilean Universities.
b. A scientist designated by the Director of the Public Health Institute of Chile.
c. A researcher nominated by the President of the Institute of Agricultural Research.
d. A scientist nominated by the President of the National Commission of Scientific and Technological Research.
e. A representative of the Guild Association of Veterinarians, the oldest association in the country.
f. A representative from nongovernmental organizations on the protection of animals, which has legal personality and is nationally represented.

This committee would define the guidelines under which the experiments on live animals may be developed in accordance with the norms of this Law. These members would serve *ad honorem* for a period of three years and may be eligible for reappointments for successive periods.

REUSE

There is no specific reference on the reuse of animals in the Mexican NOM 062.[8] However, reuse occurs more frequently in the case of large farm animals, rather than rodents, rabbits, or dogs and cats. It occurs in pigs, goats, sheep, and cattle mainly for teaching purposes.

The Brazilian Law[22] indicates in its Chapter IV, Article 14, paragraph 8, that an animal cannot be reused after the main objective of the research has been achieved. And paragraph 9 states that in teaching programs many traumatic procedures may be carried out in only one animal if all the procedures are performed under anesthesia, and that the animal be euthanized before recovering consciousness.

SETTING FREE/REHOMING

In Mexico, scientists are required to indicate how the animals will be disposed of at the end of the study. In general, rehoming of laboratory animals does not occur, although it is known that companion animals (dogs and cats), which have been socialized, are able to make the transition to a home environment. There are no programs on rehoming animals that have been used.

In Chapter IV, Article 14, paragraph 2, the Brazilian Law[22] clearly indicates that, exceptionally, animals used for demonstration purposes or research, and if not submitted to euthanasia, can be rehomed or set free after approval by the Ethics Committee and under the responsibility of an animal protection society or a responsible person.

OCCUPATIONAL HEALTH AND SAFETY

In Mexico, the NOM 062[8] has a specific section for occupational health and safety for those working with experimental animals. Section 10 of the NOM 062, "Biosafety and Health and Safety in the Workplace for Personnel Working with Laboratory Animals" provides information about regulatory requirements, dealing with biological hazards, zoonotic diseases, working with nonhuman primates safely, allergies, physical injuries, chemical hazards and radiation, and ultraviolet light.

In Brazil, Law 11794/08[22] also addresses this topic and Chapter IV, Article 14, paragraph 10 states that anyone working with animals in a closed system must follow international guidelines on safety and occupational health.

EDUCATION, TRAINING, AND COMPETENCE OF PERSONNEL

In Mexico, NOM 062[8] considers that appropriate institutional training programs are to be implemented as an essential component of any institutional animal care and use program.

Animal users (including scientists, graduate students, postdoctoral fellows, other research staff, and study directors), animal health professionals, animal care committee members, and institutional officials directly responsible for animal care and use programs should receive specific education and training according to their needs.

In Peru, the Law on the Protection of Animals,[13] Article 11, indicates that animals cannot be used for teaching purposes if this use will cause suffering to the animal.

Law n° 20.380 on the Protection of Animals, Article 7, from Chile[16] states that only qualified personnel may use laboratory animals in order to prevent their suffering. It defines qualified personnel as anyone who has studied veterinary, medical, or other related science, and is certified and recognized by a state academic institution.

Also, in Chile, the Law bans the use of animals in experiments in basic and middle levels of education. However, animals can be used in agricultural schools, as well as in higher education when their use is essential and there are no alternatives for learning.

The Brazilian Law[22] has no articles on this topic. However, education and training is one of the main concerns in this country. The new Brazilian guide, which is being prepared, emphasizes this subject. Although not required by law, there are a number of courses for technicians and researchers provided by various universities. Also, on site courses are given to personnel working at laboratory animal houses.

The available courses in Latin America can be found at: http://iclas.eventconsulting.be/wp-content/uploads/2012/08/LatinAmerica-Resources1.pdf.

TRANSPORT

There is no specific regulation in Mexico for the transport or mobilization of laboratory animals. However, the NOM 051 ZOO 1995 from the Mexican Agriculture Department[26] addresses the humane transportation of animals. The NOM 051 details requirements for the provision of food, water, rest, protection from adverse weather, use of proper containers and transport vehicles, and segregation of incompatible animals. This NOM also specifies that livestock and poultry must be accompanied by a health certificate issued or endorsed by a veterinary inspector certified by the Mexican Agriculture Department.

The Brazilian Law[22] does not mention transportation of laboratory animals, but the Brazilian Ministry of Agriculture is responsible for the humane transportation of animals. It requires following the OIE Principles for the Humane Transportation of Animals[27] and also internal rules from the Ministry of Agriculture.

HOUSING AND ENRICHMENT

In Mexico, NOM 062[22] gives specific guidelines on housing animals. Enclosure dimensions recommended by the NOM are only used as a bare minimum. The dimensions used are based on the NRC Guide for the Care and Use of Laboratory Animals.[6] Protocols which involve single housing must describe proposed measures for meeting the social requirements of the isolated animal.

As there is no mention of housing and enrichment in the Brazilian law,[22] the CEUAs are very demanding on this topic. Every project has to describe the housing and enrichment of the animals which will be used and they have to follow the specifications of the NRC Guide.[6]

HUSBANDRY AND ENVIRONMENT

In Mexico, NOM 062[8] considers general guidelines in relation to the daily care of animals in a laboratory animal facility. Moreover, it provides specific information on regulated species. This includes detailed requirements for the nutritional, minimum space requirements, and environmental conditions including some behavior recommendations.

In Mexico, NOM 062 requires that facilities for the care and use of all animals in research, all teaching and testing must be conducive to the well-being and safety of the animals, provide an appropriately-appointed and safe workplace for personnel, and establish a stable research environment. It is intended to assist the users and designers of laboratory animal facilities in achieving these objectives. The goal is to promote optimal levels of animal care and facilitate good research without curtailing new and innovative ideas for facility design. The NOM 062 includes guidelines on: spaces and areas, construction materials, sanitary requirements, localization and distribution areas, caging, bedding, nesting materials, etc.

The NOM 062 "guidelines on: spaces and areas" applies to animals such as mice, rats, gerbils, hamsters, guinea pigs, rabbits, dogs, cats, pigs, and nonhuman primates. The general principles described within the "guidelines on laboratory animal facilities" are applicable to most species maintained in captive environments for the purposes of research, teaching, and testing.

In Brazil the CEUAs also require the laboratory animal breeders and scientists to follow the NRC Guide principles on husbandry and environment.[6] This is already a solid integrated routine for almost all animal users in Brazil, and the ones who have not incorporated it yet are trying to do so.

VETERINARY CARE

In Mexico, NOM 062[8] refers that regarding laboratory animal health, a veterinarian with knowledge of laboratory animal medicine should be responsible for the daily health care. The veterinarian in charge must have not only authority and responsibility for supporting the animal care program, but also she/he must be involved in all issues and activities that relate in any way to animal care and use. Although the types of institutions and facilities widely vary, at any institution the senior director must make formal arrangements to ensure that veterinary services are readily available at all times to meet both routine and emergency needs. Animal caregivers and users must be able to report an animal health or welfare concern (e.g. injury, ill health, or death) at any time, and a veterinarian must be available to respond to the concern.

While the responsibilities relating to veterinary care for "traditional" laboratory animals is well-defined in the NOM 062, for some of the other species groups, there is no regulation at all when they are used as laboratory animals.

The Brazilian Law[22] has an annex (Annex 6) stating that any place having laboratory animals must have a veterinarian in charge of the animals. The veterinarian must have a specialization in laboratory animal medicine. As a result, this is now obliged by law.

CONDUCT OF EXPERIMENTAL PROCEDURES

In Chile, the Law[16] emphasizes the use of anesthesia for surgical interventions, and that these must be carried out by a veterinarian or other qualified professional.

The Brazilian Law[22] in its Chapter IV emphasizes the need to prevent suffering in animals (Article 4). If experiments will entail suffering or pain, analgesia or anesthesia must be used (Article 5) and projects that cause pain or suffering have to have special authorization by the CEUA (Article 6). The CEUAs are very strict on this point.

As stated before, in Mexico the IACUCs analyze the protocols and one of the main objectives is to thoroughly describe the suffering and pain caused to the animal and how to avoid it.

In the Venezuelan Law,[15] Article 54 states: "Domesticated animals which have been subjected to an experiment affecting its physical conditions in an irreversible manner or if death is inevitable, due to the research protocol, must be culled without pain."

In Cuba, there are guides for the determination of the humane endpoints in the Regulation 64/12,[28] and guidelines set forth to define the critical point or suspension of an experiment in case the animals present severe pain, suffering, are dying or have other conditions seriously affecting their welfare.

In Brazil and Mexico, the IACUCs and CEUAs use the Canadian Guide[7,7a] and the NRC Guide[6] as references and require scientists to follow these guidelines.

EUTHANASIA

In Mexico, the NOM 062[8] "guidelines on: euthanasia of laboratory animals" is based on recommendations made by the American Veterinary Medical Association (AVMA) Guidelines on Euthanasia 2007[9] with some modification to fit the Mexican context. An overview of acceptable methods of euthanasia for common species used for research, teaching, and testing primarily based on the reference documents is provided, and information is included on other methods of euthanasia that are not considered best practice, but that may be acceptable for specific purposes providing they comply with the general guiding principles and receive the approval of the IACUC reviewing the application.

In the Brazilian Law,[22] Chapter IV, Article 14, paragraph 1 requires that an animal will be euthanized when technically recommended after the end of the experiment or in any of its phases if the animal is suffering. The euthanasia must follow strict technical procedures according to the guidelines of the Ministry of Science and Technology. These guidelines are based on the AVMA Guidelines on Euthanasia 2007.

In Venezuela, the Law[15] states in its Article 6 that: "Death with no pain must be performed by a veterinarian or by a person under its supervision who must have the necessary experience, to guarantee that the death will not entail cruelty or pain to the animal."

NATIONAL LABORATORY ANIMAL SCIENCE ASSOCIATIONS

It is worth mentioning that national Laboratory Animal Science Associations in Latin America have been paramount in changing the way animals were used in experiments, testing, or teaching. Generally composed of a small number of scientists and minimal funding, they have been doing a wonderful job in organizing seminars, congresses, giving short courses, and mainly trying to call authorities' attention to the need of a law on the care and use of laboratory animals.

In some places they had success such as the Association for Laboratory Animal Science and Technology from Uruguay (AAUCyTAL; www.aucytal.org), which has had a strong role in the development of a regulatory framework for the ethical use of laboratory animals in this country.

The Brazilian Association, former COBEA, now called SBCAL (www.cobea.org.br/) also played a very important role in the development of the law and, ever since, it has been a reference for other Latin American countries.

Some, such as ABOCAL in Bolivia for instance, are in the process of organizing and achieving their legal status.

The current Laboratory Animal Science Associations in the area are:

ASOCHICAL (http://www.asochical.cl/). Chilean Association of Laboratory Animal Science.

AACyTAL (www.aacytal.com.ar). Argentinean Association for Laboratory Animal Science.

AVECAL (http://www.medicoveterinariovzla.com). Venezuelan Laboratory Animals Association.

ACCBAL Asociación Colombiana de la Ciencia y el Bienestar de los Animales de Laboratorio. CICUALES is a network of Ethics Review Committees very active in Colombia.

SCCAL Sociedad Cubana de la Ciencia del Animal de Laboratorio.

AMCAL (http://amcal-ac.blogspot.com.br). Asociación Mexicana de la Ciencia de Animales de Laboratorio.

ACCMAL Asociación Centroamericana, Caribe y Mexicana de Animales de Laboratorio.

Some of them are also members of the International Council for Laboratory Animal Science (http://iclas.org/), including those in Brazil, Chile, Uruguay, Mexico, Costa Rica, and Argentina.

An important step in Latin America has been the reactivation of the Federation of South American Societies and Associations of Laboratory Animal Science, to be the umbrella under which all the associations will work together.

REFERENCES

1. Ley Sarmiento n° 2786; Prohibición de Malos Tratos a los Animales, 1891. Argentina.
2. Decreto N° 24.645, de 10 de julho de, 1934: http://funed.mg.gov.br/wp-content/uploads/2010/05/Decreto-lei-24645-34-maus-tratos-animais.pdf. Brazil.
3. CIOMS. International Guiding Principles for Biomedical Research Involving Animals: http://www.cioms.ch/publications/guidelines/1985_texts_of_guidelines.htm.
4. Lei n° 6.638, de 8 de Maio de, 1979: http://www.lexml.gov.br/urn/urn:lex:br:federal:lei:1979-05-08;6638. Brazil.
5. *Australian code of practice for the care and use of animals for scientific purposes*. 7th ed. 2004: http://www.nhmrc.gov.au/publications/synopses/eA16syn.htm.
6. National Research Council. *Guide for the care and use of laboratory animals*. 8th ed. National Academy Press; 2011.
7. Canadian council on Animal Care. *Guide to the care and use of experimental animals*. 2nd ed. vol. 1, 1993: http://www.ccac.ca/en/CCAC_Programs/Guidelines_Policies/GUIDES/ENGLISH/toc_v1.htm.
7a. Canadian Council on Animal Care. *Guide to the care and use of experimental animals*, vol. 2, 1984. (Adopted May 1999): http://www.ccac.ca/en/CCAC_Programs/Guidelines_Policies/GUIDES/ENGLISH/TOC_V2.HTM.
8. Secretaría de Agricultura, Ganadería, Desarrollo Rural, Pesca y Alimentación. NOM-062-ZOO-1999. Especificaciones Técnicas para la Producción, Cuidado y Uso de Animales de Laboratorio. *Diario Oficial de la Federación* 22 de agosto de 2001Mexico: http://www.fmvz.unam.mx/fmvz/principal/archivos/062ZOO.PDF.
9. American Veterinary Medical Association (AVMA) Guidelines on Euthanasia, 2007.
10. Ley n° 18.611. Utilización de Animales en Actividades de Experimentación, Docencia e Investigación Científica, 2009: http://www.iibce.edu.uy/ETICA/ley-18611-oct-2-2009.pdf. Uruguay.
11. The European Parliament and the Council of the European Union. Directive 2010/63/EU of the European Parliament and of the Council of 22 September 2010 on the protection of animals used for scientific purposes. *Off J Eur Union* 2010L 276/33–79.
12. Estatuto Nacional de Protección Animal. Ley 84 de, 1989. Por la cual se adopta el Estatuto Nacional de Protección de los Animales y se crean unas contravenciones y se regula lo referente a su procedimiento y competencia: http://spac-05.tripod.com/id24.html. Colombia.
13. Ley n° 27265/2000 de Protección a los Animales Domésticos y a los Animales Silvestres Mantenidos en Cautiverio: http://www.congreso.gob.pe/ntley/Imagenes/Leyes/27265.pdf. Perú.

14. Instituto Nacional de Salud. Procedimiento para el Uso de Animales de Laboratorio en el Instituto Nacional de Salud, 2012: http://www.ins.gob.pe/repositorioaps/0/2/jer/normatividad_01/MANUAL%20PROCEDIMIENTOS%20CIEA.pdf. Perú.

15. Ley para la Protección de la Fauna Doméstica Libre y en Cautiverio. Gaceta Oficial N° 39.338 del 4 de enero de, 2010: http://es.scribd.com/doc/47545158/LEY-PARA-LA-PROTECCION-DE-LA-FAUNA-DOMESTICA-LIBRE-Y-EN-CAUTIVERIO. Asamblea Nacional de la República Bolivariana de Venezuela.

16. Ministerio de Salud, Subsecretaría de Salud Pública. Ley n° 20.380/2009 sobre Protección de Animales, 2009: http://www.leychile.cl/Navegar?idNorma=1006858. Chile.

17. Asamblea Legislativa de la República de Costa Rica. Ley de Bienestar de los Animales n° 7451, 1994: http://www.animallaw.info/nonus/statutes/stat_pdf/stcrlaw7451_welfare.pdf. Costa Rica.

18. Presidência da República, Casa Civil, Subchefia para Assuntos Jurídicos. Lei n° 9605/98: http://www.planalto.gov.br/ccivil_03/leis/L9605.htm regulated by the Decreto n° 3179/99 http://www.planalto.gov.br/ccivil_03/decreto/D3179.htm.

19. Ley General de Salud, 2007: http://www.salud.gob.mx/unidades/cdi/legis/lgs/index-indice.htm. Mexico.

20. Ley General de Bienestar Animal para el Estado de Chihuahua, 2010: http://www.congresochihuahua.gob.mx/biblioteca/leyes/archivosLeyes/689.pdf. Mexico.

21. OIE. *Terrestrial animal health code.* Chapter 7.8: Use of animals in research and education, 2012: http://www.oie.int/index.php?id=169&L=0&htmfile=chapitre_1.7.8.htm.

22. Presidência da República, Casa Civil, Subchefia para Assuntos Jurídicos. Lei n° 11.794 de 8 de Outubro, 2008 (Arouca Law): http://www.planalto.gov.br/ccivil_03/_ato2007-2010/2008/lei/l11794.htm. Brazil.

23. El Senado y la Cámara de Representantes de la República Oriental del Uruguay. Ley N° 18.611, Utilización de Animales en Actividades de Experimentación, Docencia e Investigación Científica, 2009: http://www.parlamento.gub.uy/leyes/AccesoTextoLey.asp?Ley=18611&Anchor. Uruguay.

24. Cámara de Diputados del H. Congreso de la Unión. Ley de Bioseguridad de Organismos Genéticamente Modificados Diario Oficial de la Federación 18 de marzo, 2005: http://www.conacyt.gob.mx/ElConacyt/Documentos%20Normatividad/Ley_BOGM.pdf. Mexico.

25. Lei n° 8.794, 1995 regulated by the Decreto n° 1.752, 1995. Brazil.

26. Dirección General Jurídica de la Secretaría de Agricultura, Ganadería y Desarrollo Rura. NOM 051 ZOO, 1995: http://www.cmp.org/NORMAS/NOM-051-ZOO-1995.pdf. Mexico.

27. OIE. Terrestrial Animal Health Code. Section 7, Animal Welfare, 2012: http://www.oie.int/en/international-standard-setting/terrestrial-code/access-online/?htmfile=titre_1.7.htm.

28. Ministerio de Salud Pública. Regulación 64/12, Lineamiento para la Constitución y Funcionamiento de los Comités Institucionales para el Cuidado y Uso de los Animales de Laboratorio: http://www.cecmed.cu/Docs/RegFarm/DRA/DispGen/Reg/Reg_%2064-2013.pdf2012. Cuba.

Chapter 5

The European Framework on Research Animal Welfare Regulations and Guidelines

Javier Guillén[1], Jan-Bas Prins[2], David Smith[3], and Anne-Dominique Degryse[4]

[1]AAALAC International, Pamplona, Spain, [2]Leiden University Medical Centre, Leiden, The Netherlands, [3]FELASA, Chester, UK, [4]Pierre-Fabre, Service de Zootechnie, Institute de Recherche Pierre-Fabre, Castres, France

Chapter Outline

Laboratory Animals. http://dx.doi.org/10.1016/B978-0-12-397856-1.00005-2

GENERAL FRAMEWORK

It would be an impossible task to describe a general framework for a continent of 50 countries if not for the existence of two political organizations that
group them up to some extent. These organizations are the Council of Europe
(CoE) and the European Union (EU). Both organizations have issued the documents that constitute the European framework on the protection, care, and use of
laboratory animals that will be dissected in this chapter.

The Council of Europe

The CoE (http://www.coe.int) is an intergovernmental organization that virtually
covers the entire European continent, with its 47 member countries including all

of the 28 EU Member States. It seeks to develop throughout Europe common and democratic principles based on the European Convention on Human Rights and other reference texts on the protection of individuals. The CoE has no legislative power and seeks voluntary cooperation within the member countries through Recommendations, Agreements, and Conventions. Member countries can sign and ratify the Conventions. If they do this, they are bound to implement the Convention into their national legislation. One of these Conventions is the European Convention for the Protection of Vertebrate Animals used for Experimental and other Scientific Purposes,[1] published in the European Treaty Series 123 and better known as ETS 123.

European Treaty Series 123

ETS 123 has 38 articles distributed in 12 Parts, and 2 Appendices (A and B). The first Part of ETS 123 (Articles 1–4) relates to the General Principles, where the scope and definitions are included. Part II (Article 5) focuses on General Care and Accommodation, and refers to Appendix A that contains the Guidelines for Accommodation and Care of Animals. Appendix A, revised in 2006[2] is one of the most important documents of the European framework because of the impact of its recommendations (for all common species) on housing (cage size especially) and on environmental enrichment and care, which have been incorporated in the legislation at the EU level. The minimum cage sizes recommended in the revised Appendix A are significantly larger than those in the original version, and this remains a controversial issue. Appendix A has short general recommendations on physical facilities, environment and its control, and care; and a long species-specific section where recommendations on environment, health, housing, enrichment, and care are offered for all the commonly used species.

ETS 123 has been signed and ratified by a number of member countries, which means that its provisions should have entered into force by means of national legislation. The list of countries that have done so and the dates are available at: http://www.conventions.coe.int/Treaty/Commun/ChercheSig.asp? NT=123&CM=8&DF=21/08/2012&CL=ENG.

Most of the dates of entry into force of national laws adopting ETS 123 are previous to the revision of Appendix A. The final adoption of many of the Guidelines for Accommodation and Care of Appendix A in the European Union has occurred thanks to the transposition of the new Directive 2010/63/EU which includes these standards as a mandatory requirement.

Part III of ETS 123 (Articles 6–12) discusses the conduct of procedures, stressing (implicitly) the implementation of the principles of the 3R. The articles in Part III focus on the use and promotion of alternative methods, choice of species, minimization of pain and distress, reuse, and euthanasia.

The authorization of procedures and persons carrying out procedures is required in Part IV (Article 13).

Administrative measures for breeding or supplying establishments are defined in Part V (Articles 14–17), which includes registration of establishments and records and identification of animals.

Part VI (Articles 18–24) deals with the requirements for user establishments including those related to animal needs and personnel (responsible persons, veterinarian, and training). A list of animals required to be purpose-bred (originated from breeding establishments) is included. In this original list, the nonhuman primates are not included.

The education and training of personnel is specifically addressed in Part VII (Articles 25–26), and the statistical information in Part VIII (Articles 27–28), which refers to Appendix B containing the statistical tables. Parts VIII–XII (Articles 30–38) relate to legal aspects on the implementation of the Convention.

The importance of ETS 123 relies on the fact that it was the first document trying to establish a common European framework on the protection, care and use of laboratory animals. The principles and some of the content in ETS 123 were the basis for the subsequent legislation at the EU level, Directive 86/609/EEC[3] that established the minimum legal requirements applicable in the EU for 25 years. This old Directive included the care and accommodation standards (i.e. cage sizes) of the original Appendix A of ETS 123, although only as a recommendation. Also, definitions and scope closely resembled those in ETS 123. However, this Directive failed in harmonizing the standards across the EU, as it allowed Member States to go beyond its provisions. Some Member States issued more strict rules (e.g. UK) while others just transposed the Directive literally.

The European Union

The European Union is a supra-national economic and political partnership between 28 European countries. Member States delegate sovereignty to the EU, so this can develop its activities on the treaties that are approved voluntarily and democratically by all of the Member States. The European Commission (EC) is the EU body responsible to propose new legislation to the European Parliament and the Council of the European Union (which is not the CoE described earlier) under the EU treaties. The EC is also responsible to make sure the adopted legislation is implemented correctly by the Member States. The legislation can be in the form of Regulations, which are rules that must be complied with directly by the Member States without being transposed into national legislation; and Directives, which give Member States some time to transpose and implement them.

Directive 2010/63/EU

Following the revision process of Appendix A of ETS 123, the EC started the revision of Directive 86/609/EEC with the main objectives of harmonizing the legislative framework within the EU (leveling the playing field), increasing significantly the animal welfare, and implementing explicitly the principles of replacement, reduction and refinement (3Rs principles). During this process, the EC had already issued a Recommendation to follow the guidelines of the revised Appendix A of ETS 123 with regard to the care and accommodation standards, as a first step of incorporating it in the future legislation.[4] The revision of the old 1986 Directive finalized with the publication of Directive 2010/63/EU.[5] The Directive gave 24 months to Member States for the transposition into national legislation. This period ended in November 2012, and the different transpositions at national level entered into force in January 2013.

One of the most important features of the new Directive, in contrast to the old one, is that it does not allow Member States to establish stricter measures. This provision applies one of the main objectives of the EC, which is to level the playing field in the EU with regard to the protection and use of laboratory animals. Therefore, in theory, all Member States should be applying very similar, if not equal, national laws. However, it will be discussed how this is not entirely true with regard to several of the provisions, as they can be interpreted and implemented differently in practice. Member States are allowed to communicate and maintain stricter measures that were in place before the publication of the Directive. The other main objectives of the Directive are pursued by the explicit requirement of the implementation of the 3Rs principles across the whole document.

Directive 2010/63/EU has an introductory part or preamble with 56 recitals that precedes the provisions distributed in 66 articles (in 6 different Chapters), and 8 important Annexes.

The first Chapter (Articles 1–6) relates to the General Provisions, including the scope, definitions, the application of stricter measures, the 3Rs principles, the permitted purpose of experimental procedures, and the methods of euthanasia (or "killing"). Article 6 on "methods of killing" refers to Annex IV, which lists the accepted methods by species.

Chapter II (Articles 7–11) establishes the provisions on the use of certain animals in procedures, such as endangered species, nonhuman primates, animals taken from the wild, animals that have to be purpose-bred, and stray and feral animals of domestic species. Article 10 on purpose-bred animals refers to Annex I which lists the 12 different species that must be purpose-bred to be used in procedures, and to Annex II which lists the species of nonhuman primates that are subjected to the second generation requirement. The provisions on nonhuman primates are particularly important because they are especially restrictive.

Requirements on the conduct of procedures are detailed in Chapter III (Articles 12–19). Articles refer to issues such as the choice of methods (by application of the 3Rs principle), provision of anesthesia and analgesia, the mandatory classification of the severity of procedures, the reuse of animals, the sharing of organs and tissues or the potential setting free of animals and rehoming. The requirement for the classification of severity of procedures of Article

15 is particularly important. Examples of procedure classification can be found in Annex VIII.

Chapter IV (Articles 20–45) is the most extensive. Its first section (Articles 20–33) focuses on the requirements for breeders, suppliers and users. In addition to the required administrative authorization, they refer to installations and equipment, personnel (competence and several required positions including the Designated Veterinarian), Animal Welfare Body (AWB) (the institutional oversight body), records, special considerations for nonhuman primates, cats and dogs, and care and accommodation. Articles 22 and 33 refer to Annex III for installations and equipment, and care and accommodation, respectively. Annex III evidences the importance of the Appendix A of ETS 123, because it contains part of its provisions, especially the minimum cage sizes. In contrast with the old Directive, where the provisions of Appendix A were in the form of recommendations, Directive 2010/63/EU requires the obligatory compliance with these minimum standards.

Minimum requirements for training and education of personnel are set out in Annex V. However, the EC has no competence in education, as it is at Member State level. Therefore Annex V only includes general topics which are not even assigned to the specific categories of responsible personnel.

The second Section of Chapter IV (Articles 34–45) defines the inspections and controls by the competent authorities, and the third Section focuses on the requirements for research projects. This third Section is particularly important, as it describes how the projects have to be evaluated and authorized before being initiated. Annex VI lists the elements that project applications must describe to be evaluated as required in Article 37.

The alternative approaches are addressed in Chapter V (Articles 46–49), which also includes the requirement for Member States to establish national committees for the protection of animals used for scientific purposes (Article 49). Article 48 refers to Annex VII, which defines the duties and tasks of the Union Reference Laboratory (the Commission's Joint Research Centre) to coordinate and promote the development of alternative methods.

The final Chapter VI (Articles 50–66) contains the Final Provisions, such as the requirement for reporting the use of animals in procedures, the safeguard clauses, and the definition of competent authority (CA).

The provisions of Directive 2010/63/EU have been transposed into the legislation of the EU Member States. Although there is a common legislative general framework within the EU, there may be some differences in the interpretation and implementation of some of the Directive provisions. In terms of collaborative research, it is recommended to review the specific regulations at national level.

Although the legal framework at the EU level covers a majority of the animal research conducted in Europe, there are many countries out of the EU that have other specific legislation in place or no legislation on laboratory animal protection at all. For example, countries such as Switzerland and Norway have modern regulations, while others lack them or have outdated ones.

In addition to the political organizations that frame the regulatory care and use of laboratory animals in Europe, other professional organizations have issued important guidelines that have significant influence on the ground. The most representative is the Federation of European Laboratory Animal Science Associations (FELASA), an organization that comprises many of the national/regional laboratory animal science associations in Europe. FELASA has published a number of guidelines and recommendations resulting from the collaborative work of highly respected laboratory animal science professionals.[6] Especially noteworthy are the sets of recommendations on health monitoring and on training and education. The health monitoring recommendations are widely followed by breeders and users across Europe, and the training and education recommendations have even served as the basic scheme for some regulations at national level.

All taken together, European regulations and guidelines compose a detailed framework for the protection, care and use of laboratory animals. The umbrella organizations (Council of Europe and European Union) have allowed the development of a framework based on the same principles but that still has some practical differences across countries. The areas addressed by the main European regulations and guidelines are described in the following sections.

THE PRINCIPLES

The European framework is evidently based upon the principles of Replacement, Reduction and Refinement. Although historically these principles were not mentioned explicitly in the main regulations, they were included implicitly in the various articles. ETS 123[1] is a very good example. Replacement is addressed in Article 6.1, which states that "A procedure shall not be performed for any of the purposes referred to in Article 2, if another scientifically satisfactory method, not entailing the use of an animal, is reasonably and practicably available." Also related to Replacement is Article 6.2 in which it is stated that "Each Party should encourage scientific research into the development of methods which could provide the same information as that obtained in procedures."

The principles of Reduction and Refinement can be identified in several articles, such as Article 8 (anesthesia and analgesia) and Article 7 which is the most obvious example: "When a procedure has to be performed, the choice of species shall be carefully considered and, where required, be explained to the responsible authority; in a choice between procedures, those should be selected which use the minimum number of animals, cause the least pain, suffering, distress or lasting harm and which are most likely to provide satisfactory results." Also, Article 5 and the related entire Appendix A[2] with all the guidelines for accommodation and care can be considered a call to refinement.

Similar examples could be extracted from the old Directive 86/609/EU[3] that established the basic framework in the European Union (EU) for many years.

However, the significance of the 3Rs principles is more evident in the recent legislation at the EU level. In fact, Directive 2010/63/EU[5] requires explicitly the implementation of replacement, reduction and refinement practices, with particular emphasis on replacement. This emphasis is evident from the preamble of the Directive, where it is stated that the Directive is a significant step toward achieving the final goal of full replacement of procedures as soon as it is scientifically possible (Recital 10). Recital 11 acknowledges that the care and use of laboratory animals is governed by the internationally established principles of the 3Rs, and that these principles should be considered systematically when implementing the Directive.

After the recitals, Article 4 of the Directive is explicit on the 3Rs principles, and defines specific requirements. Replacement is addressed in Article 4.1: "Member States shall ensure that, wherever possible, a scientifically satisfactory method or testing strategy, not entailing the use of live animals, shall be used instead of a procedure." Reduction is addressed in Article 4.2: "Member States shall ensure that the number of animals used in projects is reduced to a minimum without compromising the objectives of the project." In addition, Refinement is addressed in Article 4.3: "Member States shall ensure refinement of breeding, accommodation and care, and of methods used in procedures, eliminating or reducing to the minimum any possible pain, suffering, distress or lasting harm to the animals."

In addition to Article 4 that focuses specifically on the principle of the 3Rs (the title of Article 4 is: "Principle of replacement, reduction and refinement"), other clear references to the 3Rs can be found along this Directive. Article 13 ("Choice of methods"), states that "Member States shall ensure that a procedure is not carried out if another method or testing strategy for obtaining the result sought, not entailing the use of a live animal, is recognised under the legislation of the Union." It also states that the procedures to be used are those that "(a) use the minimum number of animals; (b) involve animals with the lowest capacity to experience pain, suffering, distress or lasting harm; (c) cause the least pain, suffering, distress or lasting harm; and are most likely to provide satisfactory results."

The same principles are linked to the ethical review process (ERP), which includes oversight, ethical evaluation of projects and more technical evaluation at procedure level. One of the tasks assigned to the AWB (Article 27) is to "advise the staff on the application of the requirement of replacement, reduction and refinement, and keep it informed of technical and scientific developments concerning the application of that requirement"; another task is to "follow the development and outcome of projects, taking into account the effect on the animals used, and identify and advise as regards elements that further contribute to replacement, reduction and refinement." In addition to this, the project (ethical) evaluation (Article 38) shall include "an assessment of the compliance of the project with the requirement of replacement, reduction and refinement"; and one of the areas of expertise required for the evaluation by the CA is "the areas of scientific use for which animals will be used including replacement, reduction and refinement in the respective areas." The "elements that may contribute to the further implementation of the requirement of replacement, reduction and refinement" have to be part of the retrospective assessment of projects (Article 39) and "a demonstration of compliance with the requirement of replacement, reduction and refinement" is to be part of the nontechnical summaries required for the project authorization process (Article 43). Also, the requirement of replacement, reduction and refinement is one of the elements listed in Annex V as minimum requirements for training and education of personnel.

Other articles develop requirements on specific areas related to the 3Rs, such as accommodation and care (Article 33), anesthesia (Article 14), and reuse (Article 16). Chapter V is entirely dedicated to avoidance of duplication of procedures and alternative approaches to animal use.

The same 3Rs principles govern the legislation at the national level, both within and outside of the EU, and they represent the common area of all European regulations on animal protection, care, and use. The differences between European countries are not in the principles, but may be in some of the engineering standards used for their implementation. The new regulatory framework in the EU has helped harmonizing the legislation to this respect within the EU, and other countries out of the EU may follow a similar approach.

SCOPE AND APPLICABILITY: ACTIVITIES, ANIMALS, AND INSTITUTIONS COVERED

To describe just the scope of the European legal framework is one thing. In order to fully appreciate the scope, one should include some of the motivation behind it. The motivation and rationale behind European Directive 2010/63/EU can be found in its 56 recitals.[5] As for ETS 123, its reasons and aims are described in a preamble.[1] Although they cover similar issues, the Convention's preamble uses more generic terms while the Directive's recitals are more specific. The preamble of the Convention recalls that the aim of the Council of Europe is to achieve greater unity between its members and that it wishes to cooperate with other States in the protection of the live animals used for experimental and other scientific purposes. It recognizes that man has a moral obligation to respect all animals and to have due consideration for their capacity for suffering and memory. It accepts nevertheless that man in his quest for knowledge, health and safety has a need to use animals where there is a reasonable expectation that the results will extend knowledge or be to the overall benefit of man or animal, just as he uses them for food, clothing, and as beasts of burden. It resolves to limit the use of animals for experimental and other scientific purposes with the aim of replacing such uses wherever practical, in particular by seeking alternative measures and encouraging the use of these alternative measures. Its intention is to adopt common provisions in order to protect animals used in those procedures that may possibly cause pain, suffering, distress or lasting harm and to ensure that where avoidable they shall be kept to a minimum.

Recital 1 of Directive 2010/63/EU,[5] states that it should provide more detailed rules in order to reduce disparities between Member States, i.e. establish harmonization across Member States regarding the national levels of protection of animals used for scientific procedures to ensure a proper functioning of the internal market. Since the previous Directive 86/609/EEC,[3] new scientific knowledge is available with respect to factors influencing animal welfare as well as the capacity of animals to sense and express pain, suffering, distress and lasting harm. It is therefore necessary to improve the welfare of animals used in scientific procedures by raising the minimum standards for their protection in line with the latest scientific developments (Recital 6). The Directive should be reviewed regularly in light of evolving science and animal-protection measures (Recital 10). The care and use of live animals for scientific procedures is governed by the 3Rs of laboratory animal science: replacement, reduction and refinement. To ensure the way in which animals are bred, cared for and used in procedures within the European Union is in line with that of the other international and national standards applicable outside the Union, the principles of the 3Rs should be considered systematically when implementing 2010/63/EU (Recital 11). The implementation of the 3Rs in projects is promoted by comprehensive project evaluation, taking into account ethical considerations in the use of animals, and project authorization (Recital 38). Furthermore, it is thought that the welfare of the animals used in procedures is highly dependent on the quality and professional competence of the personnel involved. Hence, it is essential and should be ensured that staff is adequately educated, trained and competent (Recital 28).

Chapter 1 of 2010/63/EU deals with the general provisions. The subject matter and scope are laid out in Article 1. First it is explicated that the Directive establishes measures for the protection of animals used for scientific or educational purposes. To that end, it lays down rules on (Article 1.1):

a. the replacement and reduction of the use of animals in procedures and the refinement of the breeding, accommodation, care, and use of animals in procedures;
b. the origin, breeding, making, care of accommodation, and killing of animals;
c. the operations of breeders, suppliers, and users;
d. the evaluation and authorization of projects involving the use of animals in procedures.

The Directive applies where animals are used or intended to be used in procedures, or bred specifically so that their organs or tissues may be used for scientific purposes. A procedure means any use, invasive or noninvasive, of an animal for experimental or other scientific purposes, with known or unknown outcome, or educational purposes, which may cause the animal a level of pain, suffering, distress or lasting harm equivalent to, or higher than, that caused by the introduction of a needle in accordance with good veterinary practice (Article 3.1). The definition of a procedure according to

Convention ETS 123 Article 1.2 is different from that of the definition in the Directive, and reads as follows: "procedure means any experimental or other scientific use of an animal which may cause it pain, suffering, distress or lasting harm, including any course of action intended to, or liable to, result in the birth of an animal in such conditions, but excluding the least painful methods accepted in modern practice (that is'humane'methods) of killing or marking animals."

Directive 2010/63/EU applies until the animals have been killed, rehomed or returned to a suitable habitat or husbandry system. The Directive applies also when anesthetic, analgesia or other methods have been applied successfully for the elimination of pain, suffering, distress or lasting harm (Article 1.2).

The following animals are included in Directive 2010/63/EU (Article 1.3):

1. live nonhuman vertebrate animals, including:
   a. independently feeding larval forms; and
   b. fetal forms of mammals as from the last third of their normal development;
2. live cephalopods.

The Directive also applies to animals in procedures, which are at an earlier stage of development than the last third of their normal development, if the animal is to be allowed to live beyond that stage of development and, as a result of the procedure performed, is likely to experience pain, suffering, distress or lasting harm after it has reached that stage of development (Article 1.4).

Directive 2010/63/EU shall apply without prejudice to the Directive of the European Council 76/78/EEC on the approximation of the laws of the Member States relating to cosmetic products (Article 1.6).[7]

The following purposes are covered and accepted by 2010/63/EU (Article 5):

1. basic research;
2. translational or applied research with any of the following aims:
   a. the avoidance, prevention, diagnosis or treatment of disease, ill-health or other abnormality or their effects in human beings, animals or plants;
   b. the assessment, detection, regulation or modification of physiological conditions in human beings, animals or plants; or
   c. the welfare of animals and the improvement of the production conditions for animals reared for agricultural purposes;
3. for any of the aims in point (2) in the development, manufacture or testing of the quality, effectiveness and safety of drugs, foodstuffs and feed-stuffs and other substances or products;
4. protection of the natural environment in the interests of the health or welfare of human beings or animals;
5. research aimed at preservation of the species;
6. higher education, or training for the acquisition, maintenance or improvement of vocational skills;
7. forensic inquiries.

Although the Directive is specifically meant for the protection of animals used for scientific purposes, Article 1.5 has been included to explicate those areas where the Directive does not apply. Those areas are covered by other pieces of (inter)national legislation. These areas are:

a. nonexperimental agricultural practices;
b. nonexperimental clinical veterinary practices;
c. veterinary clinical trials required for the marketing authorization of a veterinary medicinal product;
d. practices undertaken for the purposes of recognized animal husbandry;
e. practices undertaken for the primary purpose of identification of an animal;
f. practices not likely to cause pain, suffering, distress or lasting harm equivalent to, or higher than, that caused by the introduction of a needle in accordance with good veterinary practice.

One of the purposes of 2010/63/EU is harmonization of legislation across the Member States of the European Union. However, Article 2 allows for stricter national measures to be maintained, while observing the general rules laid down in the Treaty on the Functioning of the European Union.[8] These measures had to be in force already on November 9, 2010 and aimed at ensuring more extensive protection of animals falling within the scope of 2010/63/EU than those contained in that Directive (Article 2.1). However, Member States shall not prohibit or impede the supply or use of animals bred or kept in another Member State in accordance with 2010/63/EU, nor shall it prohibit or impede the placing on the market of products developed with the use of such animals in accordance with 2010/63/EU.

The scope and applicability of the Convention ETS 123 are very similar to that of the Directive albeit in more condensed and perhaps again more generic terms and are laid down in Article 1.1 and 2 of ETS 123. The Convention applies to any animal used or intended for use in any experimental or other scientific procedure where that procedure may cause pain, suffering, distress or lasting harm. It does not apply to any nonexperimental agricultural or clinical veterinary practice (ETS 123, Article 1.1). According to ETS 123, a procedure may be performed for any one or more of the following purposes only and subject to the restrictions laid down in the Convention:

a. Avoidance or prevention of disease, ill-health or other abnormality, or their effects, in man, vertebrate or invertebrate animals or plants, including the production and the quality, efficacy and safety testing of drugs, substances or products; diagnosis of treatment of disease, ill-health or other abnormality, or their effects, in man, vertebrate or invertebrate animals or plants;
b. Detection, assessment, regulation or modification of physiological conditions in man, vertebrate and invertebrate animals or plants;
c. Protection of the environment;
d. Scientific research;
e. Education and training;
f. Forensic inquiries.

The similarities between the Convention and the Directive are evident. However, the different wording is a potential source of differences in interpretation. Although the English versions are leading, translations are another source of interpretational differences. The EC is, therefore, maintaining an active Q&A document on the legal understanding of Directive 2010/63/EU.[9]

REQUIREMENTS FOR BREEDERS, SUPPLIERS, AND USERS

In this section, the requirements for breeders, suppliers and users to comply with Directive 2010/63/EU[1] are considered and compared with the previous Directive 86/609/EEC[3] and with the existing Council of Europe Convention for the Protection of Vertebrate Animals used for Experimental and Other Scientific Purposes (ETS 123).[1] The major difference with that of previous legislation is the harmonization of breeder and supplier requirements with the more onerous requirements of user establishments. Included in these requirements is the need for initial authorization and registration of the establishment with the CA, the nature of government inspections by the CA, and the subsequent penalties for noncompliance. The changed requirements in annual reporting of the number of animals used in procedures together with the new requirements for periodic reporting and provision of published nontechnical summaries are considered. Directive 2010/63/EU provides definitions of breeder, supplier, and user in Article 3 but these are subtly changed from those given in the original Directive 86/609/EEC and in ETS 123. Rather than referring to a breeding, supplying or user establishment they now refer to a "natural or legal person" at that establishment. The impact of this change during transposition into national law is not yet fully understood. The relevant definitions are:

"establishment"—any installation, building, group of buildings or other premises and may include a place that is not wholly enclosed or covered and mobile facilities.

"breeder"—any natural or legal person breeding animals referred to in Annex I (a list of purpose-bred animals) with a view to their use in procedures or for the use of their tissue or organs for scientific purposes, or breeding other animals primarily for those purposes, whether for profit or not.

"supplier"—any natural or legal person, other than a breeder, supplying animals with a view to their use in procedures or for their tissue or organs for scientific purposes, whether for profit or not.

"user"—any natural or legal person using animals in procedures, whether for profit or not.

Authorization of Breeders, Suppliers, and Users

Article 20 of Directive 2010/63/EU sets down the requirements for breeders, suppliers and users to gain authorization before work can commence in an establishment. Article 21 deals with the unlikely suspension and withdrawal of authorization. The requirements are now harmonized for breeders, suppliers and users. Each application for authorization must identify the person responsible for ensuring compliance with all the provisions of the Directive (for example known previously as the "Certificate Holder" in the UK, and that could be compared to the Institutional Official in the US). It must also identify the person or persons responsible for overseeing welfare, for ensuring staff have access to species-specific information and for ensuring staff are educated, competent, and continuously trained. The application should also identify a designated veterinarian with expertise in laboratory animal medicine or a suitably qualified expert where more appropriate. The duties of these responsible persons are described fully in a later section.

Further requirements of this authorization are the need to notify the CA of any changes of the person or persons referred to above or any changes to the structure or the function of the establishment. Authorizations may be granted for a limited period. Suspension or withdrawal of authorization may occur when the breeder, supplier or user no longer complies with the requirements set out in the Directive. The CA will take appropriate remedial action, or require such action to be taken, or suspend or withdraw the authorization. Where this happens the CA must ensure that the welfare of the animals is not adversely affected.

Penalties

Member States have been charged with laying down rules on penalties applicable to infringements of the national provisions adopted (Article 60). These must be effective, proportionate and dissuasive and notified to the Commission. Clearly, there is the possibility of different sanctions in each Member State but the Commission intends to review these using its regular National Contact Point meetings as a vehicle.

Inspections

Articles 34 (Inspections by the Member States) and 35 (Controls of Member State inspections) describe the requirements for inspections by competent authorities. There was a marked divergence in the frequency of inspections carried out in Member States before the revision of the Directive and the outcome was tailored to the lowest common denominator. Member States that were already performing more frequent inspections are at liberty to continue this practice in accordance with Article 2 (Stricter national measures). The Directive calls for an inspection frequency based on a risk analysis with the factors of species and number, prior history of compliance and the number and types of project to be taken into account. Inspections will be carried out on at least one third of the users each year in accordance with this risk analysis, which in practice could mean only one inspection every three years. The exception to this model is for breeders, suppliers and users of nonhuman primates who will be inspected at least once a year.

Reporting—Annual Statistics

Increased transparency of the use of animals for scientific purposes was a key objective when Directive 86/609/EEC was revised. Statistical information required by ETS 123 is limited to:

a. The numbers and kinds of animals used in procedures.
b. The numbers of animals in selected categories used in procedures directly concerned with medicine and in education and training.
c. The numbers of animals in selected categories used in procedures for the protection of man and the environment.
d. The numbers of animals in selected categories used in procedures required by law.

The statistical requirements of 86/609/EEC were disproportionately focused on regulatory toxicology with little information on the use of animals in fundamental or applied research. To this end, an Expert Working Group (EWG) was established by the Commission to consider how to present more meaningful information to the public. Article 54.2 of Directive 2010/63/EU sets out the requirements for annual reporting of the number of animals used in procedures. Member States are required to make publicly available statistical information on the use of animals in procedures, including information on the actual severity of the procedures and on the origin and species of nonhuman primates used in procedures. This information is required from November 2015 (for the year 2014) and every year thereafter. Further details of the statistical requirements have been published on the Commission website (http://ec.europa.eu/environment/chemicals/lab_animals/statistics_en.htm) and can be summarized as follows:

The revised statistical requirement represents a shift from prospective reporting to one of retrospective reporting and statistics will contain information on both the number of animals and number of procedures. The severity of each procedure needs to be recorded at its termination and reported annually. Where studies continue over two calendar years, all of the animals may be accounted for together in the year in which the last procedure ends. This system of retrospective reporting of severity has been in use in the Netherlands and Switzerland but that now required within 2010/63/EEC has some differences. The Netherlands uses a six-scale severity classification system compared to the Directive's four (nonrecovery, mild, moderate and severe). The Swiss system provides data based on the "animal" while the Directive requires both numbers of "procedures" and "animals" to be reported.

Reporting is now required for genetically altered (GA) animals used for the creation of a new line and for the maintenance of a line bred under project authorization and exhibiting a harmful phenotype. To take account of political and public concern, a one year snapshot of all GA animals not reported in annual statistics will be made every 5 years as part of the implementation report (see below). No annual statistical data collection will be made for genotyping carried out under project authorization but this will be included in the one year snapshot. Reporting on all genotyping methods will also be included in the 5-yearly implementation report.

Additional information is now required for animals that are reused. Information will be reported on each reuse if this has been previously authorized.

Animals killed for tissues, as well as sentinels, are excluded from the provision of statistical data unless the killing is performed under a project authorization using a method not included in Annex IV (Methods of killing animals) of the Directive. Surplus animals that are killed are not included apart from GA animals bred under project authorization and exhibiting a harmful phenotype.

Larval forms are to be counted once they become capable of independent feeding. Fetal and embryonic forms of mammalian species are not counted, only animals that are born, including by caesarian section, and live, are to be counted.

Despite the agreement by Member States to adopt a common statistical reporting framework, some may collect additional data if they were doing so under previous legislation.

Reporting—Implementation of the Directive

Within Article 54.1, a new requirement for periodic (5 yearly) reporting of information relating to the implementation of the Directive is included. In addition to the GA cases mentioned above, there will be a need to report on progress with a number of articles, for example, the breeding strategy for nonhuman primates (Article 28), nontechnical project summaries (Article 43), and avoidance of duplication of procedures (Article 46).

Reporting—Periodic Reports

Additional requirements for reporting are contained in Article 58 (Review). There is now a requirement for the Commission to review the Directive by November 10, 2017, taking into account advancements in the development of alternative methods not entailing the use of animals, in particular of nonhuman primates. Amendments will be proposed where appropriate. The Commission, in consultation with Member States and stakeholders, will also conduct periodic thematic reviews of the replacement, reduction and refinement of the use of animals in procedures, paying specific attention to nonhuman primates, technological developments, and new scientific and animal-welfare knowledge.

Reporting—Nontechnical Project Summaries

A new requirement for publishing nontechnical project summaries is included in Directive 2010/63/EU. This is contained in Article 43 and states that, subject to safeguarding intellectual property and confidential information, the nontechnical summary shall provide the following:

a. Information on the objectives of the project, including the predicted harm and benefits and the number and types of animals to be used.
b. A demonstration of compliance with the requirement of replacement, reduction and refinement.

Within the UK, such lay summaries have been published by the CA for a number of years without concern and have been found to be very valuable during the ERP within establishments.

NONHUMAN PRIMATES: SPECIAL CONSIDERATIONS AND RESTRICTIONS

As stated above, the revision of Directive 86/609/EEC[3] is the result of a 10-year long process during which viewpoints of various stakeholders were examined and discussed. One major matter of discussion was the continued need for use of nonhuman primates in biomedical research. Due to their genetic proximity to human beings and their highly developed social skills, their use raises specific ethical and practical challenges in terms of meeting their behavioral, environmental and social needs in a laboratory environment. Moreover, the use of nonhuman primates is of the greatest concern to the public. Limiting their use with the final goal of banning was an objective clearly stated in the *J Evans* report in 2002.[10]

In Europe, various countries have banned the use of Great Apes since more than 10 years ago. The statistical report of 2005, as published by the EU shows that, as in 2002, no Great Apes were used in experiments in 2005.[11] Also, compared to the USA, the number of nonhuman primates used in biomedical research is limited. Whereas the total number of animals used in 2005 in the EU amounts to 12.1 million animals, nonhuman primates merely represent 0.1%.

Nevertheless, after lengthy discussions and debates,[12] the allowed use of nonhuman primates was clearly defined in Article 8 of the new Directive 2010/63/EU[5]: specimens of nonhuman primates shall not be used in procedures with the exception of those procedures meeting the following conditions:

Only when there is scientific justification that the purpose of the procedure cannot be achieved by the use of species other than nonhuman primates, and on condition that procedures are undertaken with a view to the avoidance, prevention diagnosis or treatment of debilitating or potentially life-threatening clinical conditions in human beings, or for fundamental research or for research aimed at the preservation of the species is their use allowed.

The use of Great Apes is prohibited, although under exceptional circumstances a safeguard close, as defined in Article 55,[2] could provisionally allow a Member State to use Great Apes when there are justifiable grounds for believing that action is essential for the preservation of the species, or in relation to an unexpected outbreak or a life-threatening or debilitating clinical condition in human beings.

Capture of Wild-Caught Animals

Article 4.2 in Appendix A of ETS 123,[2] and thus applicable in Directive 2010/63/EU, explicitly details precautions to take into account when capturing animals from the wild. Humane methods, used by competent personnel and ensuring minimal impact on the remaining wildlife and habitat should be applied. Appropriate transport containers and means of transport should be available at capture sites in case animals need to be moved for examination or treatment. When injured or in poor health, animals should be examined by a competent person as soon as possible and, if needed, treated by a veterinarian or euthanized. Special considerations are to be given to the acclimatization, quarantine, housing, and husbandry of these animals.

The fate of wild caught animals is also to be given due consideration, taken into account possible welfare issues associated with any subsequent release into the wild.

Under the species-specific section (Section F), the requirement for group housing of nonhuman primates in an enriched environment is justified. This section also states that captive-bred nonhuman primates should be used, and where practicable, reared on site to avoid transport stress. When imported, nonhuman primates should be obtained as offspring from established colonies with high welfare and care standards.

Directive 2010/63/EU is far more restrictive, as Article 9 states that *only when* competent authorities grant exemptions can animals taken from the wild be used in procedures. Article 10 requires certain species to be purpose-bred when used in procedures, as listed in Annex I. The same article refers to Annex II for applicable dates for the use of captive-bred nonhuman primates or originating from self-sustaining colonies only. As of January 1, 2013, only captive-bred marmosets (*Callithrix jaccus*) may be used; the deadline for similar requirements for cynomolgus (*Macaca fascicularis*), rhesus (*Macaca mulatta*) and other species of nonhuman primates is set at the date of five years after the publication of a feasibility study (to be published no later than November 10, 2017), provided the study does not recommend an extended period. This feasibility study is to be conducted by the Commission in consultation with member states and stakeholders.

Use and Reuse of Nonhuman Primates

Conditions of authorized reuse of animals in procedures, as defined in Article 16, are applicable to nonhuman primates: only when the first procedure was classified as mild or moderate, after advice of the designated veterinarian and when it is demonstrated that the animal's general state of health and well-being has been fully restored, may an animal be reused in a further procedure classified as mild, moderate or nonrecovery.

In exceptional circumstances, after a veterinary examination and after derogation by the CA, reuse of an animal, provided it has not been used more than once
in a procedure classified as severe may be allowed. Whether this is a restriction
compared to ETS 123 is debatable. Article 10 of ETS 123 allowed that reuse, on
condition that the animal had returned to good health and well-being and either
was subjected to a nonsurvival procedure or to a further minor procedure only.

Specific Conditions for Establishments Using Nonhuman Primates

Article 34 requires Member States to perform annual inspections of all breeders,
suppliers and users of nonhuman primates. Records of these inspections are to
be kept for 5 years.

Article 39 requires Member States to perform retrospective assessment of all
projects using nonhuman primates.

GENETICALLY ALTERED ANIMALS: SPECIAL CONSIDERATIONS

Since its introduction, technology enabling the alteration of the genome of plants
and animals including vertebrates has attracted much interest from society at large
and has led to widespread debate. At the basis of the discussions are safety and
ethical issues concerning interference with the organism's integrity, also addressed
as manipulation of what makes an individual an individual, its genetic make-up.

Although the technology of creating GA animals is well established, specialized laboratories are needed to apply the technology successfully. Once the
GA animals are generated, the founder animals and their offspring are kept in
specialized facilities. Technical advances and availability of detailed genomic
information of an increasing number of species are contributing factors to the
increasing demand for GA animals in biomedical research.

In 1998, Directive 98/44/EC on the legal protection of biotechnological
interventions was published.[13] It was recognized then, that biotechnology and
genetic engineering are playing an increasingly important role in a broad range
of industries and that the protection of biotechnological interventions would
certainly be of fundamental importance for the Community's industrial development (Recital 1, 98/44/EC). Patentability is a major subject of this Directive.

In this context of biotechnology and its importance for industrial development, the *ordre public* and morality in particular to ethical and moral principles
are considered relevant (Recital 39, 98/44/EC). Special consideration is given to
the technology of genetic alteration of animals, which are likely to cause them
suffering without any substantial medical benefit in terms of research, prevention, diagnosis or therapy to man or animal, and to the resulting animals. Those
are excluded from patentability (Recital 45, 98/44/EC).

While the biotechnological technology and processes are subjects of 98/44/EC,
the protection of the welfare of GA animals resulting from such processes is a subject of both Directive 2010/63/EU of the European Union,[5] and in similar but more
general terms, in the European Convention for the Protection of Vertebrate Animals

Used for Experimental Purposes and other Scientific Purposes of the Council of Europe, ETS 123.[1] GA animals include transgenic, knockout and naturally occurring or induced mutant animals, and other forms of genetic alteration. For the purposes of Directive 2010/63/EU, the creation and maintenance of GA animal lines are considered procedures, hence fall within the scope of this Directive, when there is a chance of causing a level of pain, suffering, distress or lasting harm equivalent to, or higher than, that caused by the introduction of a needle in accordance with good veterinary practice (Article 3.1). With regard to new GA lines, a procedure shall be deemed to end when progeny are no longer observed or expected to experience pain, suffering or lasting harm equivalent to, or higher than, that caused by the introduction of a needle (Article 17.1).

The severity classification of procedures has been exemplified in Annex VIII of the Directive. The power of annexes to Directives is that they may be updated without having to amend the core of a Directive. In Annex VIII, four severity categories have been defined: nonrecovery, mild, moderate and severe. Breeding GA animals that are expected to have no clinically detectable adverse phenotype, may be classified as "mild". Breeding of GA animals that are expected to result in a phenotype with moderate effects, and the creation of GA animals through surgical procedures are classified as "moderate".

Through the Directive, Member States are obliged to collect and make publicly available on an annual basis, statistical information on the use of animals in procedures including information on the actual severity of the procedures; Member States shall submit that statistical information to the EC by November 10, 2015 and every year thereafter (Article 54.2). An EWG for the statistical reporting established by the EC in 2011 concluded that some further understanding was needed as to how GA animals are to be considered. In this context, the relevant articles of Directive 2010/63/EU are the above-mentioned Articles 3.1 and 17.1, and Articles 1.2 and 4.3. In Article 1.2, the elimination of pain, suffering, distress or lasting harm by the successful use of anesthesia, analgesia or other methods shall not exclude the use of an animal in procedures from the scope of the Directive. Article 4.3 stipulates that Member States shall ensure refinement of breeding, accommodation and care, and of methods used in procedures, eliminating or reducing to the minimum any possible pain, suffering, distress or lasting harm to the animals. As for GA animals, an animal with a harmful phenotype is to be understood as an animal that is likely to experience, as a consequence of the genetic alteration, pain, distress, suffering or lasting harm equivalent to, or higher than that caused by the introduction of a needle in accordance with good veterinary practice. Therefore, an EWG on severity assessment focused on GA animals. It was concluded that the creation of a new GA line is regarded as a procedure, hence requiring a project authorization until such time when the line is "established". A new strain or line of GA animals is considered to be "established" when transmission of the genetic alteration is stable, which will be a minimum of two generations, and an initial welfare assessment is completed (Table 5.1). Welfare assessment should

TABLE 5.1 Key Elements of a GA Rodent Welfare Assessment Scheme

Criteria	What to look for
Overall appearance	Is the animal morphologically "normal"? Are there any malformations or any other indicators that the phenotype has been affected? For example skeletal deformity or hydrocephalus.
Size, conformation, and growth	Are there any deviations from expected size or growth curve?
Coat condition	Is there any piloerection, areas of fur loss, loss of whiskers, barbering? Is the skin/fur in good condition?
Behavior— Posture, gait, activity, and interactions with the environment	Do they exhibit the full repertoire of behaviors appropriate for the strain/species, including social interactions, grooming, walking, running, digging, climbing? Are these normal? Is the animal hunched or reluctant to move? Is movement impaired or is there any difficulty with orientation? Any signs of rigidity or tremors? Any abnormal activity levels? Prolonged inactivity could indicate chronic stress or depression (anhedonia) and/or sickness/pain, particularly if linked with a hunched posture and/or rough or unkempt coat. Unusual activity, such as hyperactivity, could indicate stereotypy or other behavioral abnormality.
Clinical signs	For example, nasal or ocular discharge, swollen or closed eyes; increased respiratory rate; dyspnoea; seizures/twitches/tremors; increased vocalization with handling; overgrown teeth; presence of tumors, neurological or musculoskeletal abnormalities. Is metabolism impaired, for example, increased or decreased food or water intake, excessive urination? Consistency of feces.
Relative size	Any unusual changes in size of the animals should be noted, and comparisons made within the litter. It may be helpful to generate a growth curve for the line.
Numbers	Where death occurs, it is important to maintain accurate records such that any pre- or post-weaning losses can be investigated. Where appropriate (e.g. higher than anticipated mortality rate), post mortem examinations should be carried out to help determine the cause of death. A review of fertility can also be helpful in assessment of whether or not the modification is having an effect, e.g. conception rates; abortions; stillbirths.
Additional considerations for assessment in neonatal animals	
Color of pups (for neonate only)	Do any pups show evidence of abnormal skin color (e.g. anemia, poor circulation)?
Activity of pups (for neonate only)	Any abnormal activity, e.g. reduced wriggling? Righting reflex intact?
Milk spot (for neonate only)	Do any pups fail to show presence of a milk spot? Any evidence of mismothering?
Litter	Litter sizes; litter homogeneity; development and growth of pups.

include an assessment of general health, welfare and behavior together with a review of production parameters such as breeding and growth performance, which will ideally be compared with an appropriate non-GA background strain. This should be done on animals of representative age groups:

- soon after birth, around weaning and again following sexual maturity, and at additional time points as considered by a prospective review of the potential impact of the gene alteration, e.g. where there is an age dependent onset of disease;
- a minimum of seven males and seven females sampled from more than one litter;
- data from a minimum of two breeding cycles (from F2 onwards);
- comparisons made wherever possible with similar non-GA animals.

The use of animals for the maintenance of colonies of GA established lines with a likely harmful phenotype is considered a procedure, hence it requires project authorization. GA animals requiring a specific, intentional (nonaccidental) intervention to induce gene expression can be considered as having a nonharmful phenotype until deliberate induction of transgene expression. Therefore, their breeding is not regarded as a procedure and does not require project authorization. The breeding and maintenance of GA lines, which retain a risk of developing a harmful phenotype (e.g. age onset of disease or tumors; risk of infection due to compromised immune system) regardless of the applied refinement (e.g. barrier conditions, culling at early age), require project authorization as the application of refinement does not eliminate the risk. If welfare issues are later identified, these should be reviewed to consider whether the welfare problems may be attributed to the genetic alteration. If so, these should be reclassified as "harmful phenotypes" and brought under project authorization.

With regard to statistical reporting, GA animals are reported either when used for the creation of a new line; when used for the maintenance of an established line bred under project authorization and that exhibits harmful phenotype; or when used in procedures. Genetically normal animals (wild type offspring) produced as a result of creation of a new GA line should not be reported.

Genetic characterization is an essential part of GA colony management. GA animals need to be genetically characterized for scientific purposes, i.e. to settle scientific uncertainty or to confirm scientific suitability. The genetic characterization of GA animals requires sampling of the animals for genotyping. The severity classification of tissue sampling methods like fecal pellets and hair follicles could be regarded as below threshold. In contrast, more invasive methods like blood sampling, tail, ear or phalange clipping should be regarded as mild or above. By definition of a procedure as per Article 3.1, tissue sampling with methods that are below minimum threshold do not fall within the scope of the Directive. On the other hand, above threshold methods of tissue sampling carried out for the sole purposes of genetic characterization are to be considered as a procedure. Some practices for the primary purpose of identification that allow also genotyping to be performed, would not fall within the scope as per Article 1.5e.

Although GA animals are only specifically mentioned in Article 17.1 (the end of a procedure), the consequences of Directive 2010/63/EU for the breeding and maintenance of GA animals are considerable and important for protecting the welfare of these animals.

INSTITUTIONAL AND DESIGNATED PERSONNEL RESPONSIBILITIES

In this section the various institutional and personnel responsibilities are discussed. Within Directive 2010/63/EU,[5] these responsibilities are to be found spread across a number of articles. The aim in this section is to consider them together and to discuss where they may interact or overlap. It is highly likely that each Member State will allow the freedom for establishments to tailor these responsibilities according to the size and complexity of the organization. More than one responsibility can be undertaken by a single person but, for example, guidance in the UK suggests that at least three people fill the five key roles described below (roles 1–5). It is also important that scientific, financial or other conflicts of interest are avoided for those persons carrying out responsibility for animal welfare and care and for the designated veterinarian. Also, if these persons have a substantial interest in the scientific outcome of a program of work, alternative provision for the welfare and veterinary oversight should be arranged.

The Person Responsible for Overall Compliance

Directive 86/609/EEC[3] set out responsibilities for a person in overall charge of the establishment (Article 16) who was entrusted with the task of administering, or arranging for the administration of, appropriate care to the animals bred or kept in the establishment, and of ensuring compliance with Article 5 (general care and accommodation) and Article 14 (those carrying out or supervising procedures). A similar responsibility is described in ETS 123 (Article 15)[1] but there is no direct link to persons carrying out procedures. Interestingly, Directive 2010/63/EU does not have a specific article for this responsibility. It is contained within Article 20.2 (Authorization of breeders, suppliers and users) where authorization should specify the person responsible for ensuring compliance with the provisions of the Directive and the person responsible for care of the animals (Article 24.1) as well as the designated veterinarian (Article 25).

The Person(s) Responsible for Welfare and Care

The requirement to identify a person or persons who are administratively responsible for the care of the animals and the functioning of the equipment is set out in ETS 123 (Article 20a) and this wording was repeated in Directive

86/609/EEC (Article 19.2a). Directive 2010/63/EU calls for one or several persons on site that are responsible for overseeing the welfare and care of the animals in the establishment (Article 24.1a). There is no reference to the functioning of the equipment and no further description of the tasks to be undertaken.

The Designated Veterinarian

Both ETS 123 (Article 20d) and the previous Directive 86/609/EEC (Article 19.2d) mandate the requirement that a veterinarian or other competent person should be charged with advisory duties in relation to the well-being of the animals. Directive 2010/63/EU (Article 25) states that Member States shall ensure that each breeder, supplier and user has a designated veterinarian with expertise in laboratory animal medicine, or a suitably qualified expert where more appropriate, charged with advisory duties in relation to the well-being and treatment of the animals. During the revision of the Directive, and against the wishes of most stakeholders, the role of the veterinarian in the protection of animals used for scientific procedures was not strengthened.

Further discussion on the role of the veterinarian is provided in a subsequent section.

The Person Responsible for the Scientific Project

Directive 2010/63/EU places more emphasis on the requirements of the person authorized to design procedures or projects. Both ETS 123 (Article 13) and the previous Directive 86/609/EEC (Article 7) only require this person to be competent and authorized. Within Directive 2010/63/EU, the person designing procedures and projects must have had adequate education and training before authorization. They should have received instruction in a scientific discipline relevant to the work being undertaken and have species-specific knowledge. Member States have to ensure, through authorization or by other means, that these requirements are fulfilled (Article 23.2b). Annex V of the Directive sets out the elements that should have been covered in their education and training (see subsequent chapter on education and training). Two of the key responsibilities of this person are outlined in Article 24.2: They should ensure that any unnecessary pain, suffering, distress or lasting harm that is being inflicted on an animal in the course of a procedure is stopped; and they should also ensure that the projects are carried out in accordance with the project authorization and in the event of noncompliance, the appropriate correcting measures are taken and recorded. For example, additional responsibilities have been ascribed to the person responsible for the scientific project in the UK and these are outlined in the Guidance on the Operation of the Animals (Scientific Procedures) Act 1986.[14] These relate to the records to be maintained and the management and conduct of licensed personnel working on a project (personal license holders). It is likely these responsibilities will be retained after transposition.

The Person(s) Carrying Out Procedures

Requirements for those carrying out procedures were very limited in ETS 123 and Directive 86/609/EEC. They had to be competent authorized persons, or under the direct responsibility of such a person. Directive 2010/63/EU builds on these requirements. The person carrying out procedures (user) should be adequately educated and trained before they perform any task. They should also be supervised in the performance of their tasks until they have demonstrated the requisite competence. Member States have to ensure, through authorization or by other means, that these requirements are fulfilled (Article 23.2c).

The Person(s) Responsible for Education and Competence

Directive 2010/63/EU emphasizes the importance of education, training, supervision, and competence in ensuring a high standard of welfare of the animals (Recital 28). To this end, each breeder, supplier, and user should have one or several persons on site with responsibility for ensuring that the staff are adequately educated, competent and continuously trained. In addition, they should be supervised until they have demonstrated the requisite competence (Article 24.1c). Member States can determine where this responsibility lies and it could form part of an existing role, e.g. designated veterinarian, person overseeing welfare and care of animals. Ultimately, the accountability lies with the person responsible for overall compliance in the establishment. In the UK for example, this role has been formalized and the person is called the "Named Training and Competence Officer". In large establishments, such as universities and pharmaceutical companies, this role is likely to be identified as a new, stand-alone role. Tasks could include establishing and maintaining a database of education and training of individuals, recommending suitable training (including continuing education, Continuing professional development (CPD)) for persons involved in animal-related work and liaising with those designing projects on the suitability of trained persons. It is not foreseen that this individual should necessarily conduct the training.

The Person Responsible for Access to Species-Specific Knowledge

A new role has been established in Directive 2010/63/EU for a person to ensure that staff dealing with animals have access to information specific to the species housed in the establishment (Article 24.1b). Similarly to the previous role of responsibility for education and competence, this responsibility could be part of an existing person's role, e.g. person overseeing welfare and care of animals. For example, in the UK, this new role has been given the title of "Named Information Officer".

OVERSIGHT AND ERP

The oversight and ERP has been traditionally developed at the national level in Europe until the publications of Directive 2010/63/EU.[5] ETS 123[1] does not require explicitly this kind of process, and the old Directive 86/609/EEC[3] followed the scheme of ETS 123 closely, only requiring the Member States to designate the authorities responsible for verifying that the provisions of the Convention or Directive be met, which did not include requirement for ethical review. Therefore, in many European Union (EU) countries there has not been a legal mandate for ERP until the provisions of the new Directive entered into force on January 1, 2013. However, many countries or even regions within countries had established their own systems.[15] Some of these countries were not affected by the Directive because they do not belong to the EU (i.e. Switzerland), and in others the development of ERP was a voluntary process, as the old Directive did not impede the implementation of national stricter measures. The development of stricter measures happened in a number of EU Member States, such as the UK, The Netherlands, Germany, Spain, and others. Because each Member State established a particular process, there have been traditionally many different systems in place across Europe, some based more on personnel and processes at the institutional level, and others giving more predominance to the oversight by competent authorities. In all cases there was some kind of combination of institutional and government oversight activities.

Then the new Directive was published in 2010 and after its transposition into national legislation was effective in January 1, 2013, systems tended to be more harmonized within the EU. However, there may still be important differences in the way the oversight and ethical review systems are implemented because of different interpretations by Member States.

The Directive establishes many requirements to ensure oversight and ERP are conducted effectively. First, it is important to know the concept of "competent authority" (CA) because it plays an essential role in the oversight and ERP activities. The CA is defined in Article 3 as "authority or authorities or bodies designated by a Member State to carry out the obligations arising from this Directive." Although one may easily understand that the CA is the responsible person(s) at government level, this may not be entirely true in all cases, because Article 59 allows Member States to "designate bodies other than public authorities for the implementation of specific tasks laid down in this Directive, only if there is proof that the body: (a) has the expertise and infrastructure required to carry out the tasks; and (b) is free of any conflict of interests as regards the performance of the tasks. Bodies thus designated shall be considered competent authorities for the purposes of this Directive." This article means that if the requirements are met, public CA can designate institutional or external bodies or committees to be also CA with the legal competence to perform some of the activities related to the ERP, and this may be the main reason for the existence of different systems especially with regard to the project evaluation.

The first figure required by the Directive who is related to oversight at each breeder, supplier, and user is the person(s) responsible for overseeing the welfare and care of the animals in the establishment (Article 24.1a), who in addition to this general function is required to be a member of the AWB (Article 26).

The AWB plays a crucial role in oversight, is required for all establishments, and in addition to the person(s) responsible for overseeing the welfare and care of the animals in the establishment, has to include at least (in case of a user establishment) a scientific member and in all cases receive input from the designated veterinarian. The AWB could be interpreted as the equivalent to the Institutional Animal Care and Use Committee in the US, although there may be significant differences in composition and function. Especially noteworthy is that the designated veterinarian is given advisory duties only in relation to the well-being of animals (Article 25), and his/her presence as an AWB member is not mandatory (only input is needed). However, some countries consider that the position of the designated veterinarian should be stronger and require this figure to be part of the AWB.

The AWB tasks have to do with oversight and ERP, especially if we understand ERP as a set of several activities and not only the project or protocol evaluation. The tasks that as a minimum have to be performed by the AWB are: "(a) advise the staff dealing with animals on matters related to the welfare of animals, in relation to their acquisition, accommodation, care and use; (b) advise the staff on the application of the requirement of replacement, reduction and refinement, and keep it informed of technical and scientific developments concerning the application of that requirement; (c) establish and review internal operational processes as regards monitoring, reporting and follow-up in relation to the welfare of animals housed or used in the establishment; (d) follow the development and outcome of projects, taking into account the effect on the animals used, and identify and advise as regards elements that further contribute to replacement, reduction and refinement; and (e) advise on rehoming schemes, including the appropriate socialisation of the animals to be rehomed" (Article 27). Clearly the intention of the Directive is to make the AWB be the tool to create a culture of care at the institution, although it may be a matter for discussion if this can be done when institutions only comply with the minimum required composition of only two individuals at a user establishment.

The project evaluation and/or authorization functions are not assigned to the AWB, but to the CA (Articles 36 and 38). There are Member States that have kept these functions at the level of the public CA exclusively, and others have opened the possibility to assign the project evaluation (and at least in Belgium, also the authorization) to other designated CA such as institutional AWB or *ad hoc* ethics committees. In all cases, the project evaluation must be transparent and performed in an impartial manner (Article 38). The possibility of institutional committees performing the project evaluation is an object of controversy, because in addition to the transparency and impartiality, the designated CA must be free of any conflict of interest. A number of Member States consider that these requirements cannot be achieved at the institutional level and they retain the evaluation at the public CA level. On the other side, several other countries allow the delegation, which in many cases is a pragmatic decision based on the lack of resources to perform this function effectively at the government level.

At this point it is important to clarify the concept of project, which "means a programme of work having a defined scientific objective and involving one or more procedures" (Article 3.2). The Directive uses the term "project" instead of "protocol," which is more often used in other areas of the world such as the US.

Regardless of which type of CA performs the project evaluation, the project application shall include (Article 37) at least the project proposal, a nontechnical project summary (see below), and information on a number of items which are listed in Annex VI of the Directive: (1) Relevance and justification of the following: (a) use of animals including their origin, estimated numbers, species and life stages; (b) procedures. (2) Application of methods to replace, reduce and refine the use of animals in procedures. (3) The planned use of anesthesia, analgesia and other pain-relieving methods. (4) Reduction, avoidance, and alleviation of any form of animal suffering, from birth to death where appropriate. (5) Use of humane end-points. (6) Experimental or observational strategy and statistical design to minimize animal numbers, pain, suffering, distress, and environmental impact where appropriate. (7) Reuse of animals and the accumulative effect thereof on the animals. (8) The proposed severity classification of procedures. (9) Avoidance of unjustified duplication of procedures where appropriate. (10) Housing, husbandry, and care conditions for the animals. (11) Methods of killing. (12) Competence of persons involved in the project.

The nontechnical summaries (Article 43), which must be part of the application, have to be published by the Member States because one of the intentions of the European Union is to be transparent in the use of animals for research purposes. To avoid confidentiality issues, the nontechnical summaries shall be anonymous and shall not contain the names and addresses of the user and its personnel. They have to include (safeguarding intellectual property and confidential information), (a) information on the objectives of the project, including the predicted harm and benefits and the number and types of animals to be used; and (b) a demonstration of compliance with the requirement of replacement, reduction and refinement. The publication of these summaries is done for authorized projects only.

The project evaluation and authorization process has to be performed within a defined timeframe, which is 40 working days. This period can be extended with 15 extra days in the case of complex multidisciplinary projects. The decision regarding authorization must be communicated to the applicant within that period (Article 41), and very importantly, the project cannot be started until the authorization is granted and communicated, even if the CA does not respond within the established timeframe. Only projects containing procedures classified as "nonrecovery," "mild" or "moderate" and not using nonhuman primates, that are necessary to satisfy regulatory requirements, or which use animals for production or diagnostic purposes with established methods (basically regulatory testing) can be subjected to a simplified authorization administrative procedure (Article 42) if Member States decide to do so.

The project authorization has to include (a) the user who undertakes the project; (b) the persons responsible for the overall implementation of the project and its compliance with the project authorization; (c) the establishments in which the project will be undertaken, where applicable; and (d) any specific conditions following the project evaluation, including whether and when the project shall be assessed retrospectively (Article 40).

Once granted, the project authorization is valid for a period not exceeding 5 years (Article 40.3). The amendment or renewal of the project authorization is required "for any change of the project that may have a negative impact on animal welfare" and "any amendment or renewal of a project authorisation shall be subject to a further favourable outcome of the project evaluation" (Article 44).

The project evaluation may not be only performed prospectively. All projects using nonhuman primates and projects involving procedures classified as "severe," shall undergo a retrospective assessment. This retrospective assessment is the responsibility of the CA, who shall evaluate (a) whether the objectives of the project were achieved; (b) the harm inflicted on animals, including the numbers and species of animals used, and the severity of the procedures; and (c) any elements that may contribute to the further implementation of the requirement of replacement, reduction and refinement (Article 39).

Although apparently the retrospective assessment of projects is to be performed only for certain projects, in practice the actual severity of procedures inflicted to all animals have to be assessed, because this must be part of the statistical information that Member States have to submit to the EC by November 10, 2015 and yearly thereafter (Article 54). The requirement to report the actual severity of procedures forcing institutions to ensure a close oversight on the animals is implemented. This requirement may be linked to one of the tasks assigned to the AWB (following the development and outcome of projects, taking into account the effect on the animals used), but the cooperation of the researchers may be essential to perform this task effectively.

The ERP is not limited to the project evaluation. The retrospective assessment and the need to categorize the actual severity of procedures, and not only the one assigned prospectively, may be linked to the concept of post-approval monitoring. In addition to the functions (described above) assigned to the AWB, there are also other responsible persons related to oversight functions. These include the already cited person(s) responsible for overseeing the welfare and care of the animals that all institutions must have, and the persons responsible for the overall implementation of the project and its compliance with the project authorization. This last category of person has to be defined in the project authorization process.

It has been described how the EU Directive frames a complex scenario for the oversight and ERP in the European Union, in which a combination of parties play different roles. The transposition of the Directive into Member States' national legislation has led to some differences in the way they are performed across Europe. These differences are based mainly on how the balance between the functions retained by the public CA or delegated to other CA (mainly at the institutional level) is established.

Outside of the European Union the situation is diverse. There are countries with no specific requirements on ERP (for example, the Russian Federation), and others having well developed systems like Switzerland. In Switzerland, there are regional ethical committees in charge of the protocol review, which advise the Cantonal Authority whether or not the experiments should be authorized. In addition to this, a National Committee can participate in controversial cases.

The FELASA reviewed the principles and practices across Europe before the publication of the new Directive. In this report[15] (full report available on line at: http://www.felasa.eu/media/uploads/Principles-practice-ethical-review_full%20 report%20.pdf), in addition to the review of the situation of ethical review at that time across Europe, a set of 30 recommendations is offered for the conduct of effective ethical review in practice. Some of the requirements on ethical review present in the new European Directive are very similar to recommendations in the FELASA document. Because the Directive still allows room for different ways of implementation, the recommendations in the FELASA document may be a very useful guidance for AWB, ethics committees, and responsible personnel at institutional level.

REUSE OF ANIMALS

The ability to reuse animals is critical to the implementation of the reduction and refinement of procedures, two of the 3Rs that are highlighted in Directive 2010/63/ EU.[5] However, before describing the regulatory requirements concerning reuse, it is important that the two terms, "continued use" and "reuse," are clearly defined. "Continued use" is an administrative concept to describe the use of an animal in a series of related procedures or experiments for a particular scientific purpose. The use of the same animal must be essential in order to achieve the objectives of the second or subsequent experiments. An example could be the preparation of an ovariectomized mouse in one laboratory, which is transferred to another for subsequent scientific study. Reuse is a term used where, after completion of one series of procedures, an animal is used again in the same or a different experiment, where a previously unused animal would have equally sufficed to meet the objectives of the second and subsequent use. An example could be the use of a sheep to study the metabolism of a dietary supplement that may cause adverse effects followed by a further use to provide blood to make diagnostic plates for bacteriology. The two studies are not related. Any naive sheep could have been used for the second study. The second use would be reuse of that animal.

Reuse is considered in Article 16 of Directive 2010/63/EU. Paragraph 16.1 states that "Member States shall ensure that an animal already used in one or more procedures, when a different animal on which no procedure has been carried out could also be used, may only be re-used in a new procedure provided that the following conditions are met:

a. the actual severity of the previous procedures was "mild" or "moderate"
b. it is demonstrated that the animal's general state of health and well-being has been fully restored

c. the further procedure is classified as "mild", "moderate" or "nonrecovery" and
d. it is in accordance with veterinary advice, taking into account the lifetime experience of the animal."

Paragraph 16.2 states that, "in exceptional circumstances, by way of derogation from point (a) of paragraph 1 and after a veterinary examination of the animal, the competent authority may allow re-use of an animal, provided the animal has not been used more than once in a procedure entailing severe pain, distress or equivalent suffering."

Clarification on the interpretation of a number of points in this article was sought by a number of Member States and the EC has provided guidance at: http://ec.europa.eu/environment/chemicals/lab_animals/pdf/qa.pdf. A few of the questions and answers have been reproduced below:

Q1	Reuse only takes place in certain circumstances. Does it follow from this that Member States have to specifically authorize reuse?
Answer	Competent Authorities authorise "projects", not "procedures", neither "reuse". Reuse is examined during project evaluation in accordance with Article 38 (Project evaluation) to ensure that it complies with the provisions of the Directive.
Q2	Does paragraph 1 (d) require that a veterinary surgeon examines the individual animals before authorizing their reuse?
Answer	No. 1 (d) requires veterinary advice to be the basis of the decision on whether reuse should be allowed. The advice can be in a form of written instructions as to when reuse can take place. Cross reference to paragraph 16.2 that requires veterinary examination of the animal in question.
Q3	In the overall context of the Directive, the "exceptional circumstances" contemplated by the derogation must be truly exceptional i.e. that there is real urgency arising, for example, from a health emergency, such that there would be insufficient time to acquire or prepare a naive animal. Apart from this kind of situation, it is difficult to see when the derogation could be lawfully invoked?
Answer	The "exceptional circumstances" need to be justified on a case-by-case basis. The legal text does not specify the conditions to satisfy being "exceptional circumstance" and thus these are not necessarily limited to urgency or health emergencies.
Q4	Can reuse be allowed simply because it would be cheaper or easier?
Answer	No, in line with Recital 25, reuse is part of the implementation of the principle of the Three R's, in particular of Reduction. Additionally, it is worth remembering that "cheaper" or "easier" are not criteria for resorting to reuse pursuant to Article 16.

SETTING FREE/REHOMING

ETS 123[1] does not specifically mention conditions required to be fulfilled for setting free or rehoming of animals after use in a procedure, although Article 12 allows for such conditions: "Notwithstanding the other provisions of this Convention, where it is necessary for the legitimate purposes of the procedure, the responsible authority may allow the animal concerned to be set free provided that it is satisfied that the maximum practicable care has been taken to safeguard the animal's well-being. Procedures that involve setting the animal free shall not be permitted solely for educational or training purposes."

Recital 26 of Directive 2010/63/EU[5] introduces the concept of rehoming cats and dogs, taking into account the high level of public concern as to the fate of such animals. It states that, in case Member States allow rehoming, a scheme must be in place providing adequate socialization to those animals to ensure success, avoid unnecessary distress to the animals and guarantee public safety.

Articles 19 and 29 of Directive 2010/63/EU explicitly allow for a possibility for setting free or returning animals (used or intended to be used in procedures) to a suitable habitat or husbandry system, appropriate to the species. The Directive is not specific on which species are concerned. Moreover, rehoming is offered as a possibility, not a requirement. On condition that Articles 19 and 29 are met, it is left up to Member States to decide whether they want to implement these, and if yes, for which species.

Article 27 states that one of the tasks of the AWB is to advise on rehoming schemes, including the appropriate socialization of the animals to be rehomed. In case a Member State requires implementation, this could become a major responsibility for the AWB.

This task will definitely call for financial and other resources as it would not only require a careful selection of suitable animals, but also availability of appropriate means to allow for adapted socialization depending on the choice retained, careful selection of retained possibilities and, maybe more importantly, clearly defined alternative options in case of failure of the retained choice.

Conditions for setting free animals after use in procedures, as applicable to wild animals require that a program of rehabilitation be in place before they are returned to their habitat (Article 29). For nonhuman primates, this might at the time of writing imply being placed in a suitable husbandry system such as a zoo, or returned to a suitable habitat, such as release into the wild. However, once Annex II of Directive 2010/63/EU will be applicable, only allowing the use of captive-bred nonhuman primates or originating from self-sustaining colonies, the option of releasing nonhuman primates into the wild might no longer be an option as these animals might not be able to cope with those unknown conditions.

OCCUPATIONAL HEALTH AND SAFETY

In any working environment one should be safe and feel safe. The employer's responsibility is to provide a safe working place. It is the responsibility of the employee to act responsibly within this safe working place.

In February 1988, the European Parliament adopted four resolutions following the debate in the internal market on worker protection. These resolutions specifically invited the EC to draw up a framework Directive to serve as a basis for more specific Directives on minimum health and safety requirements at the work place. The important driver was the notion that the incidence of accidents at work and occupational diseases was regarded to be too high and that preventive measures had to be introduced or improved without delay in order to safeguard the safety and health of workers and ensure a higher level of protection. These issues are not specifically addressed in Convention ETS 123[1] nor in Directive 2010/63/EU.[5]

In 1989, the framework Directive on the Introduction of Measures to Encourage Improvements in the Safety and Health of Workers at Work, 89/391/EEC, was published by the Council of the European Communities.[16] According to Article 16.1 of this framework Directive 89/391/EEC, the Council, acting on a proposal from the EC, should adopt individual Directives, *inter alia*, in the areas listed in the annex of 89/391/EEC, i.e.:

- work places,
- work equipment,
- personal protective equipment,
- work with visual display units,
- handling of heavy loads involving risk of back injury,
- temporary or mobile work sites,
- fisheries and agriculture.

To date, 19 individual Directives within the meaning of Article 16.1 of 89/391/EEC have been published.[17–35]

In 1994, the Council of the European Union published Directive 94/33/EC on the Protection of Young People at Work.[36] Children and adolescents were considered specific risk groups. Following Article 15 of 89/391/EEC, particularly sensitive risk groups have to be protected against the dangers, which specifically affect them. Directive 94/33/EC applies to any person under 18 years of age having an employment contract or an employment relationship defined by the law in force in a Member State and/or governed by the law in force in a Member State. It contains articles on obligations on employers, vulnerability of young people and the consequential prohibition of work, working time, night work, rest period, breaks and annual rest.

89/391/EEC: Framework Directive

Important considerations of the framework Directive 89/391/EEC were that Member States have a responsibility to encourage improvements in the safety and health of workers on their territory. At the same time, the Member States' legislative systems covering safety and health at the work place were considered to differ widely and needed to be improved. It was anticipated that national provisions on the subject, which often include technical specifications and/or self-regulatory standards, were likely to result in different levels of safety and health protection and allowed for competition at the expense of safety and health.

Directive 89/391/EEC contains general principles concerning the prevention of occupational risks, the protection of safety and health, the elimination of risk and accident factors, the informing, consultation, balanced participation in accordance with national laws and/or practices and training of workers and their representatives, as well as general guidelines for the implementation of said principles. The Directive applies to all sectors of activity, both public and private (industrial, agricultural, commercial, administrative, service, educational, cultural, leisure, etc.). In Section II of the Directive, the employer's obligations are listed; Section III deals with the workers' obligations.

Section II: Employers' Obligations

Among the general provisions of Section II it is stated that the employer shall have a duty to ensure the safety and health of workers in every aspect related to the work. Furthermore, the workers' obligations in the field of safety and health at work shall not affect the principle of the responsibility of the employer (Articles 5.1 and 5.3). The general obligations on employers are described in Article 6: Within the context of his responsibilities, the employer shall take the measures necessary for the safety and health protection of workers, including prevention of occupational risks and provision of information and training, as well as provision of the necessary organization and means. The employer shall be alert to the need to adjust these measures to take account of changing circumstances and aim to improve existing situations (Article 6.1). The employer shall implement these measures on the basis of the following general principles of prevention (Article 6.2):

- avoiding risks;
- evaluating risks which cannot be avoided;
- combating the risks at source;
- adapting the work to the individual;
- adapting to technical progress;
- replacing the dangerous by the nondangerous or the less dangerous;
- developing a coherent overall prevention policy;
- giving collective protective measures priority over individual protective measures;
- giving appropriate instructions to the workers.

Without prejudice to other provisions of Directive 89/391/EEC, the employer shall, taking into account the nature of the activities of the enterprise and/or establishment, evaluate the risks to safety and health of workers. Subsequent to this evaluation and as necessary, the preventive measures and the working and production methods implemented by the employer must assure an improvement in the level of protection afforded to workers and be integrated into all activities of the undertaking and/or establishment and at all hierarchical levels (Article 6.3a); where the employer entrusts tasks to a worker, he should take into considerations the worker's capabilities as regards health and safety (Article 6.3b). The employer should ensure that the planning and introduction of new technologies are the subject of consultation with the workers and/or their representatives, as regards the consequences of the choice of equipment, the working conditions and the working environment for the safety and health of workers (Article 6.3c). The employer should take appropriate steps to ensure that only workers who have received adequate instructions may have access to areas where there is serious and specific danger (Article 6.3d). Where several undertakings share a work place, the employers shall cooperate in implementing the safety, health and occupational hygiene provisions and, taking into account the nature of the activities, shall coordinate their actions in matters of the protection and prevention of occupational risks, and shall inform one another and their respective workers and/or their representatives of these risks (Article 6.4). In no circumstances may measures related to safety, hygiene and health at work involve the workers in financial cost (Article 6.5). Furthermore, the employer shall: be in possession of an assessment of the risks to safety and health at work, including those facing groups of workers exposed to particular risks (Article 9.1a); decide on the protective measures to be taken and, if necessary, the protective equipment to be used (Article 9.1b); keep a list of occupational accidents resulting in a worker being unfit for work for more than three working days (Article 9.1c); and draw up, for the responsible authorities and in accordance with national laws and/or practices, reports on occupational accidents suffered by his workers (Article 9.1d). In Article 10 the obligation of informing workers is described. The consultation and participation of workers is described in Article 11 and the training of workers in Article 12.

Section III: Workers' Obligations

It is each worker's responsibility to take care as far as possible of his own safety and health and that of other persons affected by his acts or commissions at work in accordance with his training and the instructions given by his employer (Article 13.1). To this end, workers must in particular: "make correct use of machinery, apparatus, tools, dangerous substances, transport equipment and other means of production; make correct use of the personal protective equipment supplied to them and, after use, return it to its proper place; refrain from disconnecting, changing or removing arbitrarily safety devices fitted, e.g. to machinery, apparatus, tools, plant, and buildings and use such safety devices correctly; immediately inform the employer and/or the workers with specific responsibility for the safety and health of workers of any work situation they have reasonable grounds for considering represents a serious and immediate danger to safety and health and of any shortcomings in the protection arrangements; cooperate, in accordance with national practice, with the employer and/or workers, for as long as may be necessary to enable any tasks or requirements imposed by the competent authority to protect the safety and health of workers to be carried out, and to enable the employer to ensure that the working environment and working conditions are safe and pose no risk to safety and health within their field of activity" (Article 13.2).

Individual Directives

The more specific nature of the individual Directives makes them more or less relevant in the context of animals used for scientific purposes. The relevance is dependent on the type of enterprise and facilities among others. However, an attempt is made to rank them into three categories: those of general relevance, relevant for certain enterprises, irrelevant in this context.

Generally Relevant

Those of general relevance are those dealing with: the work place (Directive 89/654/EEC)[17]; the use of work equipment by workers at work (Directive 2009/104/EC)[18]; the use by workers of personal protective equipment at the work place (Directive 89/656/EEC)[19]; the manual handling of loads where there is a risk particularly of back injury to workers (Directive 90/269/EEC)[20]; the work with display screen equipment (Directive 90/270/EEC)[21]; the provision of safety and/or health signs at work (Directive 92/58/EEC)[25]; pregnant workers who have recently given birth or are breastfeeding (Directive 92/85/EEC)[26]; risks arising from vibration (Directive 2002/44/EC)[32]; risks arising from noise (Directive 2003/10/EC)[33]; young people at work (Directive 94/33/EEC).[36]

Relevant for Certain Enterprises

Relevant for certain enterprises are those dealing with: risks related to exposure to carcinogens or mutagens at work (Directive 2004/37/EC)[22]; exposure to biological agents at work (Directive 2000/54/EC)[23]; work on board fishing vessels (Directive 93/103/EEC)[29]; risks related to chemical agents at work (Directive 98/24/EEC)[30]; risks from explosive atmospheres (Directive 1999/92/EC)[31]; risks arising from electromagnetic fields (Directives 2004/40/EC; 2012/11/EC)[34]; risks arising from artificial optical radiation (Directive 2006/25/EC).[35]

Irrelevant in This Context

Irrelevant are those dealing with: safety and health requirements at temporary or mobile constructions sites (Directive 92/57/EEC)[24]; improving the safety and health protection of workers in the mineral-extracting industries through drilling (Directive 92/91/EEC)[27]; improving the safety and health protection of workers in surface and underground mineral-extracting industries (Directive 92/104/EEC).[28]

Concluding Remarks

Neither the Convention ETS 123[1] nor Directive 2010/63/EU[5] deal with occupational health and safety. The framework Directive 89/391/EEC, together with the individual Directives, form a European legislative core on these topics. The framework Directive encourages in more general terms improvements in the safety and health of workers at work. The individual Directives provide minimum health and safety requirements for specified subjects including Directive 94/33/EC on the protection of young people at work. Control of laboratory animal allergy is not among these specified subjects. At the National level, there is to varying degree, legislation in place and guidance documents being published. For example, there is the guidance note on the control of laboratory animal allergy issued by the Health and Safety Executive in the UK.[37]

EDUCATION, TRAINING, AND COMPETENCE OF PERSONNEL

In 1993, the Council of Europe adopted a Resolution on Education and Training of Persons Working with Laboratory Animals,[38] and these were embedded and expanded in the FELASA recommendations for the education and training of persons involved in animal experiments. Four categories of person were identified. Category A relates to persons taking care of the animals,[39] Category B relates to those carrying out procedures,[40] Category C relates to persons responsible for directing animal experiments,[41] and Category D relates to specialists in laboratory animal science.[42] In addition to developing the education and training recommendations, FELASA has established an accreditation system for teaching programs for the four categories.[43] This quality assurance system, introduced in 2003, is intended to assist in the development of uniform, high quality educational programs. The system is widely recognized and operates in a number of European countries.[44] CPD is a prerequisite for many professional medical and scientific disciplines but is less common for those involved in laboratory animal work. FELASA has proposed guidelines in 2010 for continuing education for all persons involved in the care and use of animals for scientific purposes and these are discussed later in this section.

Education and training requirements for those involved in animal research varies widely across Europe. In some Member States, there are mandatory courses for those carrying out procedures and designing projects while in others those applying for projects should have undertaken a course complying with the recommendations of FELASA Category C. In only one country, Switzerland, is there a regulatory requirement for continuing education.

Currently, only a few countries have mandatory requirements for those caring for animals (animal technicians and caretakers). However, structured education and training does exist in many Member States and is encouraged in academic, government, and commercial laboratories. Examples of courses developed to facilitate career progression as an animal technologist are the Institute of Animal Technology courses in the UK and the FELASA Category A levels 0 to 2.

The requirements for education and training as set forth in ETS 123[1] are very limited. Persons who carry out procedures, or take part in procedures, or take care of animals used in procedures, including supervision, shall have the appropriate education and training (Article 26). Authorization to conduct procedures should only be granted to persons deemed to be competent by the responsible authority (Article 13). The previous Directive, 86/609/EEC,[3] provided a little more definition of the actual nature of the education and training required in Article 14. In particular, persons carrying out or supervising the conduct of experiments shall have received instruction in a scientific discipline relevant to the experimental work being undertaken and be capable of handling and taking care of laboratory animals. They shall also have satisfied the authority that they have attained a level of training sufficient for carrying out their tasks.

Transposition of the revised Directive 2010/63/EU[5] offers a unique opportunity to establish a harmonized framework across Europe that assures the competence of all persons involved in the care, breeding and use of animals for scientific procedures. This would affirm the EC's commitment to uniform, high standards of animal welfare and would assist the movement of trained persons within Europe. The text of the Directive makes specific reference to education and training requirements over and above those of the previous Directive. A major challenge is to ensure consistency across Europe because each Member State will need to publish its minimum requirements.

Requirements of Directive 2010/63/EU

The Directive emphasizes the importance of training, supervision, and competence of personnel involved in the care and use of animals used for scientific procedures. Article 23 (Competence of personnel) identifies a number of requirements:

1. Member States shall ensure that each breeder, supplier, and user has sufficient staff on site.
2. The staff shall be adequately educated and trained before they perform any of the following functions:
   a. carrying out procedures on animals
   b. designing procedures and projects
   c. taking care of animals
   d. killing animals

   Persons carrying out the functions referred to in point (b) shall have received instruction in a scientific discipline relevant to the work being undertaken and shall have species-specific knowledge.

   Staff carrying out functions referred to in points (a), (c), or (d) shall be supervised in the performance of their tasks until they have demonstrated the requisite competence.

   Member States shall ensure, through authorization or by other means, that the requirements laid down in this paragraph are fulfilled.
3. Member States shall publish, on the basis of elements set out in Annex V (list of elements to be included in a training program and presented as an appendix to this section), minimum requirements with regard to education and training and the requirements for obtaining, maintaining, and demonstrating the requisite competence for the functions set out in paragraph 2.

4. Nonbinding guidelines at the level of the Union on the requirements laid down in paragraph 2 may be adopted in accordance with the advisory procedure referred to in Article 56.2.

Article 24.1c requires that each breeder, supplier, and user has one or several persons on site who shall be responsible for ensuring that the staff are adequately educated, competent, and continuously trained and that staff are supervised until they have demonstrated the requisite competence.

During 2012, the EC initiated an EWG with the objective to develop a training framework within the EU that would assure the competence of staff caring for or using animals in procedures, and facilitate the free movement of personnel within the EU. This framework included consideration of the training, supervision, competence assessment, and continuing training requirements of persons carrying out procedures, taking care of animals, killing animals and the training and continuing training requirements of those persons responsible for the design of procedures and projects. The Commission expressed its goal of moving from the current environment/time dependent, offer-driven training into a demand-driven one, which would increase flexibility. A modular structure of training provision was considered one way to facilitate this change. There is a need for more training providers to create increased accessibility, availability and affordability. This is particularly the case in Member States where training programs are not so well developed or not in place yet. Competence needs to be built around an individual's need-basis throughout the professional life of that person. Criteria need to be developed for supervision and competence assessment and these, together with a common quality standard, would provide a framework for mutual recognition and free movement of personnel between establishments and Member States. The outcome from the EWG is described below but it is important to note that this represents "work in progress" (at the time of drafting this document, February 2013).

Development of a Modular Training Structure

The EWG agreed that training should be based on a modular structure representing the minimum training necessary before persons are allowed to carry out the functions (a) to (d) stated in Article 23.2 of Directive 2010/63/EU. The content (syllabus) of each module should be described in terms of "learning outcomes," expressed in measurable verbs, as this helps to define the knowledge and skills that course participants should be able to demonstrate by the time these learning outcomes are assessed. They also help to harmonize content and guide course deliverers. The elements listed in Annex V of Directive 2010/63/EU were incorporated into the relevant modules. Four types of module were envisaged:

1. Core module—compulsory module for all functions stated in Article 23.2 and with the same learning outcomes. This would include topics such as legislation, ethics, animal welfare, and the 3Rs, basic and appropriate species-specific biology, recognition of pain, distress, and suffering.

2. Prerequisite module—a compulsory module for a specific function(s). Examples could be: conduct of minimally invasive procedures for those persons carrying out procedures and design of procedures and projects for those undertaking this responsibility.
3. National module—a compulsory module to include national/regional transposing legislation and any other legislation relevant to the use of animals for scientific purposes, e.g. transport, *cites*, occupational health and safety, GA animals.
4. Additional module—to facilitate learning specialized skills as well as lifelong learning (CPD).

The specific content of the modules for each function has yet to be finalized and agreed. A group of modules could be delivered as a single course or separately to meet individual needs for example in the case of continuing education. The importance of observational and practical skills is recognized and these can be acquired either at a teaching institute or in a work-based location. In some Member States, the use of live animals for training is not permitted except in certain circumstances and so the acquisition of technical skills has to be achieved within the scope of a scientific project. The number of animals used for educational or training purposes should be the minimum to satisfy the criteria of the CA. Supervision until competent and regular assessment of competence is integral to the training scheme and is described later in this section.

Those persons carrying out procedures will be educated and trained in a progressive manner using only the species and techniques necessary for their work and commencing with minimally invasive procedures such as blood sampling and administration of substances. Depending on the nature of the project, further modules could be undertaken to permit the use of more complex procedures, anesthesia or surgery. The requirements of a person designing procedures and projects could be very diverse and much broader than those carrying out procedures. Some of the competencies are likely to be developed through the candidate's scientific training and experience in managing staff and projects. They cannot be taught in a short course but need for these skills should be emphasized.

It has been acknowledged that the achievement of practical learning outcomes may be separate from theoretical/knowledge-based learning outcomes. When there is a likelihood of causing pain, suffering, distress, or lasting harm, the theoretical training should be completed prior to working under supervision. If this is not the case, the training could start by working under supervision and with the responsibility remaining with the supervisor. In practice, this means that someone carrying out procedures must have completed the core and prerequisite modules before working with live animals, but an animal caretaker/technician could undertake subaltern tasks within the animal facility while undergoing training.

The EWG developed principles for the assessment of learning outcomes. Assessment/pass-fail criteria should:

- be objective and transparent
- be comprehensible and clear without ambiguity
- provide reliable results
- be economically possible
- have clear pass-fail criteria
- identify critical elements that cannot fail.

Supervision and Assessment of Competence

The principle of supervision is highlighted in Directive 2010/63/EU and the person responsible for ensuring staff are adequately educated, competent and continually trained (Article 24.1c) should play a key role in the local management process. Good supervision can re-enforce and enhance learning outcomes but, equally, inappropriate supervision can have negative consequences, occasionally promulgating old or poor practices. A supervision system should therefore include a mechanism to ensure consistency. Practical training and supervision should be carried out by an experienced colleague who is:

- appropriately skilled and competent in the techniques
- sufficiently senior to command respect
- able to impart skills and knowledge to others.

The UK Laboratory Animal Science Association (LASA) has produced guiding principles on the supervision requirements for personal licensees, which can be found at: http://www.lasa.co.uk/LASA_Guiding_Principles_Supervision_for_PILs_2010.pdf. There are different levels of competence, including a level where a trainee is eventually competent as a trainer. Ideally, the person who assesses competence should not be the same person as that who did the training but this may be difficult for highly specialized skills and in small establishments. Competence and the need for retraining should be reassessed at appropriate time intervals dependent on the maintenance of skills by continued use and new developments in best practice. LASA is also producing guidance on a harmonized system for assessment of competence. Guidance will be provided by the EC in due course.

Training Records

Records should reflect the level of training and level of competence to allow skills transfer within and across the European Union (EU). Currently there is a high variability in types of records being kept and in the control of record keeping from centralized facility to individual. The culture and/or compliance history of the establishment influences record keeping. Records should be detailed down to the technique level and be species-specific. They are an essential part of the training scheme and should incorporate professional education and competencies acquired prior to employment. Records may be kept centrally by individuals and either stored electronically or in a paper-based system. The records should be available for audit by the CA. The LASA document on supervision referred to above contains some examples of suitable records. Draft proposals for training record templates have been produced by the EC.

Quality Assurance

The quality of training needs to be assured to allow transfer of staff. Perhaps one of the most difficult issues to address is how to apply some form of "quality standard" to the course in terms of its content, delivery, and outcome. Within the UK, three accrediting bodies are approved by the CA to oversee their mandatory courses. A similar scheme is in operation in Switzerland where the CA itself manages the scheme. FELASA introduced an accreditation program in 2003[43] and has considerable experience of accrediting laboratory animal science courses which use the FELASA recommendations for education and training (FELASA categories A, B, C, and D). In 2012, there were 28 accredited courses within 13 countries. The accreditation scheme is operated on behalf of the FELASA Board of Management by the Accreditation Board for Education Teaching and Training, which comprises experienced educators and trainers from Member States. Courses meeting the criteria laid down by FELASA for each of the four categories can be described as offering "FELASA accredited training" for the duration of the validity of the approval. Further details can be found on the FELASA website: http://www.felasa.eu/accreditation-boards/accreditation-board-for-education-and-training1/.

The EC, through its EWG, is developing proposals for a framework for mutual acceptance. It should cover all courses, including in-house courses, and be of "light-touch."

Continuing Professional Development

Article 23.3 requires that staff shall maintain competence through a process of continuing education (CPD). This addresses both professional development and the need to implement best practices as these are identified. This is a new requirement for many Member States. Currently, only Switzerland has such a mandatory requirement. In 2010, FELASA proposed Guidelines for Continuing Education for all Persons Involved in the Care and Use of Animals for Scientific Purposes and details can be found at: http://www.felasa.eu/recommendations/guidelines/guidelines-for-continuing-education-for-persons-involved-in-animal-experime/. The EC has endorsed the principles contained in the FELASA guidelines and these are listed below:

• Personnel working with animals should maintain state of the art knowledge and skills
• CPD should be available and organized in a flexible way

- CPD should commence when a person starts working with animals and continue through the working career
- The system is based on the award of credits over a period of time
- There should be a process for review and endorsement of CPD activities, but up to 50% of credits can be achieved from activities recognized by the institution
- The operation of the scheme should be reviewed
- Joint courses should be encouraged between countries.

In practice, FELASA encourages the uptake of the scheme by Member Associations in their respective country. Member Associations facilitate the introduction of the CPD scheme and adapt it to local needs. They endorse courses and provide guidance and information. The institute, together with the individual, determines his/her CPD needs. Compliance with the scheme is by self-regulation with oversight by the CA during inspection.

APPENDIX

ANNEX V of Directive 2010/63/EU
List of Elements Referred to in Article 23(3)

1. National legislation in force relevant to the acquisition, husbandry, care, and use of animals for scientific purposes.
2. Ethics in relation to human-animal relationship, intrinsic value of life and arguments for and against the use of animals for scientific purposes.
3. Basic and appropriate species-specific biology in relation to anatomy, physiological features, breeding, genetics, and genetic alteration.
4. Animal behaviour, husbandry, and enrichment.
5. Species-specific methods of handling and procedures, where appropriate.
6. Animal health management and hygiene.
7. Recognition of species-specific distress, pain, and suffering of most common laboratory species.
8. Anesthesia, pain-relieving methods, and killing.
9. Use of humane end-points.
10. Requirement of replacement, reduction, and refinement.
11. Design of procedures and projects, where appropriate.

Transport

Appendix A of ETS 123[2] states that "for animals, transportation is a stressful experience which should be mitigated as far as possible. Animals should be transported in accordance with the principles of the ETS on the Protection of Animals during International Transport (ETS N°65[45] and ETS N°193[46]), having regard to the Resolution on the Acquisition and Transport of Laboratory Animals, adopted by the May 1997 Multilateral Consultation of the Parties to Convention ETS N°123. Both sender and recipient should agree on the conditions of transport, departure and arrival times to ensure that full preparation can be made for the animals' arrival." Appendix A of ETS 123 further lists conditions of animals to be taken into account to allow for transport and requires that, in case of transport of sick or injured animals, a competent person to confirm these animals are fit for transport. The route should be planned in order to ensure that the transport is carried out efficiently to minimize journey time, from loading to unloading, and to avoid delays in order to limit any stress and suffering of the animals. Care is needed to ensure that animals are maintained under suitable conditions for the species, and that measures are taken to minimize sudden movements, excessive noise, or vibrations during transport. ETS 123 specifies conditions to be fulfilled for transport of nonhuman primates, amphibians, fish, and reptiles.

Directive 2010/63/EU[5] however only consecrates one article (Article 23) to transport of animals: "Member States shall ensure that animals are transported under appropriate conditions." Annex III does mention transport, but only refers to transport container requirements.

Whereas the vast majority of animals used in research in Europe are purpose-bred in facilities within Europe, very few institutions only use animals bred on site; some animal models (some transgenic lines, nonhuman primates) are only available in non-European countries. Transport thus is a very important issue and all steps should be taken to ensure that the time taken to transport animals between breeders and users is kept to a minimum, thus avoiding unnecessary stress.

Transportation of animals is governed by a number of international bodies and regulation. ETS 193 clearly states that animals shall be transported without delay to their place of destination. However, that condition can only be fulfilled if the various aspects of the process are clearly understood. Animal transportation will not only require careful planning but also ensure that all the required conditions are fulfilled. It includes checking authorizations for transporters, knowledge of design and construction of means of transporters, fulfilling the various conditions of the preparation for transport (planning, presence of attendants, certificates of fitness for transport), inspections needed and health certificates required, possibilities for loading and unloading and transport practices among which special provisions for transport by rail, by road, by water, in road vehicles or rail wagons on roll-on/roll-off vessels, or by air.

In the European Union, all these requirements have been specifically defined in the Council Regulation (EC) 1/2005 on the protection of animals during transport and related operations.[47] At the European Union level, Regulations do not need transposition into the national legislation (Directives are different and need to be transposed), and therefore, they enter into force automatically in all Member States. This Regulation 1/2005 applies to the transport of live vertebrate animals in general, thus it also applies to laboratory animals. It includes specific requirements on documentation, transporters (must be authorized), inspections, emergency measures, penalties, and training of personnel. It also includes a number of technical rules (Annex I) related to fitness of animals, means of transport, transport practices, transport vessels, watering, long journeys, space allowances, logs (Annex II), forms (Annex III), and training (Annex IV).

There are two important requirements in Regulation 1/1005 that may specifically impact the transport of laboratory animals. One refers to the obligatory successfully completed training courses (Annex IV) that road drivers and attendants have to follow. The courses have to include technical and administrative aspects of Community legislation concerning protection of animals during transport and more particularly items such as animal physiology, animal behavior, practical aspects of handling of animals, impact of driving behavior on welfare, emergency care, and safety consideration for personnel handling animals. Many animals are routinely transported by "common" couriers, which may have problems implementing these requirements. Another important requirement is in Annex I, Chapter V, where it is stated that for several species (equidae, bovine, ovine, caprine, and porcine), when transport time exceeds the maximum for each species, animals must be unloaded, fed and watered, and be rested for at least 24 h. In the case of pigs, the maximum transport period is 24 h, which may be exceeded in the transportation of specific pathogen free (SPF) animals (i.e. mini-pigs). Indeed, to follow the Regulation in cases of long road transportation, animals would have to be unloaded and the retention of the SPF condition of the animals would be at stake.

The International Air Transport Association (IATA) updates the Live Animals Regulations annually, and IATA member airlines and many countries agree to comply with these regulations to ensure safe and humane transport of animals. In September 2008, an interesting initiative was taken by the World Organisation for Animal Health (OIE) which signed a joint IATA–OIE agreement[48] to cooperate, through formal and informal consultations, on issues of common interest including the transport of live animals. IATA and OIE have also started to dialogue on problems affecting the international transport of research animals by air and possible solutions.

Air transportation of animals not only promotes good research; it also promotes good animal welfare. The IATA recently stated, "In today's modern world, carriage of live animals by air is considered the most humane and expedient method of transportation over long distances." IATA ensures that both safety and animal welfare are addressed in all regulatory issues pertaining to transportation of live animals by air.

However, over the last years, ongoing campaigns by animal rights activists have targeted the transport of live animals for biomedical research, with the objective to stop research. As a result, a significant number of European and International airline companies, together with ferry operators, no longer transport live animals for research purposes. Undoubtedly this will result in undermining animals' welfare, either because of prolonged transport or because the studies will be conducted outside of European authorities jurisdiction and scrutiny.

At the time of writing (February 2013), several initiatives are ongoing in the biomedical community to raise awareness at the level of legislators and airlines CEO's, as thousands of research programs at institutions worldwide are dependent on the shipment of laboratory animals.

HOUSING AND ENRICHMENT

The natural histories of the species that are used for experimental purposes are fascinating. What is the historical background of our laboratory mouse (*Mus musculus*) before it ended up in the cages in our animal facilities? What are the characteristics of the species? How is that for species more recently introduced as experimental animals, like the Zebra fish (*Danio rerio*)? One thing is certain: Their captive environment in the laboratory does not compare to anything in their natural habitat. Some suggest that the animals born and bred in our facilities for generations do not know any better, but the generally accepted obligation is to ensure the optimal well-being of the animals in our care. Proper housing conditions and the application of enrichment are among the basic requirements enforced through (inter)national legislation. In 1986, the Council of Europe published ETS No. 123, the European Convention for the Protection of Vertebrate Animals used for Experimental and Other Scientific Purposes[1]; an amended text was published in 2005.[2] Appendix A of this Convention contains guidelines for the accommodation and care of these animals.[2] These guidelines served as the basis for the requirements on accommodation and care as stipulated in the European Directive 2010/63/EU of the European Parliament and of the Council on the Protection of Animals Used for Scientific Purposes.[5]

European Treaty Series 123

Article 5.1 deals with the animal's housing and environment: "Any animal used or intended for use in a procedure shall be provided with accommodation, an environment, at least a minimum degree of freedom of movement, food, water and care, appropriate to its health and well-being. Any restriction on the extent to which an animal can satisfy its physiological and ethological needs shall be limited as far as practicable. In the implementation of this provision, regard should be paid to the guidelines for accommodation and care of animals set out in appendix A to this convention."

ETS 123, Appendix A

In the introduction to Appendix A it is stated that it provides guidelines for the accommodation and care of animals, based on present knowledge and good practice. The object is thus to help authorities, institutions and individuals in their pursuit of the aims of the Council of Europe in this matter. A general section and species-specific sections follow the introduction.

The general section provides guidelines on accommodation, housing, and care relevant to all animals used for experimental and other scientific purposes. Supplementary guidance concerning commonly used species is presented in specific sections. Where no information is included in these specific sections the provisions of the general section apply. The species-specific sections are based on proposals made by expert groups on rodents, rabbits, dogs, cats, ferrets, nonhuman primates, farm species, mini-pigs, birds, amphibians, reptiles, and fish. In addition to these proposals, the expert groups also submitted background information to support their proposals, based on scientific evidence and practical experiences. This background information is the sole responsibility of the respective expert groups and is separately available. For some groups of species, namely amphibians, reptiles, and fish, these explanatory documents also provide additional information on less commonly used species not referred to in the species-specific provisions.[49]

In case behavioral or breeding problems occur or further information on specific requirements for other species is required, there is the obligation to seek advice from experts and care staff specialized in the species concerned, to ensure that any particular species' needs are adequately addressed.

ETS 123 Appendix A includes advice about the design of appropriate animal facilities and provides recommendations and guidance about how the welfare provisions contained within the Convention can be met. It is important to understand that the standards of space and dimensions given represent minimum allowances. These may have to be increased in some circumstances, as environmental requirements for individual animals might vary according to, for example, species, age, physiological conditions, stocking density, and whether the animals are kept as stock, for breeding or experimental purposes, whether for long or short-term.

Environmental enrichment is also an important aspect contributing to the well-being of the animals. It has been suggested to change the term "environmental enrichment" to "environmental refinement" when applied to laboratory animals since "environmental enrichment" is used in neuroscience in the context of novelty-induced stimulation and regularly changing items, primarily to measure the effects of brain neuronal plasticity. In contrast, "environmental enrichment" in laboratory animal science is coined as a term for the enhancement of welfare by means of appropriate enrichment focused on animal needs.[50] However, "environmental enrichment" will still be used here since that is the term used in legislative documents as well as in most guidelines.

The General Section of ETS 123, Appendix A

The general section starts with the functions and general design of the physical facilities. It covers not just construction and maintenance, but also security (item 1.1). It deals with these issues down to the level of the holding rooms (item 1.2), the general and special purpose procedure rooms (item 1.3), and the service rooms (item 1.4).

The general section continues with "the environment and its control" (item 2). Topics are: ventilation (item 2.1); temperature (item 2.2); humidity (item 2.3); lighting (item 2.4); noise (item 2.5); and alarm systems (item 2.6).

Next are "education and training" (item 3) and "care" (item 4). Under care, item 4.5 is on "housing and enrichment." The introduction to this item describes the preconditions set for housing, or rather "lodging," animals: "all animals should be allowed adequate space to express a wide behavioral repertoire. Animals should be socially housed wherever possible and provided with an adequately complex environment within the animal enclosure to enable them to carry out a range of normal behaviors. Restricted environments can lead to behavioral and physiological abnormalities and affect the validity of scientific data. Consideration should be given to the potential impact of the type of accommodation, and that of the environmental and social enrichment programmes, on the outcome of scientific studies, in order to avoid the generation of invalid scientific data and consequential animal wastage. The housing and enrichment strategies used in breeding, supplying and user establishments should be designed to fulfill the needs of the species housed and to ensure that the animals can make the best use of the space available. Their design should also take into account the need to observe the animals with minimum disruption and to facilitate handling. Suggested minimum animal enclosure sizes and space allowances are included in the subsequent individual species sections. Unless otherwise specified, additional surface areas provided by enclosure additions, such as shelves, should be provided in addition to the recommended minimum floor areas" (item 4.5.1).

Housing is addressed in item 4.5.2: "Animals, except those, which are naturally solitary, should be socially housed in stable groups of compatible individuals. Single housing should only occur if there is justification on veterinary or welfare grounds. Single housing on experimental grounds should be determined in consultation with the animal technician and with the competent person charged with advisory duties in relation to the well-being of the animals. In such circumstances, additional resources should be targeted to the welfare and care of these animals. In such cases, the duration should be limited to the minimum period necessary and, where possible, visual, auditory, olfactory, and tactile contact should be maintained. The introduction or re-introduction of animals to established groups should be carefully monitored by adequately trained staff, to avoid problems of incompatibility and disrupted social relationships. The possibility of social housing should be promoted by purchasing compatible individuals when procuring animals of gregarious species."

Enrichment is addressed in item 4.5.3: "All animals should be provided with sufficient space of adequate complexity to allow expression of a wide range of normal behaviour. They should be given a degree of control and choice over their environment to reduce stress-induced behaviour. This may be achieved by using appropriate enrichment techniques, which extend the range of activities available to the animal and increase their coping activities. In addition to social activities, enrichment can be achieved by allowing and promoting physical exercise, foraging, manipulative, and cognitive activities, as appropriate to the species. It is advisable to allow the animals to exercise at every possible opportunity. Environmental enrichment in animal enclosures should be appropriate to the species-specific and individual needs of the animals concerned. Forms of enrichment should be adaptable so that innovation based on new understanding may be incorporated. The enrichment programme should be regularly reviewed and updated. The staff responsible for animal care should understand the natural behaviour and biology of the species, so that they can make sensible and informed choices on enrichment. They should be aware that all enrichment initiatives are not necessarily to the advantage of the animal and therefore should monitor their effects and adjust the programme as required."

And animal enclosures in item 4.5.4: "Animal enclosures should not be made out of materials detrimental to the health of the animals. Their design and construction should be such that no injury to the animals is caused. Unless they are disposable, they should be made from materials that will withstand cleaning and decontamination techniques. In particular, attention should be given to the design of animal enclosure floors, which should be appropriate to the species and age of the animals and be designed to facilitate the removal of excreta."

The Species-Specific Section of ETS 113, Appendix A

The species-specific section consists of 11 subsections each addressing to a specific order of animals. The subsections contain considerations valid for the order, and where appropriate, additional order or species-specific provisions are included:

A. species specific provisions for rodents;
B. species specific provisions for rabbits;
C. species specific provisions for cats;
D. species specific provisions for dogs;
E. species specific provisions for ferrets;
F. (a) species specific provisions for nonhuman primates with additional provisions for housing and care of (b) marmosets and tamarins; (c) squirrel monkeys; (d) macaques and vervets; (e) baboons;
G. (a) species specific provisions for farm animals and mini-pigs with additional provisions for housing and care of (b) cattle; (c) sheep and goats; (d) pigs and mini-pigs; (e) equines, including horses, ponies, donkeys and mules;
H. (a) species specific provisions for birds with additional provisions for housing and care of the (b) domestic fowl; (c) domestic turkey; (d) quail; (e) ducks and geese; (f) pigeons; (g) zebra finch;
I. species specific provisions for amphibians;
J. species specific provisions for reptiles;
K. species specific provisions for fish.

The items addressed in the general section are filled in with more specific information in the species-specific subsections. Appendix A includes performance standards for each species. For example: acceptable temperature and humidity ranges are specified; in more general terms requirements for light intensities and color, noise levels and type of noise are given.

In each subsection, housing, enrichment, and care are dealt with under one subheading. Emphasis is given to the enclosures, their dimensions, and flooring. Recommended minimal enclosures seizes are listed in table format. As an example the table presenting the minimum enclosure dimensions and space allowances for mice (ETS 123, Appendix A, Table A.1) is copied here (Table 5.2).

In 2007, the FELASA published the *Euroguide*.[51] It is an abbreviated version of the revised Appendix A of ETS 123 published in 2005. The primary

TABLE 5.2 Minimum Enclosure Dimensions and Space Allowances for Mice, ETS 123 Appendix A

	Body weight (g)	Minimum enclosure size (cm^2)	Floor area per animal (cm^2)	Minimum enclosure height (cm)
In stock and during procedures	Up to 20	330	60	12
	Over 20–25	330	70	12
	Over 25–30	330	80	12
	Over 30	330	100	12
Breeding		330 For a monogamous pair (outbred/inbred) or a trio (inbred). For each additional female plus litter, 180 cm^2 should be added		12
Stock at breeders*				
Enclosure size 950 cm^2	Less than 20	950	40	12
Enclosure size 1500 cm^2	Less than 20	1500	30	12

*Post-weaned mice may be kept at these higher stocking densities, for the short period after weaning until issue, provided that the animals are housed in larger enclosures with adequate enrichment. These housing conditions should not cause any welfare deficit, such as increased levels of aggression, morbidity or mortality, stereotypies and other behavioral deficits, weight loss, or other physiological or behavioral stress responses.

consideration was to produce a concise, "user-friendly" reference for anyone working with laboratory animals and also for those who provide services related to the animals' housing and care. It should be understood, however, that the approved Appendix A is the definitive document that should be consulted for points of certitude on the accommodation and care of animals.

Directive 2010/63/EU

Part of the guidelines of Appendix A of ETS 123 (mainly the tables with the enclosures' dimensions) has been included as requirements in Annex III— "requirements for establishments and for the care and accommodation of animals"—of Directive 2010/63/EU of the European Parliament and of the Council on the Protection of Animals Used for Scientific Purposes.[5] Like Appendix A of ETS 123, Annex III has a general section and a species-specific section. Tables with minimum enclosure dimensions and space allowances have been copied in. However, the tables have been complemented with an additional column in which the date of 1st of January 2017 is listed. This is the date by which the Member States of the European Union shall ensure that the care and accommodation standards as set out in Annex III are applied (Article 33.2). The extensive performance standards as described in Appendix A of ETS 123 have not been adopted in Directive 2010/63/EU.

The final provisions of 2010/63/EU ensure adaptation of annexes to technical or scientific progress (Article 50). The EC may adopt, by means of delegated acts, modifications to that effect. This could mean that the information presented in Annex III may be updated according to technical or scientific advances while Appendix A of ETS 123 remains unchanged. The dates referred to in the species-specific section of Annex III may be subject to change, but shall not be brought forward. In line with the intentions of the Convention, the European Directive 2010/63/EU in Article 33.1 states that the Member States shall, as far as the care and accommodation of animal is concerned, ensure that:

1. all animals are provided with accommodation, an environment, food, water, and care which are appropriate to their health and well-being;
2. any restrictions on the extent to which an animal can satisfy its physiological and ethological needs are kept to a minimum;
3. the environmental conditions in which animals are bred, kept or used are checked daily;
4. arrangements are made to ensure that any defect or avoidable pain, suffering, distress or lasting harm discovered is eliminated as quickly as possible; and
5. animals are transported under appropriate conditions.

Member States may allow exemptions from the requirements of paragraph 33.1a or paragraph 33.2 for scientific, animal-welfare, or animal-health reasons.

Zebra Fish

Both Appendix A of ETS 123 and Directive 2010/63/EU include a more general section on fish without addressing the more species-specific requirements. Since zebra fish are gaining popularity in different areas of research, there is an increasing interest in establishing and maintaining productive zebra fish housing facilities. Recommendations and guidance notes are available for further reference. Among those is the Guidance on the Housing and Care of Zebra fish.[52] Also the Guide for the Care and Use of Laboratory Animals provides species specific information.[53] A joint working group of FELASA and the European Network on Fish Biomedical Models (EuFishBioMed) is expected to report its recommendations on zebra fish housing, husbandry and health monitoring in 2013. The report will be made available through the FELASA website (www.felasa.eu).

Conclusion

Minimum cage sizes or space allocations to be provided per animal are generally based on professional judgment. This does not necessarily mean that they are backed up by scientific evidence nor that animal well-being is guaranteed when legal requirements are met. Further research is needed to elucidate the effect of captivity and space confinement under varying experimental conditions on the welfare of the different animal species, strains, including genetically modified animals, in our care. The most important remains careful observation and evaluation of the animals' well-being on a day-to-day basis.

HUSBANDRY AND ENVIRONMENT

There is a close relationship between ETS 123[1] and Directive 2010/63/EU[5] with regard to guidelines or requirements on husbandry and environment. Both documents have an appendix or annex dealing with these topics, among other important issues such as housing and enrichment. In the ETS, it is the Appendix A, entitled "Guidelines for Accommodation and Care Of Animals,"[2] and in the Directive it is Annex III, entitled "Requirements for Establishments and for the Care and Accommodation of Animals." The article in ETS 123 referring to the Appendix A is Article 5 of Part II (General Care and Accommodation), and the article in the Directive referring to Annex III is Article 33, which mandates, among other things, that Member States ensure that "all animals are provided with accommodation, an environment, food, water and care which are appropriate to their health and well-being" (Article 33.1a); and "the environmental conditions in which animals are bred, kept or used are checked daily" (Article 33.1.c). The close relationship between the husbandry and environment guidelines of ETS 123 Appendix A and the requirements of the Directive Annex III is that these last ones have been extracted from those in the Appendix A. In most cases, the contents in the Directive Annex III are only a short summary of those in ETS 123 Appendix A, which generally contains more performance standards. The main difference is that ETS 123 Appendix A is more intended as a guideline (the term "should" is used), while in the Directive the more mandatory term "shall" is used.

Husbandry (sanitation, feeding, watering, and bedding) is addressed very briefly. There is little more emphasis on the environmental conditions, such as temperature, relative humidity, ventilation, lighting, and noise. As the requirements in the Directive are similar, although shorter, to the guidelines of ETS 123 Appendix A, the description below focuses first on the Directive requirements, which have been transposed into the legislation of the European Union (EU) Member States, and secondly includes some of the performance standards given in ETS 123 Appendix A.

Sanitation is addressed very shortly and generically in the General Section of Annex III of Directive 2010/63/EU. First, when dealing with "Holding Rooms" (1.2), establishments are explicitly required to have a regular and efficient cleaning schedule for the rooms and to maintain satisfactory hygienic standards. The need to have walls and floors (the Appendix A also refers to ceilings) surfaced with materials resistant to sanitation is also included. Other minor references to sanitation address the existence of cleaning and washing areas large enough and the separation of the flow of clean and dirty equipment to prevent cross-contamination; the mandate that all feed hoppers, troughs or other utensils used for feeding be regularly cleaned and, if necessary, sterilized; and the need to clean enrichment devices and sleeping areas. There are no more detailed or specific requirements on sanitation. ETS 123 Appendix A contains some additional guidelines in a performance-based approach (General Section 4.9): "A very high standard of cleanliness and order should also be maintained in holding, washing and storage rooms. Adequate routines for the cleaning, washing, decontamination and, when necessary, sterilisation of enclosures and accessories, bottles and other equipment should be established and carried out." It is also recommended that cleaning and disinfection regimens should not be detrimental to animal health or welfare, be recorded, and follow well-established procedures. The importance to address behavioral needs such as odor-marking when establishing cleaning frequency is recognized and a performance approach is given with regard to frequency of sanitation: "decisions on frequency of cleaning should be based on the type of animal enclosure, the type of animal, the stocking density, and the ability of the ventilation system to maintain suitable air quality."

With regard to feeding, the General Section of the Directive Annex III (subsection 3.4) requires the form, content and presentation of the diet meet the nutritional and behavioral needs of the animal, as well as be palatable and non-contaminated. Practices to minimize or avoid contamination or deterioration during production, transport and storage are required. In addition to the sanitation of the feed utensils, the access of all animals to food must be ensured. ETS 123 Appendix A contains a few more guidelines addressing opportunities for foraging, information on the feedbags (production and expiry dates), and conditions for storage (cool, dark, dry, vermin-proof, and also refrigeration for perishable feed). However, there are no specifications on nutritional quality, sterilization treatments, or other practices.

Watering is briefly addressed too in subsection 3.5 of Annex III. The only requirements for drinking water are that: "(a) Uncontaminated drinking water shall always be available to all animals; and (b) When automatic watering systems are used, they shall be regularly checked, serviced and flushed to avoid accidents. If solid-bottomed cages are used, care shall be taken to mini-mise the risk of flooding." There are almost no extra guidelines in ETS 123 Appendix A with regard to drinking water either. There are no indications on specific quality parameters, frequency of changing bottles, or treatment regimens of water.

There is only one reference to bedding, in subsection 3.6 of Directive Annex III, where bedding materials adapted to the species are required. ETS 123 Appendix A extends more on the functions of the bedding: "to absorb urine and faeces, and thus facilitate cleaning; to allow the animal to perform certain species-specific behaviour, such as foraging, digging or burrowing; to provide a comfortable, yielding surface or secure area for sleeping; and to allow the animal to build a nest for breeding purposes." Also on the recommended charac-teristics for the bedding materials: "dry, absorbent, dust-free, non-toxic and free from infectious agents or vermin and other forms of contamination. Materials derived from wood that has been chemically treated or containing toxic natural substances as well as products which cannot be clearly defined and standardised should be avoided."

The European documents are not very thorough with regard to environmental conditions either. However, the requirements, although with a positive perfor-mance approach, are extremely succinct. For ventilation, subsection 2.1 of the General Section states that "Insulation, heating and ventilation of the holding room shall ensure that the air circulation, dust levels, and gas concentrations are kept within limits that are not harmful to the animals housed." There is nothing more on the subject. On the other hand, the ETS 123 Appendix A does contain some good performance standards: "The purpose of the ventilation system is to provide sufficient fresh air of an appropriate quality and to keep down the levels and spread of odours, noxious gases, dust and infectious agents of any kind. It also provides for the removal of excess heat and humidity." This performance approach is merged with the recommendation that 15–20 air changes would be normally adequate, but that a lesser number (i.e. 8–10) may be enough depend-ing on the density of animals.

The only requirement in the Directive for temperature and relative humid-ity in the holding rooms is that they have to be adapted to the species and age groups housed, and be measured and logged on a daily basis. There are no specific ranges required. ETS 123 Appendix A recommends temperature ranges for adult, normal animals, and considers that: "new-born, young, hairless, newly-operated, sick or injured animals will often require a much higher temperature level." It also acknowledges the importance of potential changes in animal's thermal regulation and the effect of temperature on behavior and metabolism of animals and therefore in experimental results. In practice, European institutions follow the recommended ranges in the species-specific section of Appendix A (Table 5.3). No recommended ranges for farm animals (except for pigs and mini-pigs) are offered, as it is explained they may be dependent on a number of factors including, for example, breed, age, caloric intake, weight, stage of lactation, and type of environment. Ranges for some avian species are also given depending on the age of the animals, and for some amphibians and reptiles. With regard to relative humidity, only a specific comment in the General Section of ETS 123 Appendix A states that: "For some species, such as rats and gerbils, the relative humidity may need to be controlled within a fairly narrow range to minimise the possibility of health or welfare problems, whereas other species, such as dogs, tolerate well wide fluctuations in humidity levels." However, recommended ranges are also given for some species in the species-specific section (Table 5.3).

Lighting is limited in the Directive to a few requirements on the need of an appropriate dark/light cycle (that may be based on natural light), regular photoperiods and intensity of light (no ranges given), and the need to consider sensitivity of albino animals. These requirements are clearly extracted from the same recommendations in ETS 123 Appendix A.

The reference to noise is even shorter in both the Directive and ETS 123. The few requirements/recommendations focus on avoiding effects on animal

TABLE 5.3 ETS 123 Appendix A Recommended Temperature and Relative Humidity Levels for Several Species

Species	Temperature (°C)	Relative humidity (%)
Rodents	20–24	45–65 (35–55 for gerbils)
Rabbits	15–21	≥45
Cats/Dogs	15–21	No need to control
Ferrets	15–24	No need to control
Marmosets and tamarins	23–28	40–70
Squirrel monkeys	22–26	40–70
Vervets	16–25	40–70
Macaques	21–28	40–70
Baboons	16–28	40–70

welfare, and isolating as much as possible animal areas from noise, and also from that noise above the hearing range of humans, which Appendix A states is above 20 kHz.

The last page of the species-specific section of the Directive Annex III (subsection 11) includes the requirements for the fish environment. Without detailing ranges on water parameters, it is required to control the water quality, and keep parameters within acceptable levels that sustain normal activity and physiology for a given species and stage of development. The water parameters mentioned are oxygen, nitrogen compounds, pH, and salinity. Other environmental factors mentioned very briefly are temperature, lighting, and noise. Some more information on the importance and effects of these parameters (including CO_2) is given in the species-specific section of ETS 123 Appendix A.

VETERINARY CARE

Veterinarians play an essential role in ensuring that high animal welfare standards are met. Often decisions concerning animal welfare are based on expert judgment especially in areas where clear clinical signs are not always available, such as assessment of pain, distress and suffering. Veterinarians are specifically trained in diagnosing and treating diseases in animals. As such, they are recognized by the CA, and have not only legal rights but also duties and responsibilities in relation to public health and the animals they care for.

Whereas ETS 123[1] clearly identifies requirements for the care and use of animals in procedures, respecting the Five Freedoms (freedom of movement, hunger, thirst, disease and ill-health, behavioral needs), the education and training requirements of the person responsible for ensuring/supervising the implementation of this requirement are not spelled out, and arrangements are left to each Party.

Although wording used in ETS 123 to describe the role of the veterinarian might at first glance seem limited, specific and major responsibilities in decision taking and care of animals are attributed to the veterinarian in Articles 11 and 20. Article 11 specifies that the decision whether an animal shall be kept alive or killed by a humane method at the end of a procedure and the supervision of the care that animal (when kept alive) is to receive is the responsibility of a veterinarian or other competent person. Article 20 requires user establishments to ensure that a veterinarian *or other competent person* be charged with advisory duties in relation to the well-being of animals.

Directive 2010/63/EU[5] confirms the essential role of the veterinarian in the protection of animals used for scientific procedures. Indeed, Article 25 states "Member States shall ensure that *each* breeder, supplier and user has a *designated veterinarian with expertise in laboratory animal medicine, or a suitably qualified expert where more appropriate*, charged with advisory duties in relation to the well-being and treatment of the animals."

This provision "where more appropriate" lays out two conditions that are explained in the questions and answers website of the European Commission[9]: the expert has to be suitably qualified and more appropriate to the institution than a designated veterinarian. Such could be the case when the user/breeder/supplier is dealing with more rare species, and veterinarians specialized in these specific species are not available.

The primary role of the designated veterinarian is thus to oversee the well-being and clinical care of animals used for scientific procedures. That responsibility extends to promoting and monitoring animal well-being at all times and during all phases of animal use and the animal's life, as also described in the FELASA/ECLAM/ESLAV Guidelines for the Veterinary Care of Laboratory Animals[54]:

"Traditional veterinary care such as acute medical treatment, while still important, takes up a relatively small proportion of the laboratory animal veterinarian's time. Adequate veterinary care in laboratory animal science encompasses several aspects which all reflect the complexity of the veterinarian's role in this specialty. These include the following:

- All activities directly related to the animals to promote their welfare, such as during transportation, health monitoring and health management, husbandry, selection of environmental enrichment, surgery, anesthesia, analgesia and euthanasia.
- Scientific activities, often as a scientific collaborator and advisor in laboratory animal science.
- Activities related to regulatory and administrative compliance. The veterinarian must be knowledgeable about relevant legislation, including any appropriate ERP.
- Education and training of personnel and guidance of administrative staff, animal care staff and scientists to the benefit of the animals, the science and the institution."

Article 24.1a of Directive 2010/63/EU requires establishments to identify a person(s) responsible for overseeing the welfare and care of the animals. In addition to this general function, and according to Article 26, this person is required to be a member of the AWB, which the tasks are outlined in this chapter in the section on oversight. This function is separated from the designated veterinarian, although apparently there is no prohibition for the same person to hold both roles.

Depending on the complexity of the establishment, the species housed and the techniques used, the designated veterinarian can be a major asset and an added value to the AWB. Indeed, in his/her role as designated veterinarian, he/she will have the best inside knowledge on the adequacy of veterinary care provided, the pain/distress/welfare of animals involved in experiments, the competence of personnel, and the type and severity of procedures used at the institution. Moreover, as the project evaluation requires expertise in areas of his/her competence and as the process should allow for complete transparency and be performed in an impartial manner (Article 38), the designated veterinarian could be the crucial link.

Some countries have therefore opted to require the designated veterinarian to be an integral part of the AWB. In other countries, institutions are free to choose that option. In any case, the AWB is to seek input from the designated veterinarian or the expert referred to in Article 25, record the advice and decisions taken regarding that advice, and keep those records available for competent authorities for at least three years.

The designated veterinarian/AWB in their day-to-day responsibilities will evaluate/advise compliance with the Directive and implementation of the 3Rs. They could therefore also act as the contact person for the institution and have a privileged liaison with the inspectors. Procedures may be very specific to an institution; the designated veterinarian/AWB may give the necessary insight into questions raised during inspections; the same may be true for deviations with, e.g. housing requirements. Exchange of such specific information could allow for the building of a relationship of mutual trust and respect.

Harmonized inspections and establishing guidance for these is one of the aims of the EC. Veterinarians working in the field of Laboratory Animal Science and Medicine are often subject to different certification/accreditation of the Animal Care and Use programs they supervise. Their willingness to share their experience and participate in the preparation of Guidelines and training of inspectors has recently been communicated to the EC.

Compared to the US applicable policies and guidelines,[53,55] Directive 2010/63/EU does not impose explicit authority of the designated veterinarian. Although Recital 30 states that "…appropriate veterinary care should be available at all times…," Article 25 gives the designated veterinarian "advisory duties in relation to the well-being and treatment of the animals." With regard to the AWB, Article 26 indicates that "The animal welfare body shall also receive input from the designated veterinarian…" but does not require explicitly that he/she be a member. While some Member States have maintained the specific role and duties of the attending veterinarian as already existing in their legislation (e.g. UK), other Member States have used the occasion of the transposition of the Directive into national legislation to reinforce the role of the designated veterinarian. References to the need of veterinary intervention can be found in Article 9 related to animals taken for the wild: "Any animal found, at or after capture, to be injured or in poor health shall be examined by a veterinarian…"; Article 16 on reuse, where the veterinary advice and examination is required to allow animal reuse; and Article 17: "At the end of a procedure, a decision to keep an animal alive shall be taken by a veterinarian."

The recommendations and requirements of ETS 123 and Directive 2010/63/EU do not include a detailed or organized program of veterinary care as described in other international documents. Areas related to veterinary care such as anesthesia and analgesia, euthanasia, etc. are referred to in different sections of the documents. For example, Article 14 of the Directive focuses on anesthesia and analgesia to ensure that pain, suffering, and distress are kept to a minimum. Euthanasia is addressed in Article 6 and Annex IV and is discussed in other sections of this chapter. More detailed European guidelines for the veterinary care were produced by FELASA, ECLAM and ESLAV.[54]

CONDUCT OF EXPERIMENTAL PROCEDURES

The European Convention for the Protection of Vertebrate Animals used for Experimental and Other Scientific Purposes (ETS123)[1] sets down the requirements for the conduct of procedures in Part III, Articles 6 to 12. The provisions in Directive 2010/63/EU[5] are largely similar but have been extended to include additional refinement needs and the introduction of a system of classification of severity of procedures.

Choice of Methods

The emphasis on the 3Rs in Directive 2010/63/EU is contained in Article 4 (Principle of replacement, reduction and refinement) and this Article has, in the choice of methods, to be implemented in accordance with Article 13. Member States shall ensure that, wherever possible, a scientifically satisfactory method or testing strategy, not entailing the use of live animals, shall be used instead of a procedure (Article 4.1). They shall also ensure that the number of animals used in projects is reduced to a minimum without compromising the objectives of the project (Article 4.2). Member States shall ensure refinement of breeding, accommodation and care, and methods used in procedures, eliminating or reducing to the minimum any possible pain, suffering, distress or lasting harm to the animals (Article 4.3). The requirement to use the least number of animals can, at times, seem to conflict with the principle of refinement and the EC has provided an answer to this specific question posed by a Member State. After all means to replace are exhausted, the Directive does not make a hierarchy between reduction and refinement. This is part of the case-by-case project evaluation that needs to take into account the principle of the 3Rs as well as ethical considerations (Article 38.2d).

Article 13 (Choice of methods) states that "a procedure cannot be carried out if another method or testing strategy for obtaining the result sought, not entailing the use of a live animal, is recognised under the legislation of the Union."

This is more restrictive than previous legislation that forbade an experiment where one not entailing the use of an animal was reasonably and practicably available. This restriction has led to questions of clarification being sought from the EC.[9] One typical example is reproduced below:

Q   If a third country requires the use of an animal method, even when a nonani-
     mal method is recognized by the legislation of the Union, would it be allowed
     to be performed under this Directive?

Answer The requirement to use a nonanimal method is conditional on various factors,
     amongst others:
     1. That the result sought by the procedure is achieved by the nonanimal
        method or testing strategy.
     2. That the nonanimal method or testing strategy is recognized under Union
        legislation.
     All the above is overridden by any national legislation prohibiting certain
     types of methods.

Therefore, if a nonanimal method is recognized under Union legislation but a third country requires an animal method to be performed, the only situation where the animal method could be authorised would be that where the non-animal method does not provide "the result sought," provided national legislation of the Member State authorizing the procedure does not prohibit it. In this context, we believe that the concept of "result sought" should be interpreted as meaning "end-point": obtaining information from the animal test could not be argued to be a result in itself.

As an example, a replacement alternative recognized by the EU legislation that identifies a toxic hazard but is not able to differentiate between different levels of that toxicity (for example as required under a legislation for Classification and Labeling depending on the endpoint); If the third country regulatory requirement is to be able to identify only the existence of that toxicity, the respective animal method could no longer be used. However, if the third country requirement is to discriminate between different levels of that toxicity, the animal method could still be used.

Anesthesia

The requirements for anesthesia and analgesia are contained in Article 14 of Directive 2010/63/EU:

1. Member States shall ensure that, unless it is inappropriate, procedures are carried out under general or local anesthesia, and that analgesia or other appropriate method is used to ensure that pain, suffering, and distress are kept to a minimum. Procedures that involve serious injuries that may cause severe pain shall not be carried out without anesthesia.

2. When deciding on the appropriateness of using anesthesia, the following should be taken into account:
   a. whether anesthesia is judged to be more traumatic to the animal than the procedure itself, and
   b. whether anesthesia is incompatible with the purpose of the procedure.
3. Member States shall ensure that animals are not given any drug to stop or restrict their showing pain without an adequate level of anesthesia or analgesia. In these cases, a scientific justification shall be provided, accompanied by details of the anesthetic or analgesic regimen.
4. An animal, which may suffer pain once anesthesia has worn off, shall be treated with pre-emptive and post-operative analgesics or other appropriate pain-relieving methods provided that is compatible with the purpose of the procedure.
5. As soon as the purpose of the procedure has been achieved appropriate action shall be taken to minimise the suffering of the animal.

Classification of Severity of Procedures

A new requirement to classify the severity of procedures is set down in Article 15 of Directive 2010/63/EU. Before the transposition of the Directive, a number of Member States together with Switzerland (in this case the Directive makes no change) had a national requirement to prospectively estimate the severity of procedures during the preparation of the project proposal. However, there was no consistency in the process across those countries that have such a requirement. The Netherlands had a six-scale classification system while Switzerland and other countries a four-scale system. In those countries with a four-scale system there were variations in the definitions of the lowest severity category.

Now, the Article 15 requires that:

1. Member States shall ensure that all procedures are classified as "nonrecovery," "mild," "moderate" or "severe" on a case-by-case basis using the assignment criteria set out in Annex VIII.
2. Subject to the use of the safeguard clause in Article 55.3, Member States shall ensure that a procedure is not performed if it involves severe pain, suffering or distress that is likely to be long-lasting and cannot be ameliorated.

Annex VIII has three sections, a description of the severity categories, the criteria to be considered during the assignment process and a set of examples illustrative of each of the severity of categories:

Section I: Severity Categories

Nonrecovery

Procedures which are performed entirely under general anesthesia from which the animal shall not recover consciousness shall be classified as "nonrecovery".

Mild

Procedures on animals as a result of which the animals are likely to experience short-term mild pain, suffering or distress, as well as procedures with no significant impairment of the well-being or general condition of the animals shall be classified as "mild."

Moderate

Procedures on animals as a result of which the animals are likely to experience short-term moderate pain, pain, suffering or distress, or long-lasting mild pain, suffering or distress as well as procedures that are likely to cause moderate impairment of the well-being or general condition of the animals shall be classified as "moderate."

Severe

Procedures on animals as a result of which the animals are likely to experience severe pain, suffering or distress, or long-lasting moderate pain, suffering or distress as procedures, that are likely to cause severe impairment of the well-being or general condition of the animals shall be classified as "severe."

Section II: Assignment Criteria

The assignment of the severity category shall take into account any intervention or manipulation of an animal within a defined procedure. It shall be based on the most severe effects likely to be experienced by an individual animal after applying all appropriate refinement techniques. When assigning a procedure to a particular category, the type of procedure and a number of other factors shall be taken into account. All these factors shall be considered on a case-by-case basis. The factors related to the procedure shall include:

1. type of manipulation, handling
2. nature of pain, suffering, distress or lasting harm caused by (all elements of) the procedure, and its intensity, the duration, frequency, and multiplicity of techniques employed
3. cumulative suffering within a procedure
4. prevention from expressing natural behavior including restrictions on the housing, husbandry, and care standards

In addition to the factors above which are based on the procedure, for the purposes of the final severity classification, the following factors should also be taken into account on a case-by-case basis:

1. Type of species and genotype.
2. Maturity, age, and gender of the animal.
3. Training experience of the animal with respect to the procedure.
4. If the animal is to be reused, the actual severity of the previous procedures.

5. The methods used to reduce or eliminate pain, suffering, and distress, including refinement of housing, husbandry, and care conditions.
6. Humane end-points.

Section III: Examples of Different Types of Procedure

This section contains at least 10 examples of procedures in each of the mild, moderate and severe categories. The full list can be seen at: http://eur-lex.europa.eu/LexUriServ/LexUriServ.do?uri=OJ:L:2010:276:0033:0079:EN:PDF. Unfortunately, the examples lack sufficient detail, particularly of adverse effects, to facilitate a consistent approach to classification. Additionally, there are no examples of models involving pain research or CNS models in the severe category. To address this shortfall, a European Commission Expert Working Group has developed some further examples that will be published on the EC's website. This work will be further supplemented with additional examples from a FELASA/ESLAV/ECLAM working group that will demonstrate the continuous process of prospective assignment of severity, management of the in-life phase using welfare assessment tools and, finally, conducting the retrospective assessment severity. Further details of this work can be found on the FELASA website (http://www.felasa.eu/media/uploads/Classification_Severity_TOR_Final_v2.pdf) and the full report will be published in due course.

End of the Procedure

Article 17 of Directive 2010/63/EU sets down the requirements when an animal has come to the end of its procedure:

1. A procedure shall be deemed to end when no further observations are to be made for that procedure or, as regards new genetically modified animal lines, when the progeny are no longer observed or expected to experience pain, suffering, distress or lasting harm equivalent to, or higher than, that caused by the introduction of a needle.
2. At the end of a procedure, a decision to keep the animal alive shall be taken by a veterinarian or by another competent person. An animal shall be killed when it is likely to remain in moderate or severe pain, suffering, distress or lasting harm.
3. Where an animal is to be kept alive, it shall receive care and accommodation appropriate to its state of health.

EUTHANASIA

Euthanasia is an ethically contentious subject that places moral responsibilities on all those breeding, caring for and using laboratory animals. There is no ideal way of ending an animal's life, and the method adopted is a compromise between the requirements of the experiment, the facilities and expertise available, and minimizing distress to the animal.

The word "euthanasia" is derived from the Greek "eu" meaning good and "thanatos" meaning death. In this context, "good" should be interpreted as causing minimal pain, distress or suffering to the animal.

Laboratory animals may be euthanized for various reasons. For example when a procedure is finished and there is no justification for continuing to house them in laboratory conditions, rehoming is not permitted nor an option, when tissues need to be removed for further analysis or when adverse effects such as pain, distress or suffering can no longer be justified or when animals are no longer suitable for breeding. Euthanasia is difficult to justify ethically when the animals cannot be used in procedures because of inappropriate genotypic and/or phenotypic characteristics including the wrong sex. In those and related cases it is more appropriate to use the terms "humane death" or "humane killing."[56]

In the European Convention for the Protection of Vertebrate Animals used for Experimental and Other Scientific purposes (ETS 123),[1] it is stated that an animal shall not be kept alive if it is likely to remain in lasting pain or distress (Article 11.1). The decision on whether the animal shall be kept alive or not has to be taken by a competent person or the person how is responsible for, or has performed the procedure (Article 11.2). The Appendix to ETS 123[2] specifies, that the killing shall be done by a humane method, in line with the principles set out in the European Commission Recommendations for the Euthanasia of Experimental Animals (Part 1 and Part 2).[57,58]

In a recent review, some key issues of euthanizing experimental animals were listed[56]:

- All those engaged in animal experimentation must recognize the ethical sensitivity and close regulatory environment in which euthanasia takes place.
- Staff carrying out euthanasia should be trained, competent, experienced, and have access to well-maintained and adequate facilities and equipment.
- When choosing the method one should consider welfare, research needs, safety, and aesthetical and ethical issues. The method of choice should assure a rapid, stress-free loss of consciousness without subsequent recovery.
- Unnecessary distress should be avoided by using the home cage and established social groupings whenever possible, careful handling, skillful manipulation under circumstances where alarm communication between animals is not possible.
- Establishments should develop a broad strategy to minimize the number of animals euthanized, for example by encouraging good experimental design, efficient breeding policies, and germ-line banking.

Some of these considerations have found their way into Directive 2010/63/EU more specifically in Recital 15:

"The use of inappropriate methods for killing an animal can cause significant pain, distress and suffering to the animal. The level of competence of the person carrying out this operation is equally important. Animals should therefore be killed only by a competent person using a method that is appropriate to the species."

In Directive 2010/63/EU, killing is listed as an act that needs specific rules as a measure for the protection of animals used in scientific or educational purposes. Article 6 and Annex IV specifically deal with the issue of methods of killing. The animals listed are: fish; amphibians; reptiles; birds; rodents; rabbits; dogs, cats, ferrets and foxes; large mammals; and nonhuman primates. Member States shall ensure that animals are killed with minimum pain, suffering, and distress. They shall ensure that animals are killed in the establishment of a breeder, supplier or user, by a competent person. However, in the case of a field study an animal may be killed by a competent person outside of an establishment. The issue of competence is addressed in Article 23.2: the killing should only be done by adequately educated and trained persons. Those persons shall have received relevant instructions and shall have species-specific knowledge. Also, these persons shall be supervised in the performance of their tasks until they have demonstrated the requisite competence.

In relation to the animals covered by Annex IV of Directive 2010/63/EU, the appropriate method of killing as set out in that Annex shall be used. Competent authorities may grant exemptions from these requirements:

a. to allow the use of another method provided that, on the basis of scientific evidence, the method is considered to be at least as humane; or
b. when, on the basis of scientific justification, the purpose of the procedure cannot be achieved by the use of a method of killing set out in Annex IV.

These requirements shall not apply where an animal has to be killed in emergency circumstances for animal-welfare, public-health, public-security, animal-health or environmental reasons.

Annex IV of Directive 2010/63/EU lists the methods accepted for killing animals as there are: anesthetic overdose, captive bolt, carbon dioxide, cervical dislocation, concussion or percussive blow to the head, decapitation, electrical stunning, inert gases, and shooting with a free bullet with appropriate rifles, guns and ammunition (Annex IV-1). A table with acceptable methods per animal class/species is included (Annex IV-3) (Table 5.4).

The killing of animals shall be completed by one of the following methods:

1. confirmation of permanent cessation of the circulation;
2. destruction of the brain; dislocation of the neck;
3. exsanguination;
4. or confirmation of the onset of *rigor mortis.*

The table of Annex IV is the result of literature review and experts advice. The Federation of European Laboratory Science Associations (FELASA; http://www.felasa.eu) together with the European College of Laboratory Animal Medicine (ECLAM; http://www.eclam.org) and the European Society of Laboratory Animal Veterinarians (ESLAV; http://www.eslav.org) published

TABLE 5.4 Euthanasia Methods by Species

Animals-remarks/methods	Fish	Amphibians	Reptiles	Birds	Rodents	Rabbits	Dogs, Cats, Ferrets and foxes	Large mammals	Nonhuman primates
Anesthetic overdose	(1)	(1)	(1)	(1)	(1)	(1)	(1)	(1)	(1)
Captive bolt	No	No	(2)	No	No		No		No
Carbon dioxide	No	No	No		(3)	No	No	No	No
Cervical dislocation	No	No	No	(4)	(5)	(6)	No	No	No
Concussion/percussive blow to the head				(7)	(8)	(9)	(10)	No	No
Decapitation	No	No	No	(11)	(12)	No	No	No	No
Electrical stunning	(13)	(13)	No	(13)	No	(13)	(13)	(13)	No
Inert gases (Ar, N₂)	No	No	No	No	No	No	No	(14)	No
Shooting with a free bullet with appropriate rifles, guns and ammunition	No	No	(15)	No	No	No	(16)	(15)	No

Requirements: (1) Shall, where appropriate, be used with prior sedation. (2) Only to be used on large reptiles. (3) Only to be used in gradual fill. Not to be used for fetal and neonate rodents. (4) Only to be used for birds under 1kg. Birds over 250 g shall be sedated. (5) Only to be used for rodents under 1 kg. Rodents over 150 g shall be sedated. (6) Only to be used for rabbits under 1 kg. Rabbits over 150 g shall be sedated. (7) Only to be used for birds under 5 kg. (8) Only to be used for rodents under 1kg. (9) Only to be used for rabbits under 5 kg. (10) Only to be used on neonates. (11) Only to be used for birds under 250 g. (12) Only to be used if other methods are not possible. (13) Specialized equipment required. (14) Only to be used on pigs. (15) Only to be used in field conditions by experienced marksmen. (16) Only to be used in field conditions by experienced marksmen when other methods are not possible.
Source: From Annex IV of Directive 2010/63/EU.

guidelines on the veterinary care of laboratory animals which includes the act of euthanasia.[54]

The power of annexes to Directives is that they may be updated without having to amend the core of a Directive. Methods of killing animals are subject to active debate and ongoing research. Novel insight and technological advances will lead to refinement. The inclusion of the various methods in an annex to 2010/63/EU allows for relatively easy updating of the methodology of killing animals.

EQUIPMENT AND FACILITIES

The scheme for equipment and facilities follows the same pattern as for other described areas: Directive 2010/63/EU[5] incorporates as requirements ("shall") a summary of the recommendations ("should") of ETS 123[1] and more specifically of its Appendix A.[2] In summary, the legal requirements are very brief and general, with very little specifications.

In the Directive, Article 22 is specifically dedicated to "Requirements for installations and equipment." This article requires in general terms that all establishments "have installations and equipment suited to the species of animals housed and, where procedures are carried out, to the performance of the procedures." Facilities and equipment are connected in this article to the 3Rs principles, because it also requires their design, construction and functioning have to ensure procedures be performed effectively, "and aim at obtaining reliable results using the minimum number of animals and causing the minimum degree of pain, suffering, distress or lasting harm." To achieve this, the brief Article 22 refers to the requirements for establishments and the care and accommodation of animals described in Annex III. The text of Article 22 of the Directive is almost copied literally from Article 19 of ETS 123. Moreover, in ETS 123, the guidelines for equipment and facilities are described in its Appendix A, which serves as the basis for the requirements in the Directive Annex III.

References to equipment and facilities (mainly to facilities) follow the same scheme in both documents, which address functions and general design; holding rooms; general and special purpose procedure rooms; and service rooms. The only two pages that include recommendations about these areas in ETS 123 Appendix A are transformed in one page of very general requirements in the General Section (subsection 1) of Directive Annex III.

The only requirements in the Directive Appendix A concerning functions and general design (subsection 1.1) are (a) "all facilities shall be constructed so as to provide an environment which takes into account the physiological and ethological needs of the species kept in them. Facilities shall also be designed and managed to prevent access by unauthorised persons and the ingress or escape of animals"; and (b) "establishments shall have an active maintenance programme to prevent and remedy any defect in buildings or equipment."

With regard to the animal holding rooms (subsection 1.2), requirements are focussed on having good and regular sanitation practices, and constructions surfaces and materials that facilitate sanitation and are not detrimental to animal health. Species that are incompatible (e.g. predator and prey) have to be provided with independent environments. ETS 123 Appendix A has some additional recommendations on observation windows, drains (if existing should be covered and filtered), and the possibility of having facilities to perform minor procedures and manipulations inside the holding rooms.

The Directive Annex III (subsection 1.3) indicates that general and special purpose procedure rooms shall be available for situations where it is undesirable to carry out the procedures or observations in the holding rooms. Also, there must be areas for separation of sick or injured animals. This is the only potential relation in the Directive between facilities and quarantine; although the concept of quarantine is present, there is no more specific requirement for a quarantine area. This relation is more clear in ETS 123 Appendix A, where it is stated that facilities for newly acquired animals should be provided.

Requirements for service rooms (subsection 1.4) include food and bedding store rooms which have to be vermin-proof, cleaning and washing areas which must be large enough for the site needs and allow separation of dirty and clean materials, and surgical areas (when aseptic technique is required) with facilities for postoperative care. These requirements that take only a few sentences mimic the recommendations in ETS 123 Appendix A.

Facilities are expected to have appropriate heating and ventilation systems to ensure appropriate environment (subsection 2.1) as well as illumination (subsection 2.2) with controlled photoperiod, and alarm systems for all environmental control systems including essential services such as heating and ventilation, and illumination.

The few other references to equipment or facilities are scattered throughout the text of the Directive Annex III and ETS 123 Appendix A, such as the indication in Annex III of the need for the floors to be resistant to removal of excreta; or that protection has to be provided to equipment to avoid being damaged by animals or they do not cause damage to the animals themselves.

In summary, it is evident that the regulations and guidelines on equipment and facilities are brief and general. There are no mention of specialized equipment and facilities such as individually ventilated cages, isolators, imaging facilities, etc. There is not a description of all areas that a facility must/should have or of the different type of facilities depending on its function (breeders, users, etc.), species used, or type of research. On the other side, there is always a connection between the concept of equipment/facilities and the 3Rs principles, and more specifically with behavioral needs of animals. The proof of that is the first sentence of the General Section of both Directive Annex III and ETS 113 Appendix A. In this last one, it reads, "All facilities should be so constructed as to provide a suitable environment for the species to be kept, taking into account their physiological and ethological needs." In the Directive Annex III, it reads, "All facilities shall be constructed so as to provide an environment which takes into account the physiological and ethological needs of the species kept in them." This is not only the proof of the main focus of the European framework but also evidently of the influence of ETS 123 in the Directive, as has been shown in other sections of this chapter.

REFERENCES

1. Council of Europe. European convention for the protection of vertebrate animals used for experimental and other scientific purposes. *Eur Treaty Ser* 1986;**123**.
2. Council of Europe. Appendix A of the European convention for the protection of vertebrate animals used for experimental and other scientific purposes (ETS No. 123). Guidelines for accommodation and care of animals (Article 5 of the convention). Approved by the multilateral consultation. *Cons* 2006;**123**:3.
3. EEC. Council Directive of 24 November 1986 on the approximation of laws, regulations and administrative provisions of the Member States regarding the protection of animals used for experimental and other scientific purposes (86/609/EEC). L358. *Off J Eur Communities* 1986:1–29.
4. Commission. Commission recommendation of 18 June 2007 on guidelines for the accommodation and care of animals used for experimental and other scientific purposes. L 197. *Off J Eur Union* 2007:1–89.
5. The European Parliament and the Council of the European Union. Directive 2010/63/EU of the European Parliament and of the Council of 22 September 2010 on the protection of animals used for scientific purposes. *Off J Eur Union* 2010;L 276/33–79.
6. Guillen J. FELASA guidelines and recommendations. *J Am Assoc Lab Anim Sci* 2012;**51**(3): 311–21.
7. OJ L 262, 27.9.1976, p. 169. Directive recast by Regulation (EC) No 1223/2009 of the European Parliament and the Council of 30 November 1009 on cosmetic products (OJ L 342, 22.12.2009, p. 59), which applies from 11 July, 2013.
8. Consolidated versions of the Treaty on European Union and the Treaty on the Functioning of the European Union. *Off J Eur Union* 2010;**53**:1–388.
9. http://ec.europa.eu/environment/chemicals/lab_animals/pdf/qa.pdf.
10. Evans J. 2002. Report on Directive 86/609 on the protection of animals used for experimental and other scientific purposes (2001/2259 INI). http://www.europarl.europa.eu/sides/getDoc. do?pubRef=-//EP//TEXT+REPORT+A5-2002-0387+0+DOC+XML+V0//EN.
11. Report from the Commission to the Council and the European Parliament—Fifth Report on the Statistics on the Number of Animals used for Experimental and other Scientific Purposes in the Member States of the European Union (SEC(2007)1455).
12. Scientific Committee on Health and Environmental risks SCHER; The need for non-human primates in biomedical research, production and testing of products and devices. http://ec.europa.eu/environment/chemicals/lab_animals/pdf/scher_o_110.pdf.
13. Directive 98/44/EC of the European Parliament and of the council on the legal protection of biotechnological interventions. *Off J Eur Communities* 1998;L 213/13.
14. Guidance on the Operation of the Animals (Scientific Procedures) Act 1986. Ordered to be printed by the House of Commons 23 March 2000. http://www.official-documents.gov.uk/document/hc9900/hc03/0321/0321.pdf.

15. Smith JA, van den Broek FAR, Cantó Martorell J, Hackbarth H, Ruksenas O, Zeller W. Principles and practice in ethical review of animal experiments across Europe: summary of a report of a FELASA working group on ethical evaluation of animal experiments. *Lab Anim* 2007;**41**:143–60.
16. The Council of the European Communities. Council Directive 89/391/EEC on the introduction of measures to encourage improvements in the safety and health of workers at work. *Off J Eur Communities* 1989;L 183/1.
17. 89/654/EEC concerning the minimum safety and health requirements for the workplace (first individual directive within the meaning of article 16(1) of Directive 89/391/EEC).
18. 2009/104/EC concerning the minimum safety and health requirements for the use of work equipment by workers at work (second individual directive within the meaning of article 16(1) of Directive 89/391/EEC) (codified version).
19. 89/656/EEC on the minimum health and safety requirements for the use by workers of personal protective equipment at the workplace (third individual directive within the meaning of article 16(1) of Directive 89/391/EEC).
20. 90/269/EEC on the minimum health and safety requirements for the manual handling of loads where there is a risk particularly of back injury to workers (fourth individual directive within the meaning of article 16(1) of Directive 89/391/EEC).
21. 90/270/EEC on the minimum safety and health requirements to work with display screen equipment (fifth individual directive within the meaning of article 16(1) of Directive 87/391/EEC).
22. 2004/37/EC on the protection of workers from risks related to exposure to carcinogens or mutagens at work (sixth individual directive within the meaning of article 16(1) of Directive 89/391/EEC).
23. 2000/54/EC on the protection of workers from risks related to exposure to biological agents at work (seventh individual directive within the meaning of article 16(1) of Directive 89/391/EEC).
24. 92/57/EEC on the implementation of minimum safety and health requirements at temporary or mobile constructions sites (eighth individual Directive within the meaning of article 16(1) of Directive 89/391/EEC).
25. 92/58/EEC on the minimum requirements for the provision of safety and/or health signs at work (ninth individual directive within the meaning of article 16(1) of Directive 89/391/EEC).
26. 92/85/EEC on the introduction of measures to encourage improvements in the safety and health at work of pregnant workers who have recently given birth or are breastfeeding (tenth individual directive within the meaning of article 16(1) of Directive 89/391/EEC).
27. 92/91/EEC concerning the minimum requirements for improving the safety and health protection of workers in the mineral-extracting industries through drilling (eleventh individual directive within the meaning of article 16(1) of Directive 89/391/EEC).
28. 92/104/EEC on the minimum requirements for improving the safety and health protection of workers in surface and underground mineral-extracting industries (twelfth individual directive within the meaning of article 16(1) of Directive 89/391/EEC).
29. 93/103/EC concerning the minimum safety and health requirements for work on board fishing vessels (thirteenth individual directive within the meaning of article 16(1) of Directive 89/391/EEC).
30. 98/24/EC on the protection of the health and safety of workers from the risks related to chemical agents at work (fourteenth individual directive within the meaning of article 16(1) of Directive 89/391/EEC).

31. 1999/92/EC on minimum requirements for improving the safety and health protection of workers potentially at risk from explosive atmospheres (fifteenth individual directive within the meaning of article 16(1) of Directive 89/391/EEC).
32. 2002/44/EC on the minimum health and safety requirements regarding the exposure of workers to the risks arising from physical agents (vibration) (sixteenth individual directive within the meaning of article 16(1) of Directive 89/391/EEC).
33. 2003/10/EC on the minimum health and safety requirements regarding the exposure of workers to the risks arising from physical agents (noise) (seventeenth individual directive within the meaning of article 16(1) of Directive 89/391/EEC).
34. 2004/40/EC on minimum health and safety requirements regarding the exposure of workers to the risks arising from physical agents (electromagnetic fields) (eighteenth individual directive within the meaning of article 16(1) of Directive 89/391/EEC) and Directive 2012/11/EU amending Directive 2004/40/EC.
35. 2006/25/EC on the minimum health and safety requirements regarding the exposure of workers to the risks arising from physical agents (artificial optical radiation) (nineteenth individual directive within the meaning of article 16(1) of Directive 89/391/EEC).
36. The Council of the European Union. Council Directive 94/33/EC on the protection of young people at work. *Off J Eur Communities* 1994; L216/12.
37. http://www.hse.gov.uk/pubns/eh76.pdf.
38. Council of Europe. 1993. Multilateral consultation of the parties to the Council of Europe Convention ETS no. 123 on 3 December, 1993.
39. Weiss J, Bukelskiene V, Chambrier P, Ferrari M, van der Meulen M, Moreno M, et al. FELASA recommendations for the education and training of laboratory animal technicians: Category A. Report of the Federation of European Laboratory Animal Science Associations Working Group on Education of Animal Technicians (Category A) accepted by the FELASA Board of Management. *Lab Anim* 2010;**44**:163–9.
40. Nevalainen T, Dontas I, Forslid A, Howard BR, Klusa V, Käsermann HP, et al. FELASA recommendations for the education and training of persons carrying out animal experiments. Report of the Federation of European Laboratory Animal Science Associations Working Group on Education of Persons Carrying out Animal Experiments (Category B) accepted by the FELASA Board of Management. *Lab Anim* 2000;**34**:229–35.
41. Wilson MS, Berge E, Maess J, Mahouy G, Natoff I, Nevalainen T, et al. FELASA recommendations on the education and training of persons working with laboratory animals: Categories A and C. Reports of the Federation of European Laboratory Animal Science Associations Working Group on Education accepted by the FELASA Board of Management. *Lab Anim* 1995;**29**:121–31.
42. Nevalainen T, Berge E, Gallix P, Jilge B, Melloni E, Thomann P, et al. FELASA guidelines for education of specialists in laboratory animal science (Category D). Report of the Federation of European Laboratory Animal Science Associations Working Group on Education of Specialists (Category D) accepted by the FELASA Board of Management. *Lab Anim* 1999;**33**:1–5.
43. Nevalainen T, Blom HJM, Guaitani A, Hardy P, Howard BR, Vergara P. FELASA recommendations for the accreditation of laboratory animal science education and training. Report of the Federation of European Laboratory Animal Science Associations (FELASA) Working Group on Accreditation of Laboratory Animal Science Education and Training. *Lab Anim* 2002;**36**:373–7.
44. Van der Valk J, et al. Education and training in the 3Rs. *Altex* 2010;**27**:169–75.
45. European Convention for the Protection of Animals during International Transport. European Treaty Series 65. http://conventions.coe.int/Treaty/en/Treaties/Html/065.htm.

46. European Convention for the Protection of Animals during International Transport (revised). European Treaty Series 193. http://conventions.coe.int/Treaty/en/Treaties/Html/193.htm.

47. Council Regulation (EC) No 1/2005 of 22 December, 2004 on the protection of animals during transport and related operations and amending Directives 64/432/EEC and 93/119/EC and Regulation (EC) No 1255/97.

48. http://www.oie.int/fr/a-propos/principaux-textes/accords-de-cooperation/accord-entre-lorganisation-mondiale-de-la-sante-animale-et-lassociation-internationale-du-transport-aerien-iata/.

49. http://www.felasa.eu/about-us/library/.

50. Baumans V, Augustsson H, Perretta G. Animal needs and environmental refinement. Chapter 4. In: Howard BR, Nevalainen T, Perretta G, editors. *The cost manual of laboratory animal care and use: refinement, reduction, and research.* USA: CRC-press, Taylor & Francis Group; 2011.

51. Forbes D, Blom H, Kostomitsopoulos N, Moore G, Perretta G. *Euroguide on the accommodation and care of animals used for experimental and other scientific purposes.* London, UK: Published by Royal Society of Medicine Press limited; 2007; 978-1-85315-751-6.

52. Reed B, Jennings M. *Guidance on the housing and care of Zebrafish (Danio rerio).* Southwater, Horsham, UK: RSPCA; 2010. http://www.rspca.org.uk/researchanimals.

53. Committee for the Update on the Guide for the Care and Use of Laboratory Animals. *Guide for the care and use of laboratory animals.* 8th ed. Washington, D.C., USA: Institute for Laboratory Animal Research, National Research Council. The National Academies Press; 2011.

54. Voipio HM, Baneux P, Gomez de Segura IA, Hau J, Wolfensohn S. Guidelines for the veterinary care of laboratory animals: Report of the FELASA/ECLAM/ESLAV Joint Working Group on Veterinary Care. *Lab Anim* Jan 2008;**42**(1):1–11.

55. Unites States Public Health Service, 2002. Public Health Service Policy on Humane Care and Use of Laboratory Animals.

56. Antunes L Euthanasia. In: Howard B, Nevalainen T, Perretta G, editors. *Chapter 8 of the COST manual of laboratory animal care and use: refinement, reduction and research.* Boca Raton, FL, USA: CRC press, Taylor and Francis Group; 2011.

57. Close B, Banister K, Baumans V, Bernoth E-M, Bromage N, Bunyan J, et al. Recommendations for euthanasia of experimental animals: Part 1. *Lab Anim* 1996;**30**:293–316.

58. Close B, Banister K, Baumans V, Bernoth E-M, Bromage N, Bunyan J, et al. Recommendations for euthanasia of experimental animals: Part 2. *Lab Anim* 1997;**31**:1–32.

Chapter 6

Israeli Legislation and Regulation on the Use of Animals in Biological and Medical Research

Rony Kalman[1], Alon Harmelin[2], Ehud Ziv[3], and Yacov Fischer[4]

[1]*Hebrew University, Jerusalem, Israel,* [2]*Department of Veterinary Resources, The Weizmann Institute of Science, Rehovot, Israel,* [3]*Diabetes Research Unit, Hadassah University Hospital, Jerusalem, Israel,* [4]*Israeli Council for Animal Experimentation; and Ministry of Health, Jerusalem, Israel*

GENERAL FRAMEWORK

Animal care and use has been fully regulated in Israel since 1994 when the Animal Welfare Law—Animal Experiments[1] (Law) was legislated; until then,

Laboratory Animals. http://dx.doi.org/10.1016/B978-0-12-397856-1.00006-4

animal care and use was carried out on a voluntary basis according to the NRC *Guide for the care and Use of Laboratory Animals*[2] (*Guide*) standards. At the heart of the Law is the establishment of a National Council (Council) (Figure 6.1). The 23 member Council represents different stake holders including government, professional medical and veterinary organizations, industry, animal welfare groups, and the National Academy of Science. The Council operates in the Ministry of Health and all members are appointed for a 4 year term by the Minister of Health. The Law regards the Council as a professional body and the chairman is appointed by the Minister of Health from the National Academy of Sciences' members of life-science and medicine disciplines. The Council has all legal authority regarding animal care and use, and is entrusted with regulation and supervision duties.

The essence of the Law is the establishment of an Institutional Animal Care and Use Program (Program) in each and every institution that wishes to carry out research with animals. Each institution must be authorized by the Council prior to the introduction of animals to the facility or obtaining research permits. Council authorization indicates the existence of a Program including an appropriate infrastructure, and the appointment of an attending veterinarian (AV). All introductions of animals to the facilities require an Animal Experiment Permit—(Permit), which can be obtained from a designated (by the Council) Institutional Animal Care and Use Committee (IACUC). Permit applications (Applications) can be submitted only by principal investigators (PI) who have passed appropriate training for the purpose of the Law.

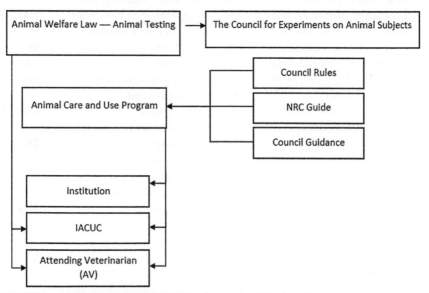

FIGURE 6.1 General schema of the legislative and supervising bodies involved in the regulation of the use of animals for biological and medical research in Israel.

An IACUC can be established only following Council authorization. The IACUC is independent in its deliberations, and the Applications are submitted on a Council-approved National Standard Form, which applies for the entire country and includes all aspects of animal care and use.

Applications can be submitted by an appropriately trained PI. The Council places particular importance on the issue of training, and each of the training courses is approved by the Council. All courses are built on a unified national framework of two levels: a basic training part, and a specific supplementary practical part for each animal species not covered by the basic course.

The Institutional Program is built on three arms: the Institute, the IACUC, and the AV. Council guidance on the establishment of the Program is comprised of: Council Rules, Council Guidance, and the instructions of the latest edition of the *Guide*, which has an official status in Israel.

The Law places special importance on the role of the AV, whose post is a legal requirement, and who is entrusted, among other things, with supervisory duties at the institution. Moreover, the AV has to be specifically authorized for this duty by the National Director of Veterinary Services and Animal Health—Ministry of Agriculture or be a Diplomat in Laboratory Animal Medicine (LAM), and the institute has to employ them according to a specific Council guidance.

Institutions are required to report to the Council on all approved permits, on the actual progress of research projects, and on the general veterinary conditions. Institutions are also supervised by Council members and the professionals of the Council (veterinarians and the academic advisor) who have the legal authority to inspect and even recommend to the Council the revocation of permits when severe irregularities are identified.

THE PRINCIPLES

The principles by which animal care and use are regulated in Israel are based on the 3Rs and are explicitly written in paragraphs eight (regarding reduction and refinement) and nine (regarding replacement) of the Law:

- The number of animals in any experiment shall be limited to the minimal number required for the performance of the experiment.
- Experiments on animals shall be carried out with close attention to minimizing the pain and suffering caused to them.
- No permit shall be granted for the conduct of any experiment on animals, if the objective of the experiment can be attained by reasonable alternate means.

In addition the Rules give a legal status to the *Guide* in Israel, which also follows the same principles.

The above legal instructions find their expression in each and every activity that follows the legislative act. The Animal Welfare Rules—on animal experiments[3] (Rules) published in 2001 state the necessary elements that need to be part of the training process for prospective animal users. These include detailed ethical rules on the use of animals—the contribution to scientific research, the fundamental principles of minimizing the number of animals in an experiment, finding alternatives to experiments with animals, moderation of experimental methods, legal principles, and the ethical system in the country and in the institution.

SCOPE/APPLICABILITY

The Law enables the use of animals (only live vertebrates, invertebrates, and fetuses are not covered) for research purposes in order to achieve one of the following goals:

1. The advancement of health and of medicine and the prevention of suffering.
2. The advancement of scientific research.
3. Testing or producing materials or objects.
4. Education and teaching (this is further discussed under the section "Education and Teaching with Animals").

Achieving the above goals must be done only after careful deliberations and taking into account numerous limitations and restrictions, including the 3Rs.

The bodies responsible for implementing the Law in Israel are the Council and IACUCs empowered by the Council.

All the activities related to animal experimentation in Israel are under the Council's authority. The Council grants authorization to institutes to perform experiments, and to establish IACUCs and training courses. Additionally, the Council receives reports from the IACUCs, from the Council's members[4] and from the Council's supervising veterinarians, as to the activities in these institutes.

An important role in the Council's activity is performed by the Council's supervising veterinarians whose role and authorities are specified in paragraph 15 of the Law.

The Council oversees all research activities with animals in Israel including, but not limited to, academic institutes—public or private, commercial institutions, public health institutions, and governmental agencies.

Requirements for institutional permits are described in paragraph 12 of the Law. Council authorization confirms the existence of a Program, the existence appropriate infrastructure, and the appointment of an AV.

NONHUMAN PRIMATES

According to the Law, the care and use of nonhuman primates (NHP) must follow the Guide recommendations. In addition, since the large majority of NHP experiments in Israel are electrophysiological studies, and because these studies are long term (usually 3–4 years) and have animal welfare aspects, the Council published special guidelines on harmonization of electrophysiological studies with NHP in Israeli institutions.[5] The purpose of these guidelines is to update users on available means of conducting the research and to harmonize the way electrophysiological studies with NHP are performed.

These guidelines emphasize the importance of specific points related to studies with NHP such as veterinary care, behavioral studies, enrichment and behavioral welfare, task learning by the monkeys, requirements for personnel, and the training of the research team at all levels. All the points listed above must be addressed in each new application for an experiment with NHP before approval is granted. To meet the goals of this harmonization, each institute is expected to acquire knowledge on all the subjects listed below, including the help of behavioral research groups, either locally or abroad. One of the innovations of the harmonization guidelines is the requirement for institutions that conduct research with NHP to establish an institutional animal welfare position.

The guidelines apply to:

1. The number of animals allocated to the project. When examining the number of NHP used in the research project, the IACUC should evaluate, among other things, if the minimum possible number of monkeys are allocated to the study while considering:
   a. The study objectives, taking into account the validity and significance of the potential findings.
   b. Potential loss of NHP due to health problems and/or failure of training.
   c. The social needs of NHP, who should be housed in direct proximity with at least one more individual when they are not involved in experimental procedures.
2. Personnel. The IACUC must condition the approval of an application to the constant presence of appropriate personnel, in quality and quantity, to ensure animal suffering is minimized and animals are maintained in the study for the minimum period of time necessary to achieve the study objectives. The IACUC must ensure that:
   a. There are enough personnel to fulfill the tasks outlined in the planned protocol of the experiment at all times.
   b. The absence of the PI or other researchers will be backed up by suitable replacements so that the experiment will not be unnecessarily delayed.
   c. The IACUC will condition its approval to perform any invasive and/or surgical procedure that have to be carried out under anesthesia by the presence of a trained veterinarian, and a person competent in performing the procedure.
   d. Institutions must have an institutional behavioral animal welfare position that addresses the NHP social and behavioral needs, establishes learning paradigms and the environmental enrichment program, as well as addresses the teaching and training of the research team on the use of NHP.
3. NHP training. The IACUC must verify that the research team has the appropriate knowledge regarding the training of NHP (or uses an external expert for this purpose) and that a training program is in place. This training program will address the following aspects:

    a. The researchers must have thorough acquaintance with the NHP.

    b. In order to allow NHP to appropriately perform the required experimental procedures, nonaggressive training methods shall be preferentially selected using positive reinforcement paradigms. Negative reinforcement paradigms of aggressive measures should be avoided as much as possible.

    c. If the training paradigms are not successful in the preselected NHP, the paradigms shall be revised and adapted.

    d. All members of the research team (including supporting staff) must be updated on the training objectives.

4. Education. IACUC approval will be granted only after verification that all members of the research team are qualified to work with NHP, and have followed and acquired the relevant theoretical and practical training and knowledge. This training must be implemented under the guidance of a professional and experienced instructor in the field.

5. NHP not in experiment. The IACUC needs to verify the existence of an environmental enrichment program appropriate to the NHP daily routines and needs, especially for periods prior to the experiment, between the various parts of the study, and until the NHP leaves the facility upon the completion of the study.

GENETICALLY ALTERED ANIMALS

The Israeli legislation does not have special regulations and requirements regarding genetically modified animals (GMA). As GMA are becoming an increasingly large proportion of the animals used in research in Israel, the IACUC and the AV have the mutual responsibility of assessing and monitoring the protocol and the experiments.

GMA require special genetic management procedures including maintenance of detailed pedigree records and genetic monitoring, which is the responsibility of the scientist. However, generation of animals with multiple genetic alterations often involves crossing different GMA lines and can lead to unexpected phenotypes. Any new phenotypes that negatively affect the animal's well-being should be reported to the IACUC and managed in a manner that ensures the animals' health and well-being. The issue is further addressed in the mandatory section of Humane End Points of the research permit application from.

INSTITUTIONAL AND DESIGNATED PERSONNEL RESPONSIBILITIES

Several designated Institutional bodies and professionals are specifically mentioned in the Law and in Council's regulations. These have a major role in approving and supervising animal experimentation, in order to guarantee their adherence to the legal requirements.

The Institutional Animal Care and Use Committee

According to the Law, research institutions may establish their own Institutional Animal Care and Use Committees (IACUCs) (subject to Council authorization) or make use of a National Animal Care and Use Committee (National ACUC) formed by members of the Council. The National ACUC provides service to institutions that do not have their own IACUC and its authority is limited to the review and approval of research permit applications.

The IACUC has a dual responsibility. It is the highest authority regarding the care and use of animals in research and, at the same time, it is the delegate of the Council. As such, it is responsible for the application of the Law and National Regulations at the institutional level.

The IACUC's roles as outlined in the Law (paragraph 14) indicate that the IACUC in public institutes must include at least three members: one veterinarian (according to the definition of veterinarian in the Law), one researcher from the life sciences or medicine disciplines, and one member who is not from the disciplines of the life sciences or medicine.

In private institutions the committee must also include a fourth member who is a public, nonaffiliated member (nominated by the institution and approved by the Council).

The IACUC's activities include:

- The examination and evaluation of all the animal related activities in the institution.
- The approval of protocol applications and amendments to already approved protocols.
- The control of all the animal facilities and the research activities.
- Reporting of the regular activities and any irregularities to the Council. Regular activities include:
  - Online reporting of all the approved applications.
  - Annual reporting on the project progress for each approved protocol.
  - Annual summary of institutional animals used by species.
  - Details of the program and facilities, including personal details of the AV.
- Approval of the syllabus of the institutional courses for personal accreditation of researchers, students, and technicians.
- Processing complaints from institutional personnel and the public about concerns regarding the use of animals in research.

The Attending Veterinarian

In order to become the AV (or as described in the Law "the veterinarian") a veterinarian must, as stated in the Law, be either a diplomat in LAM (by the Israeli college or any other recognized college) or personally accredited by the National Director of Veterinary Services and Animal Health—Ministry of Agriculture.

In September 2008[6] the Council instructed the minimum time dedication for AVs at institutions. The percentage of dedicated time depends on the size and complexity of the animal research at the specific institution:

- Facilities housing over 1000 rodent cages and also other species different than rodents—at least two full-time veterinarians.
- Facilities housing over 1000 rodent cages and without other species—at least one full-time veterinarian.
- Facilities housing 500–1000 rodent cages and also other species different than rodents—at least one full-time veterinarian.
- Facilities housing 500–1000 rodent cages and without other species—at least 50% part-time veterinarian (at least three weekly visits).
- Facilities housing less than 500 rodent cages and without other species—at least 25% part-time veterinarian (at least two weekly visits).
- Facilities housing up to 50 "large animals" (nonrodent species that are of a higher phylogenetic level)—at least 50% part-time veterinarian (at least three weekly visits).

In addition, the Council passed a resolution[6] that as of 2015, all newly appointed AV's will be Diplomats in LAM, and that by 2020 all AV's will be Diplomats in LAM.

The Law (paragraph 12) describes in general terms the roles of the AV.

In practice, AVs in Israel have many additional roles as indicated in the *Guide*, and therefore the AVs' authorities are broad, according to what is written in the *Guide*. These roles include membership in IACUCs; conduction of courses for new researchers, students, and technicians; advise on technical methods (i.e. analgesia and anesthesia) in research models; purchase and breeding of animals; and defining environmental enrichment programs.

OVERSIGHT AND ETHICAL REVIEW PROCESS

Animal experiment permit applications are reviewed by the designated IACUC. In principle, IACUCs conduct the review process according to the *Guide*, and applications are submitted on a standard national Council approved form, which requires the following information:

- Scientific abstract of the research.
- Duration of the experiment.
- Personnel authorized to perform the experiment.
- The goals of the experiment (according to the Law).
- Justification for the use of animals in the experiment.
- Number of animals required and justification for the species, strain, and number.
- Description of the procedures to be performed.
- Anesthetics and analgesics to be used.
- Follow up on the animals and Humane End Points.

- Fate of the animals at the end of the experiment and the method of euthanasia, if performed.
- Severity level (these are described in the "Conduct of Experimental Procedures" Section).
- Signed declaration of the PI acknowledging awareness of the Law.

There are several exceptions where specific guidance and preruling is required from the relevant authorities before IACUC deliberations can take place.

- Applications that propose the use of LD50 protocols have to gain specific Council authorization for each suggested use before IACUC deliberations occur.
- Animal testing for cosmetic substances is prohibited in Israel, unless it is intended for medicinal purposes. Therefore any application with such intent has to be cleared by the regulatory authority for pharmaceuticals as one that satisfies the regulatory definition for cosmetics for medicinal purposes, which is then communicated to the IACUC by the Council.
- In applications intended for the purpose of teaching the Council requires that the evaluation of alternative methods will not be the sole responsibility of the PI and asks for additional preview by the institutional teaching committee. The approval by this committee regarding the justification of using animals for the purpose of teaching is required before the IACUC commence its independent deliberations. Additionally, permits for teaching purposes are limited in time for 1 year only.

Each Permit is submitted upon IACUC approval to the Council, where they have a legally protected status for the purpose of supervision and statistical requirements of the Council. The representatives of the Council (professional civil servants) will conduct a review of the permits. In some cases this will lead to a further exchange between the Council and the IACUC for discussions and clarifications, which may result in protocol modifications. In extreme cases, where common ground cannot be found, or when there is a significant deviation from requirements, Council representatives may instruct cancellation of the permit.

The IACUC oversight does not terminate in the approval of applications. The IACUC is the highest authority at institutional level concerning the use of animals for research. As such, the IACUC supervises and controls all the animal related activities such as animal facilities, research laboratories, and training programs.

VETERINARY CARE

According to the Law, the *Guide* standards (pages 106–124), apply with regard to the Veterinary Care of Laboratory Animals in Israel.

The AV's roles, according to the Law, are indicated in the *Guide*, which broadens the AV's authority, making him the final voice concerning the animal's health and fate. The AV is required to oversee animal health and welfare, provide medical care, prevent diseases, reduce the suffering of animals before, during, and after the experiments, euthanize animals where necessary, and instruct staff members on these issues.

The AV must be a diplomat in Laboratory Animal Medicine (LAM) or a veterinarian whom the National Director of Veterinary Services and Animal Health—Ministry of Agriculture authorized for the purposes of the Animal Welfare Law. The Council in collaboration with the National Director of Veterinary Services and Animal Health—Ministry of Agriculture set accreditation criteria for veterinarians who are not diplomats in LAM, and decided that this authorization shall be limited for a period not exceeding 3 years from the first approval. After such period, renewal applications shall be submitted to the accreditation committee.

REUSE

The reuse of animals is a delicate issue that demonstrates the possible tension between the Rs of Reduction and Refinement. The Council has given its attention to this point in the Rules Article 6 indicating that a reuse can be considered by the IACUC only if one of the following conditions exist:

- The Committee is convinced that the experiment already carried out involved little suffering, though no association to a defined severity level is required.
- The animal will be anesthetized at the beginning of the additional experiment and will be euthanized at its conclusion, without regaining consciousness in its course.

The reuse may be approved only following a specific request of the PI to the IACUC. Such request should include information regarding both the experiment that was already performed as well as the requested additional experiment. All these approvals must be specifically marked by the IACUC and receive special attention by the Council.

In addition, the Rules clearly state that reuse of animals cannot be justified for economic reasons. There must be a scientific (or other) justification.

SETTING FREE/REHOMING

Research applications in Israel must include a paragraph describing the fate of the animal at the end of the experiment. Although there is no legal requirement to set free or rehome animals, it is the common understanding that farm animals, dogs, cats, and NHP should not be euthanized at the end of the experiment, unless there exists a scientific justification. The Council supports the rehoming of cats and dogs, the return to nature of appropriate wild animals, the return to the herd of farm animals, and the rehabilitation of NHP.

The setting free of NHP is currently a well-established solution. Research institutes can set free NHP at the end of the experiment to a specific NHP shelter. This shelter is supported and controlled by the Ministry of Environmental Protection. According to the Council's data over 50% of NHP that ended their participation in experiments were set free.

OCCUPATIONAL HEALTH AND SAFETY

Principles

Legally required institutional Safety Units monitor the health of faculty, staff, students, and any other persons on campus exposed to hazardous materials and situations, in accordance with Israeli laws and Institutions policy. All new employees must undergo a health evaluation, and persons whose work involves hazardous substances or procedures are required to undergo special medical tests at regular intervals, in accordance with Ministry of Labor guidelines.[7] Special medical examinations are required if a significant risk of excessive exposure to a hazard exists, and for a work-related health problem. The supervisor and the institutional Safety officer, in agreement with an Industrial-Hygiene Medical Doctor, decide on the specific medical examinations. Any person suspecting adverse health effects from a substance or procedure in the work environment is entitled to consult the doctor, who will decide if further testing is required. Noncompliance with the medical examinations is a disciplinary transgression and can bring disciplinary punishment.

The institutional Safety Unit maintains health records and monitors the medical surveillance program. The individuals are summoned to the examinations by the Personnel Department or the Safety officer, who coordinate the process with the health authorities. The process may include any measurements taken to monitor exposures such as questionnaires, medical consultations or physical examinations. The Industrial-Hygiene Medical Doctor enters the results to the workers medical file, and deals with any cases requiring a second test, medical treatment, or change in work conditions. The employee may ask to see and receive a copy of their personal records.

Experiments with special agents such as biological hazards, chemicals, etc., require biohazard specialized facilities or other adequate methods of containment.

The following are principles of biosafety that are used in Israel[8]:

The term containment describes safe methods, or properly managing infectious agents in the laboratory environment where they are being handled or maintained. The purpose of containment is to reduce or eliminate the exposure to potentially hazardous agents of laboratory workers, other persons, and the outside environment.

Primary containment, which is the protection of personnel and the immediate laboratory environment from exposure to infectious agents, is provided by both good microbiological technique and the use of appropriate safety equipment. The use of vaccines may provide an increased level of personal protection.

Secondary containment, which is the protection of the environment external to the laboratory from exposure to infectious materials, is provided by a combination of facility design and operational practices.

To review, the three elements of containment include:

- laboratory practice and technique;
- safety equipment;
- facility design.

The risk assessment of the work to be done with a specific agent will determine the appropriate combination of these elements.

All regulations, publications, and guidelines of the Israeli Institute for Occupational Safety and Hygiene, Ministry of Labor are online at: http://osh.org.il/site/english_main.html.[9]

EDUCATION, TRAINING, AND COMPETENCE OF PERSONNEL

The issue of education and training has been extensively dealt with by the Law, Rules, and Council decisions.[10,11] Instructions detail both the requirement on how to obtain a personal license as well as the relevant content of the accreditation courses to be taken for this purpose.

In order to conduct any experiment on animals, the PI and all the personnel who handle the animals (students, technicians, and animal care-takers) must obtain, prior to the experiment, a personal license as indicted in the Law (Paragraph 11).

In the 2001 Rules (Paragraph 9), the Council issued the general outline for such accreditation courses including topics to be covered, minimal content, as well as the qualification and supervision methods to be taken by the Council in order to ensure the academic level of such courses.

Nevertheless, as courses are not available continuously, and in order to enable students to start working under the supervision of qualified researchers even before they take the courses, the Council, in its 2001 Rules, enabled all personnel to obtain a one-time temporary-supervised permit for up to 6 months. Such temporary permits are supervised by the IACUC and must be followed by a permanent qualification to continue and conduct experiments with animals. The training should include at least the following:

- Explanations on the correct handling of laboratory animals, the use of alternatives, and minimization of the suffering of animals used in experiments.
- A practical demonstration of the experiment on animals in which the employee will engage.
- The conduct of at least two of those experiments on animals by a qualified researcher in the employee's presence.

In its circular dated September 2006,[11] the Council issued further instructions regarding qualification courses. The circular indicated that qualification courses must open with a basic theoretical part common to all the species in research. Participating in this basic part is a prerequisite to participate in the second-practical part. The practical part is specific for each species and each person will be qualified only to work with the species he or she has passed in the practical part. In the same circular, the courses in Israel of which the syllabus has been approved by the Council are listed.

EDUCATION AND TEACHING WITH ANIMALS

Animals can be used for two main teaching purposes: personnel qualification and academic education. These purposes were regulated by the Council in its circular dated October 2007.[10] The Purposes are:

- Qualification courses—(see Education, Training, and Competence of Personnel).
- Courses for teaching students known biological principles. The regulations for such courses include:
  - Each course has to be approved by the institutional academic teaching committee, which has to evaluate the need to use live animals—and not alternatives—for that specific course, prior to being finally reviewed by the IACUC.
  - The approval has to be renewed every academic year.
  - The use of animals in the course must be known to the students during their registration.

TRANSPORT

Transportation of animals is governed by a number of Israeli regulatory agencies and international bodies. The International Air Transport Association (IATA) updates the Live Animals Regulations annually and IATA member airlines and many countries agree to comply with these regulations to ensure the safe and humane transport of animals by air (IATA 2009). The Ministry of Agriculture with its Veterinary Services regulates the permit and vaccination programs if needed for the importation and exportation of animals, enforces regulations to prevent the introduction, transmission, or spread of communicable diseases, and regulates the importation of any animal or animal product capable of carrying a zoonotic disease. The Ministry of Environmental Protection regulates importation/exportation of wild vertebrate and invertebrate animals and their tissues through the Wild Life Protection Law[12] and the national authority of the Convention on International Trade in Endangered Species of Wild Fauna and Flora (CITES). The National Parks Service, operating under the Minister of Environmental Protection, regulates movement of CITES-listed species that are captive bred, including NHP. Institutions should contact appropriate authorities to ensure compliance with any relevant statutes and other animal transportation requirements that must be met for animals to cross international boundaries.

Animal transportation may be intra-institutional, inter-institutional, or between a commercial or noncommercial source and a research facility. For wildlife, transportation may occur between the capture site and field holding facilities.

According to the Law and Rules, the *Guide* standards (pages 107–109) apply to transport of animals to and from Israel.

HOUSING AND ENRICHMENT

According to the Law and Rules, the *Guide* standards (pages 42–88), apply to the housing and environmental enrichment of Laboratory Animals in Israel. This includes the minimum cage sizes for all species except rabbits. For rabbits, the Council made an exception and decided they should be housed according to the European Convention ETS 123 Appendix A.[13]

HUSBANDRY AND ENVIRONMENT

All the parameters related to husbandry and environment are equivalent to those indicated in the current edition of the *Guide*. The 2001 Rules specifically indicate that the leading document in Israel regarding environment and animal management is the *Guide* (the current updated version applies).

CONDUCT OF EXPERIMENTAL PROCEDURES

The need for refining experimental procedures is one of the basics 3Rs of conducting animal procedures. It is referred to in the Law and in all the Council's publications.

In 2006 the Council instructed that all research projects must be assigned a severity level ranging from 1 to 5. The severity level must be indicated in the research protocol and reported to the Council.

The five severity levels are:

1. Collection of organs from animals that did not go under any experimental procedure and were euthanized using a method acceptable for organ collection.
2. Experiments that cause slight temporary discomfort or stress. Examples: intravenous, intramuscular, intraperitoneal, and subcutaneous injections; behavioral experiments that do not cause stress (but not including water maze or predator experiments); infliction of slight pain that the animal can avoid; withdrawing blood from peripheral vessels up to the quantity of blood that does not require anesthesia (10% of the total blood volume or 1% of the body weight); feeding experiments that do not cause clinical manifestations; tail tip sampling.
3. Experiments that cause slight stress or short term pain. Such experiments should not cause significant changes to the animal's appearance, to physiologic

parameters such as heart rate or respiratory rate or to social behavior. During and after such experiments animals should not exhibit signs of self-injury, anorexia, dehydration, anxiety, excessive recumbence, vocalization, over aggressiveness, or tendency for isolation. Examples: nonsurvival major surgery; cannulation; minor survival surgery; blood withdrawal under anesthesia from the retro-orbital sinus or from the heart; restraint for short periods; water or food restriction for less than 12 h a day.

4. Experiments that cause medium pain or distress that is alleviated by analgesics. Examples: major survival surgeries where animals receive analgesics; local nonmetastatic tumors where animals receive analgesics; restraining animals for over 60 min; restriction of water or food for over 12 h during the animal's activity phase; significant changes in environmental parameters (temperature, lighting); procedures that cause sensory or motor damage or severe and constant anatomical and/or physiological changes; use of Complete Freund's Adjuvant.

5. Experiments that cause severe and lasting pain or distress that are not alleviated by analgesics. Examples are metastatic tumors or experiments in which the end point is death. In all such experiments the researcher is requested to justify why analgesics are not used.

EUTHANASIA

Both the Law and the Rules indicate the need to include in the qualification courses the adequate euthanasia methods and materials for each species.

The adequate methods and materials required by the Israeli regulation are detailed in the AVMA guidelines on euthanasia.[14]

EQUIPMENT AND FACILITIES

According to the Law and Rules, the *Guide* standards (pages 133–151), apply with regard to equipment and facilities for laboratory animals in Israel.

AUTHORIZATION AND INSPECTION

The Council must approve facilities in each institution before animals can be housed and before any experiment can take place. The Council's supervising veterinarian must approve that the facility of an institution complies with the *Guide* standards. Once the facility is approved, periodic monitoring by the Council veterinarians is performed.

The Council's supervising veterinarians and the Council's academic advisor may enter, at any time, any institution and any animal facility, on the condition that he takes the necessary steps to prevent interference with an experiment. In addition, any Council member can enter any institution after informing the Council chair of their intent to visit, and after receiving the Chair's approval.

Such person may read any document in order to check whether the provisions of the Law are complied with. Reports to the Council on the state of the veterinary program at the institutions are submitted on a quarterly basis and a more detailed report is submitted annually. The Annual Report provides a description of any changes in the facilities and the veterinary program in the institution, and should report among other things special problem or irregularities that occurred.

REFERENCES

1. Animal Welfare Law—Animal Care and Use 5754-1994, www.weizmann.ac.il/vet/Law.htm.
2. NRC. *Guide for the care and use of laboratory animals.* Washington: National Academies Press; 2011. http://www.aaalac.org/resources/Guide_2011.pdf.
3. Animal Welfare Rules—on Animal Care and Use 576-20011. www.weizmann.ac.il/vet/Law.htm.
4. Visits by Council members, Council circular 26.12.2004. http://www.old.health.gov.il/Download/pages/VisitCouncilMem.pdf.
5. *Council's periodic report.* Feb 2011, 19–21. http://www.old.health.gov.il/Download/pages/d14022011.pdf.
6. Veterinarian role, Council circular 14.9.2008. http://www.old.health.gov.il/Download/pages/InsVet-nov09.pdf.
7. Ministry of Labor Guidelines. http://www.osh.org.il/uploadfiles/gov_hok_takanot.html.
8. Israeli Regulations for Safe Work with Biohazard Agents. http://www.osh.org.il/uploadfiles/takanot_gehut_maabadot.pdf.
9. Israel Institute for Occupational Safety and Hygiene website. http://osh.org.il/site/english_main.html.
10. Use of animals for teaching, Council circular 23.10.2007. http://www.old.health.gov.il/Download/pages/TeachingAndEducation.pdf.
11. Format of qualification courses, Council circular 4.9.2006. http://www.old.health.gov.il/Download/pages/Training.pdf.
12. Israeli Wild Life Protection Law. http://www.parks.org.il/ConservationAndheritage/LawAnd Enforcement/Regulation/%D7%97%D7%95%D7%A7%20%D7%9C%D7%94%D7%92%D7%A0%D7%AA%20%D7%97%D7%99%D7%99%D7%AA%20%D7%94%D7%91%D7%A8%20%D7%AA%D7%A9%D7%98%D7%95%20-%201955.pdf.
13. Appendix A of the European Convention ETS 123. http://conventions.coe.int/Treaty/EN/Treaties/PDF/123-Arev.pdf.
14. AVMA. *AVMA guidelines on euthanasia.* Schaumburg IL: AVMA; 2007. https://www.avma.org/KB/Policies/Documents/euthanasia.pdf.

Chapter 7

Animal Experimentation in Africa: Legislation and Guidelines

Prospects for Improvement

Amanda R. Hau[1], Faisal A. Guhad[2], Margaret E. Cooper[3], Idle O. Farah[4], Ouajdi Souilem[5], and Jann Hau[6]

[1]Faculty of Law, University of Lund, Lund, Sweden, [2]Jigjiga Export Slaughter House (JESH) PLC, Somali Regional State, Federal Democratic Republic of Ethiopia, [3]Faculty of Veterinary Medicine, University of Nairobi, Kenya; and DICE, The University of Kent; and Wildlife Health Services, UK, [4]National Museums of Kenya, Nairobi, Kenya, [5]Laboratory of Physiology and Pharmacology, National School of Veterinary Medicine, Sidi Thabet, Tunisia, [6]Faculty of Health and Medical Sciences, University of Copenhagen; and Department of Experimental Medicine, University Hospital, Copenhagen, Denmark

INTRODUCTION

Animal experimentation is common in Africa, a region where—according to Kimwele and co-workers—economic development and social development take priority over the issue of animal protection.[1] Animal welfare has become a

Laboratory Animals. http://dx.doi.org/10.1016/B978-0-12-397856-1.00007-6

major issue worldwide, and it is of interest to examine the state of the situation in a region of the world (among others) where animal experimentation is common, but little or no legislation exists.

The aim of this chapter is to present an investigation and review of the shortfalls in African legislation and guidelines concerning animal welfare in Africa in general and laboratory animal experimentation in particular. The investigation focused on three primary questions:

- What is the present situation?
- What are the main issues?
- What are the prospects for improvement?

The method used in the process of writing the current chapter was to study various documents found on the subject: Acts of Parliament, ordinances and proclamations, university guidelines, and scientific papers. In particular, the work of Kimwele and co-workers, "*A Kenyan perspective on the use of animals in science education and scientific research in Africa and prospects for improvement*"[1] and the World Organisation for Animal Health (OIE) summaries of "*Animal welfare in OIE member countries & territories in the SADC region,*"[2] have been very helpful. Summaries and statements from NGOs and other organizations were also reviewed. Correspondence with experts in the field has also been helpful throughout the process.

After our initial expectation of studying legislation and guidelines from as many African countries as possible, it soon became apparent that due to the scarcity of legislation in most countries, the main focus of the chapter would be on Kenya, Tanzania, and South Africa. These countries have produced the four most comprehensive pieces of legislation, namely The Animal Welfare Act of Tanzania,[3] the Prevention of Cruelty to Animals Act, Cap 360 of the Laws of Kenya,[4] and the two main laws regulating the welfare of animals in South Africa, the Animals Protection Act 71 of 1962,[5] and the Performing Animals Protection Act 24 of 1935.[6]

THE PRESENT SITUATION

Empirical data regarding the existence or adequacy of national or institutional policies and guidelines on the use of animals in research in Africa are limited. However, the limited availability of evidence indicates that most African countries lack relevant legislation and guidelines.[7]

General Framework

Masiga and Munyua[8] report that "currently in West, East, and Southern Africa, animal welfare issues are addressed under several Acts of Parliament, ordinances, and proclamations including the following (the titles of these acts vary from country to country, so some of the alternative names have been provided in brackets):

1. The Prevention of Cruelty to Animals Act (Animal Welfare Act),
2. The Branding Act (Animal Identification Act),
3. The Animal Diseases and Pest Control Act,
4. The Wildlife Conservation and Management Act (Wildlife Act),
5. The Meat Control Act (Meat Inspection Act, Veterinary Public Health Act), and
6. The Veterinary Surgeons Act (Veterinary Professionals and Paraprofessionals Act)."[8]

Additional acts include the following:

1. The Animal (Control of Experiments) Act, and
2. The Scientific Experiments on Animals Act.[9]

However, in this study, specific legislation concerning the welfare of animals *used in research and education* has only been found in a limited number of cases, namely in the animal welfare laws of Kenya, Tanzania, the Seychelles,[10] Zimbabwe, and South Africa.

Britain's Protection of Animals Act, which went through Parliament in 1911 and was enforced on January 1, 1912, has been the model for most of the legislation in African countries that were under British influence. Some nations such as Kenya, South Africa, and Tanzania have enacted their own animal welfare legislation,[3–6,9–11] however, even in these cases, implementation and enforcement mechanisms have, according to Kimwele and co-workers, been largely ineffective.[1]

Kenya's Prevention of Cruelty to Animals Act[4] is a comprehensive piece of legislation regarding animal welfare issues in general. There are three other acts containing comprehensive legislation: the Animal Welfare Act of Tanzania[3] and the two main laws regulating the welfare of animals in South Africa, the Animals Protection Act 71 of 1962[5] and the Performing Animals Protection Act 24 of 1935.[6] However, the use of animals in research, explicitly in South African legislation, is only covered in the South African Medical Research Council Act, 58 of 1991.[11]

Some countries—Angola, Democratic Republic of Congo, Madagascar, and Mozambique—are reported to have literally no significant animal welfare legislation.[2] The authors have reviewed all relevant documents accessible; yet, it is important to bear in mind that there is the possibility that guidelines and other regulations may exist in some African countries, but they are not available in the public domain.[1]

The Principles

There are certain principles generally accepted by the scientific community that are followed when using animals in research. One of the most well known and internationally accepted is the principle of the "Three Rs" (Replacement, Reduction, and Refinement); a guiding principle for the use of research animals in many countries, and now also explicitly endorsed in the text of the recently revised European Directive.[12]

The Three Rs were first described by Russell and Burch in 1959[13] and represent a practical strategy for researchers to apply when considering experiments with laboratory animals. *Replacement*, which is the first "R," refers to striving to avoid the use of animals completely, if possible. Replacement also encompasses replacing animals higher on the phylogenetic scale with those animals that are lower on the phylogenetic scale. *Reduction* refers to strategies to allow researchers to obtain comparable levels of information from the use of as few animals as possible, or to obtain more information from the same number of animals.[13] And, *Refinement* concerns the objective to use research methods that alleviate or minimize potential pain, distress, or suffering of the animals that are used. The enhancement of animal welfare is also included.

The Five Freedoms[14] are another set of guiding principles used by animal researchers, among others, and endorsed by many international guidelines on animal welfare in general. They have been the basis of much legislation in many parts of the world and were originally developed in 1965 in the UK Government Brambell Report on livestock husbandry.[15] This set of principles entails the freedom from hunger, thirst, and malnutrition; freedom from fear and distress; freedom from physical discomfort; freedom from pain, injury, and disease; and freedom to express normal patterns of behavior. The Freedoms define ideal states, rather than standards for acceptable animal welfare yet form a logical and comprehensive framework for the study of welfare in almost all situations.[14]

The Universal Declaration on Animal Welfare[16] is another source of guiding principles in the animal welfare field. It is a draft agreement built on feedback from United Nation (UN) member states, international organizations, and nongovernmental organizations. If endorsed by the UN, the Declaration would encompass a nonbinding set of animal welfare principles affirming, among other things, that animals are sentient beings and acknowledging that their welfare must be respected. The Declaration would provide a solid uniform basis for member states to improve already existing animal welfare legislation or introduce new national animal welfare legislation built on the agreed-upon principles.[16]

Principles of animal welfare can be found, both implicitly and explicitly, in various relevant documents in African legislation. The most obvious reference to fundamental animal welfare principles can be found in the Animal Welfare Act, 2008, of the Laws of Tanzania.[3] According to Part I section 4 (a) in the aforementioned Act, "every person exercising powers under, applying or interpreting this Act shall have regard to ensuring that animals are cared for according to their universally adopted five freedoms." Additional fundamental principles of animal welfare to be regarded are listed in Part I section 4 (b), "recognizing that an animal is a sentient being; animal welfare is an important aspect of any developed society, which reflects the degree of moral and cultural maturity of that society; animal welfare enhances livestock productivity, and that a human being has a moral obligation to care, respect, and protect an animal." The concept of the Three Rs is mentioned explicitly in the Act, Part I section 4 (c), "ensuring that animal experimentation promotes the reduction in number of animals used, refinement of experimental methods, and replacement of animal with non-animal use techniques."

Principles regarding the ethics of animal experimentation are also found in less formal forms, namely in the guidelines and codes of conduct governing the use of animals in research and teaching at universities in Africa. A telling example is the guide presented by the Animal Ethics Committee of the University of KwaZulu-Natal, South Africa.[17] The guide describes the concept of the Three Rs and also lists principles concerning the humane treatment of animals by referring to the principles described by the *Canadian Council on Animal Care*.[18]

Nonhuman Primates and Genetically Altered Animals

The use of nonhuman primates in scientific research is a controversial subject and is, therefore, a part of animal welfare that often entails additional attention. Internationally, the use of nonhuman primates and genetically altered animals in research often requires special considerations and regulations. In the European Union (EU), for example, several member states such as Austria, Britain, the Netherlands, and Sweden have introduced special regulations and/or some form of prohibition on the use of nonhuman primates in research, thus applying additional considerations beyond the requirements set by the 1986 European Directive controlling animal experimentation.[19,20] In Africa, legislation covering the subjects is generally lacking.

Nonhuman primate models are used in several research institutions in Africa. For instance, the National Museums of Kenya's Institute of Primate Research (IPR), which is located in Nairobi, Kenya, is a well-known research center focused on primatology. Animal models, in particular vervet monkeys and baboons, are used in studies of human reproductive disorders and tropical infectious diseases. Many scientific publications have been published from IPR, including a number that focus on animal welfare and the Three Rs.[21,22]

IPR is also engaged in nonhuman primate conservation,[23] and research ethics and animal welfare issues are taken into account at IPR. According to their webpage, "Ethical and animal welfare concerns form a strong component of the Department's animal husbandry and research activities." Before any experimental procedure is carried out, review committees examine all of the proposals for scientific merit and welfare concerns.[24] Nyika reported that in order to ensure that animals are used humanely, the institute has an Animal Resources Department equipped with modern diagnostic, therapeutic, and surgical facilities, which takes care of all animals used in research at the institution.[7] The IPR is currently leading a multi-institutional effort to develop comprehensive guidelines for laboratory animals used in research and in education. IPR has partnered with the National Council for Science and Technology, Kenya, and the Consortium for National Health Research (a local funding agency for Health Research) for this task.

Research using nonhuman primates, in particular vervet monkeys and baboons, as animal models is also conducted in other African countries including South Africa and Ethiopia.[25,26]

The present inadequate production of captive-bred primates in the United States and Europe necessitates importations from source countries in Asia, Africa, and South America. Long distance transportation to North America, Japan, and Europe, induces considerable stress in the animals. The change in temperate climate, new patterns of light cycles and dark cycles, and a different diet may result in a permanent reduction in the quality of life for the animals. For animal welfare reasons, as well as scientific and financial reasons, it is has thus been argued that it is important to develop primate research centers in source countries.[27] Due to political pressure and the introduction of the new EU Directive,[12] biomedical research using nonhuman primates as animal models is increasingly difficult to carry out in Europe. European scientists are seeking collaboration with nonhuman primate centers outside of Europe. Nonhuman primates such as vervet monkeys and baboons are considered agricultural pests in many parts of Africa. Considering that nonhuman primates remain vital as animal models in biomedical research,[28,29] many countries in Africa thus have unique possibilities to develop infrastructures focusing on studies of nonhuman primates under conditions approximating those of free-ranging members of the same species.[30,31] However, in order to successfully accomplish this, it seems essential that effective legislation and guidelines on the humane care and the use of animals are established in the countries where these possibilities are being investigated.

Special considerations and restrictions regarding genetically altered animals in the reviewed documents are also limited. However, according to the Tanzanian Animal Welfare Act, a special permit is necessary for practices involving "the alteration of the genetic material of an animal in a manner which infringes the natural barriers of sexual reproduction and of recombination; the application of biotechnical technology or the administration of substances to animals or embryos that modify genetic makeup of animals."[32]

In South Africa, there is a specific Act that addresses genetically altered animals, namely the Genetically Modified Organisms Act, 15 of 1997.[33] The objective of the Act is to provide for measures to promote the responsible development, production, use, and application of genetically modified organisms. This is implemented by the establishment of a council to advise the Minister on all aspects concerning genetically altered organisms and to ensure that all activities with regard to the development, production, use, and application of genetically modified organisms are performed in accordance to the Act.[33]

Transport, Housing, and Enrichment

The transport of laboratory animals, the housing of laboratory animals, and the enrichment of the animal's environment are areas of animal welfare typically covered in legislation in some other parts of the world.

Among all of the reviewed documents, only the Animal Welfare Act of Tanzania covers transportation of animals. Even in the Tanzanian Act, it is not apparent whether the transportation of laboratory animals is included specifically. The Act deals with the transportation of animals, animals unfit for transportation, revocation of movement permits, and certificates and documents required for transporting animals.[34]

Although legislation in the area of animal transportation may be limited, guidelines within the various institutions can be found. One example is the guide to the care and use of animals in research and teachings used at the University of KwaZulu-Natal in South Africa, a guide containing standards to be met when transporting animals.[17]

The Office International des Épizooties—or OIE, which is the French acronym for the World Organisation for Animal Health—has also set standards for the transport and use of animals in research and education.[35]

In the reviewed documents, husbandry and environment are covered only on a general level, and no specific requirements for cage sizes, ventilation, temperature, etc. can be found. Confining an animal in a manner which causes that animal unnecessary suffering and starving, underfeeding, or denying water to an animal are considered offenses, according to the animal welfare acts of both Kenya and South Africa. In this context, it is interesting to consider whether the aforementioned regulations have their basis in the principle of the Five Freedoms, mentioned in section 2.2.

Euthanasia

Euthanasia is another example of a frequently and specifically regulated procedure. Euthanasia is the act of inducing humane death in an animal by a method that induces rapid loss of consciousness and death with a minimum of pain, discomfort, or distress.

Material and information concerning euthanasia of research animals in the various African countries was very difficult to obtain. No approved methods are stated, and only general guiding principles regarding slaughter, such as "A person who, whether in a slaughterhouse or abattoir or in any other place, and whether for human consumption or not, slaughters any animal in such a manner as to cause it more suffering than is necessary (...) shall be guilty of an offence (...),"[36] can be found in the reviewed documents. However, in Kenya, reported methods of euthanasia have according to Kimwele and co-workers[1] raised serious welfare concerns. For example, the use of chloroform has been criticized based on concerns about human and animal welfare. Various methods of euthanasia have been reported being used in Kenyan institutions. Inhalant chloroform and diethylether were most commonly used, indicating a need to up-date guidelines introducing more modern and internationally recommended methods. Other reported methods included cervical dislocation, barbiturate overdose, and slaughter.[1]

Education and Awareness

Levels of public awareness about animal welfare are generally low in most of the African countries, according to the OIE. An exception is South Africa, where awareness is high in some sectors of the society. However, according to OIE's paper on Animal Welfare in OIE Member Countries & Territories in the Southern African Development Community (SADC) region, there are "considerable activities being carried out on humane education and or animal welfare classes in schools by NGOs such as in Tanzania, South Africa, and the Seychelles." Courses outlined after the guidelines for the Federation of Laboratory Animal Science Associations category C have been organized on several occasions in countries including Tunisia and Kenya.[37] The OIE also reports that "In some countries, humane education is included in the national environmental education programs (NEEPs) and is widely defused and resourced with assistance from civil societies and animals welfare networks. South Africa, Tanzania, and Zimbabwe have included animal welfare in their veterinary curricula." Very few veterinary authorities seem to be conducting joint activities on animal welfare education and awareness.[10]

In some countries, e.g. Botswana veterinarians are trained outside the country, as there is no university-level veterinary educational establishment in the country. In Botswana, the veterinarians are trained at the University of the West Indies, in Trinidad. In other countries, e.g. Malawi, there is reportedly a complete lack of training in professional and occupational courses in veterinary education. However, in countries such as South Africa, Kenya, Tanzania, Zambia, and Zimbabwe, faculties of veterinary science or veterinary educational establishments exist whose mission is to provide undergraduate and postgraduate professional veterinary training of an international standard.[2]

Acts of Parliament regarding the registration of veterinarians exist in a very similarly worded fashion in Kenya, Tanzania, Uganda, Zambia, and Zimbabwe.[2]

Inspection, Oversight, and Enforcement

There is no uniform approach for dealing with inspection, oversight, and enforcement in the African countries. In Tanzania, Kimwele and co-workers reported "Researchers must obtain a permit issued by the relevant minister or by the Director of Animal Welfare under advice of the Animal Welfare Council. Some guidelines are also provided, such as the need for veterinary supervision. However, implementation has been delayed, and the various statutory organs stipulated in the Animal Welfare Act have yet to be established."[1]

In Kenyan law, a license to perform animal experimentation may be granted only to a person who is registered under the Veterinary Surgeons Act or the Medical Practitioners and Dentists Act (or under the supervision of a person duly registered under one or another of those Acts).[38] A majority of the institutions in Kenya do not have established committees to review experimental protocols or to provide oversight regarding the use of animals in education and research.

Kimwele and co-workers purported that "Only two of the institutions with an established animal care and use committee refer to documented guidelines, one of which makes reference to external guidelines, specifically the United States National Research Council's Guide for the Care and Use of Laboratory Animals and the United Kingdom Animals (Scientific Procedures) Act 1986."[1] There are no requirements by law for protocol review by an ethics committee.[39]

In the Seychelles, only persons licensed by the Minister under the Animals (Control of Experiments) Act may carry out experiments on animals. The Minister may grant a license for animal experimentation to any person to perform any experiment for any purpose specified with such a license during such period subject to such conditions, in addition to the considerations specified in Animals (Control of Experiments) Act, as he may think fit. The Minister may grant a teaching permit to perform specified experiments for teaching purposes and may authorize experiments on animals without the use of anesthetic.[40]

In South Africa, animal research is guided by the South Africa Medical Research Council Act, 58 (1991), which are guidelines on the use of animals in research in 2004.[41] The affairs of the Council are managed and controlled by a board that regulate and control research on, or experimentation with, animal material.[8]

According to Masiga and Munyua,[8] many African countries have adopted initiatives that support research and development work; however, legislation is often lacking to support these initiatives. In former British colonies such as Kenya, the legal system criminalized animal abuse without any provisions for community awareness or education about animal welfare. This is not unique to Africa as few countries, including Britain, at that time did more than make animal cruelty and unnecessary suffering an offense; it is an issue nonetheless. Furthermore, Masiga and Munyua[8] considered that "internal committees at the various research institutions, which are often subservient to the directorate, do not have the capacity and/or authority to perform self-inspections or enforce regulations, which makes them essentially powerless." For the African countries that are affected by this situation, it would be beneficial if provisions for regular external monitoring and evaluation of the current systems were incorporated in legislation so that a legal framework exists to support the new initiatives.[42]

DISCUSSION

Compared with other parts of the world, policies, legislative frameworks, and guidelines on the use of animals in experimentation are generally poorly developed in Africa.

Main Issues

The reasons as to why legislation and guidelines concerning animal experimentation are poorly developed in Africa are numerous.

The first issue encountered when reviewing relevant material for this chapter, was that although many of the African countries had laws aimed at protecting the welfare of animals, most of these laws covered the welfare of animals in general, without specifically addressing animals used in research or teaching. This is a major problem because there is a large difference between regulating the welfare of animals on a general level (animal abuse, animal hunting, etc.) and the rather complex and multifaceted level of animal experimentation. Preferably, laws and guidelines covering the various sectors—including universities, breeding facilities, and pharmaceutical companies—should be specialized and detailed. To regulate and legislate animal welfare on such a general level, as is commonplace in most African countries, will not suffice in protecting animals in research and education.

The second obstacle the authors faced during this process was simply retrieving relevant information on the subject of animal experimentation. The inaccessibility of legislation is an issue in Africa; not only concerning animal experimentation and welfare. It is possible that some African countries may have developed guidelines that are not available in the public domain and are, therefore, not known to experts in the field. Animal welfare is a worldwide issue, and the dissemination and inaccessibility of relevant information to potential users is a major problem. Subsequently, efforts toward alignment with modern standards will not serve the purpose for which they were developed.

Enforcement is another important issue. Many of the reviewed Acts lacked adequate enforcement mechanisms. The success of animal experimentation legislation depends strongly on ensuring effective enforcement; all efforts to regulate the welfare of research animals become meaningless if there is no authority to enforce regulations. The absence of legal and ethical frameworks and committees to review protocols that involve animals in research and teaching also leaves major gaps in the protection of the animals involved.

It is, perhaps, not surprising that legislation regulating animal experimentation is not a top priority in many African countries. Africa is the world's poorest and most underdeveloped continent. Poverty, illiteracy, malnutrition, and poor health affect a large proportion of the people who reside in the continent. Animal welfare involves the physical and psychological well-being of animals and the human concern for the welfare of them. It is neither surprising nor unexpected that, in regions where the prevalence of respect for even basic human rights is questionable, more important issues relating to human welfare overshadow the concern for the welfare of animals. In countries with considerable social welfare challenges, meeting animal welfare standards is neither a priority nor, in some cases, a possibility.

Prospects for Improvement

Policies, guidelines, and legislation regarding the use of animals in research need to be established and enforced. Improvements must be made.

The first issue considered was the fact that many of the African countries have laws aimed at protecting the general welfare of animals, without specifically addressing animals used in scientific research or education. This existing legislation urgently needs to be adapted and developed to include codes on animal experimentation specifically. The necessary regulations can initially be implemented within the current legislative framework while more effective legislation is being drafted.

The problem of dissemination and accessibility of relevant codes and regulations also needs to be remedied. One approach to this issue would be to either create a regional organization dealing with the assembly of legislation, or to develop already existing bodies for that purpose. It is naturally problematic to decide which body would be an appropriate candidate for the responsibility of assembling the legislation and guidelines; however, there do exist various government organizations and NGOs, both locally in Africa and internationally, that might be appropriate to undertake the task.

Improving enforcement is critical. Enforcement mechanisms should be incorporated into the animal welfare acts. However, the addition to relevant laws is not enough to ensure implementation of regulations; ethical review committees and oversight boards should be established to oversee and exercise institutional control over research involving animals.

The main prospect for improvement begins with changing and increasing education and awareness on the subject. Levels of public awareness on animal welfare are generally low and, although awareness has increased over the years, humane education on animal welfare is lacking at most of the institutions of higher learning. This is unacceptable by modern standards, considering how commonly animals are used in academic research in Africa. Education concerning the welfare of animals in research should be a major part of every animal care, animal science, and veterinary curriculum. The simplest way to facilitate increased awareness and education would be to replicate well-established modern standards and incorporate them into national legislation.

Internationally, the accepted guiding principles include the universally recognized Five Freedoms, the Three Rs, and the fact that the use of animals carries with it a duty to ensure the welfare of such animals to the greatest extent achievable. These principles need to be specifically integrated into all African legislation concerning animal experimentation. Potentially appropriate legislation and guidelines from Europe, the US, Canada, New Zealand, and Australia could be reviewed, edited, and adopted, along with input from international animal welfare organizations and the draft Universal Declaration on Animal Welfare.

CONCLUSION

African legislation on animal experimentation is in urgent need of review. In order for the relevant legislation to meet modern standards, significant improvements must be made. A much greater awareness is warranted, and a vast improvement in animal welfare education is necessary.

One approach to resolve this major problem with immediacy would be to modify current legislation concerning animal welfare, in general, to also cover the welfare of animals in research. This would constitute a short-term solution; a provisional solution while new, more comprehensive and effective legislation is being drafted. In order for new legislation to be efficient and to suit the right purpose, it is important that modern standards are clearly incorporated into the statutes. Appropriate legislation and guidelines from other parts of the world could conveniently be used as a foundation for local and/or regional legislation. However, based on experience from other parts of the world, it is important to emphasize that regardless of how comprehensive and inclusive the novel legislation would be, the acts and ordinances will be of little impact without effective enforcement mechanisms. Consequently, new legislation should be created hand in hand with new systematic methods of enforcement and inspection. Provisions should be made in the statutes for regular internal monitoring, as well as external monitoring, and evaluation of the current systems and animal care and use programs. Systems should be developed for regulatory authorities to ensure routine inspections of animal facilities and review research protocols and practices to ensure that animal welfare issues are addressed appropriately in all institutions and facilities responsible for the care and use of laboratory animals. Ethical committees and inspection agents should be given the mandate and power to execute the relevant provisions stated in the legislation.

REFERENCES

1. Kimwele C, Matheka D, Ferdowsian HA. Kenyan perspective on the use of animals in science education and scientific research in Africa and prospects for improvement. *Pan Afr Med J* 2011;**9**:45.
2. World Organisation for Animal Health, Animal Welfare in OIE member countries & territories in the SADC region, Gaborone, Botswana, April 2011.
3. Tanzania Animal Welfare Act, Part V 2008, 19 of 2008.
4. Kenya Prevention of Cruelty to Animals Act of 1983, Chapter 360.
5. South Africa Animals Protection Act, 71 of 1962.
6. South Africa Performing Animals Protection Act, 24 of 1935.
7. Nyika A. Animal research ethics in Africa: an overview. *Acta Tropica* 2009;**112**.
8. Masiga WN, Munyua SJM. Global perspectives on animal welfare: Africa. *Rev Sci Tech Off Int Epiz* 2005;**24**(2):579–86.
9. Zimbabwe Scientific Experiments on Animals Act, Chapter 19:12.
10. Seychelles Animals (Control of Experiments) Act, Chapter 8.
11. South African Medical Research Council Act, 58 of 1991.
12. EU Directive 2010/63/EU of the European Parliament and of the Council of 22 September 2010 on the protection of animals used for scientific purposes.
13. Russell WM, Burch RL. *The principles of humane experimental technique*. Springfield, II: Charles C Thomas; 1959.
14. Farm Animal Welfare Council, Five Freedoms. http://www.defra.gov.uk/fawc/about/five-freedoms/. Accessed October 30, 2012.
15. Report of the Technical Committee to Enquire into the Welfare of Animals Kept under Intensive Livestock Husbandry Systems, the Brambell Report, December 1965 (HMSO London, ISBN 0 10 850286 4).

16. Universal Declaration on Animal Welfare.
17. University of KwaZulu-Natal, Guide to the care & use of animals in research & teaching.
18. *CCAC guide to care and use of experimental animals.* 2nd ed. vol. 1, 1993. *CCAC guide to the care and use of experimental animals.* vol. 2, 1984.
19. EU Directive 86/609/EEC on the protection of animals used for scientific purposes.
20. Langley G. Next of kin: a report on the use of primates in experiments, British Union for the Abolition of Vivisection.
21. Suleman MA, Wango BVM, Sapolsky RM, Odongo H, Hau J. Physiologic manifestations of stress from capture and restraint of free-ranging male african green monkeys (*Cercopithecus aethiops*). *J Zoo Wildl Med* Mar 2004;**35**(1):20–4.
22. Farah IO, Kariuki TM, King CL, Hau J. An overview of animal models in experimental schistosomiasis and refinements in the use of nonhuman primates. *Lab Anim* Jul 2001;**35**(3):205–12.
23. Moinde NN, Suleman MA, Higashi H, Hau J. Habituation, capture and relocation of Sykes monkeys (*Cercopithecus mitis albotorquatus*) on the coast of Kenya. *Anim Welfare* Aug 2004;**13**(3):343–53.
24. Institute of Primate Research, Kenya. http://www.primateresearch.org//index.php?option=com_content&task=view&id=39&Itemid=41. Accessed November 5, 2012.
25. Gregory MA, Mars M. A model for determining baseline morphometrics of the capillary bed in skeletal muscle. *Cardiovasc J S Afr* 2006 Jul–Aug;**17**(4):172–7.
26. Hailu A, Negesse Y, Abraham I. Leishmania aethiopica: experimental infections in non-human primates. *Acta Tropica*; **59**(3):243–250.
27. National Research Council. *International perspectives: the future of nonhuman primate resources.* Washington, DC: National Academy Press; 2003.
28. Carlsson HE, Schapiro SJ, Farah IO, Hau J. The use of primates in research: a global overview. *Am J Primatol* Aug 2004;**63**(4):225–37.
29. Hau J. Nonhuman primates must remain accessible for vital biomedical research. *Atla* 1999;**27**(special issue):221.
30. Hau J, Schapiro SJ. Nonhuman primates in biomedical research. *Scand J Lab Anim Sci* 2006;**33**:9–12.
31. Hau J, Schapiro SJ. The welfare on nonhuman primates. In: Kaliste E, Phillips Clive, editors. *The welfare of laboratory animals.* In: Kaliste Eila, editor. *Series: Animal welfare*, vol. 2. Cambridge: Springer; 2004.
32. Tanzania Animal Welfare Act, 2008, Part V section 39(1)(a–b).
33. South Africa Genetically Modified Organisms Act, 15 of 1997.
34. Tanzania Animal Welfare Act, 2008, Part III, section 22–25.
35. http://www.oie.int/animal-welfare/animal-welfare-key-themes/.
36. Kenya Protection of Cruelty to Animals Act 1983, Part I, section 8.
37. Wilson MS, Berge E, Maess J, Mahouy G, Natoff I, Nevalainen T, et al. FELASA recommendations on the education and training of persons working with laboratory animals: category A and category C. *Lab Anim* 1995;**29**:121–31.
38. Kenya Veterinary Surgeons Act, Chapter 366, (Repealed by No. 29 of 2011); Tanzania Veterinary Act 2003, (No. 16 of 2003); Zambia Veterinary Surgeons Act, Chapter 243; Zimbabwe Veterinary Surgeons Act, Chapter 27:15.
39. Prevention of Cruelty to Animals Act, Chapter 360, Kenya, Part III, section 16.
40. Seychelles Animals (Control of Experiments) Act, section 7.
41. http://www.kznhealth.gov.za/research/ethics3.pdf.
42. South Africa Medical Research Council Act, 58 of 1991, section 6 and section 17.

Chapter 8

Laws, Regulations, and Guidelines Shaping Research Animal Care and Use in India

Syed SYH Qadri[1] and Christian E. Newcomer[2]

[1]*National Institute of Nutrition (ICMR), Hyderabad, Andhra Pradesh, India,* [2]*Association for Assessment and Accreditation of Laboratory Animal Care International, Frederick, MD, USA*

GENERAL FRAMEWORK AND PRINCIPLES

The Constitution of India (Part IV) in the Directive Principles of State Policy; Article 48, says that the State shall endeavor to organize agriculture and animal husbandry on modern and scientific lines and shall take steps to preserve and improve breeds, prohibit slaughter of cows and calves and other milch and draft cattle. Article 48A directs the state to protect and improve the environment and to safeguard the forests and wildlife of the country. In Part IV A, under Fundamental Duties, a citizen is duty bound to protect and improve the natural environment

Laboratory Animals. http://dx.doi.org/10.1016/B978-0-12-397856-1.00008-8

including forests, lakes, rivers, and wildlife, and to have compassion for living creatures.[1]

The Indian Penal Code, 1860 (45 of 1860), Chapter II, Section 47, defines the word "Animal" as any living creature, other than a human being. Chapter III of the code deals with the punishments for mischief by killing or maiming animals.[2]

The use of Laboratory animals (especially rodents) in research started in India in 1876 at the Madras Veterinary College (presently the Tamil Nadu University of Veterinary and Animal Sciences) during the rule of the British, followed by the Indian Veterinary Research Institute and the Haffkine's Institute in 1889. The first laws that governed animal welfare were passed by the Indian Parliament, and were called the Prevention of Cruelty to Animals Act 1960 (Act 59 of 1960).[3] This particular act, under definitions, describes an "animal" as any living creature other than a human being. Chapter II of this Act deals with the establishment of an animal welfare board functioning to promote animal welfare and to protect animals from being subjected to unnecessary pain or suffering. Chapter III of the Act lists and describes cruelties, penalties, and the procedure to destroy suffering animals.

Chapter IV of the Act[3] deals entirely with Experimentation on Animals and specifies the legal constraints on the sale and acquisition of animals for research purposes.[4] Section 14 of the Act permits experiments (including experiments involving surgery) on animals for the purpose of advancing new discovery of physiological knowledge which will be useful for saving or for prolonging life, alleviating suffering or combating disease, whether of human beings, animals or plants. The Act (Chapter IV, Section 15) also instigated the constitution of a committee, called the Committee for the Purpose of Control and Supervision of Experiments on Animals (CPCSEA). From 1960 until 1998 the CPCSEA was chaired by the Director General of Indian Council of Medical Research and remained dormant. It was revived in 1998 and the new "Breeding of and Experiments on Animals Rules 1998," were promulgated by the Indian Parliament.[5] These rules were again amended in 2001.[6] Under these rules registration of animal establishments for breeding or for experimentation was made mandatory. Registration of the facility necessitates the submission of the design and floor plan of the animal facility, and the requirements of the Institution. Approval of a facility is subject to an inspection of the facility by a person nominated by the Member Secretary of the CPCSEA and to the establishment of the Institutional Animal Ethics Committee (IAEC) (the counterpart of the IACUC in the USA).

The IAEC (a mandatory provision in the law) is constituted of eight members. Five of the members of the committee are from the Institution (a biological scientist; the scientist in charge of the animal facility; the named veterinarian; two scientists from a different biological discipline), and the other three members (the main nominee; scientist from outside the institution; and a nonscientific socially aware member) are appointed by the CPCSEA. The CPCSEA also designates a link nominee who may serve in the capacity of the main nominee if that individual conveys his/her unavailability in writing to the IAEC Chairman. The act permits all diploma and degree holders in biological sciences to conduct experiment on animals. However, the named or attending veterinarian is vested with the legal authority to supervise the conduct of animal experiments. The presence of the named veterinarian in the IAEC has therefore been mandatory. In an institution, it is mandatory to hold an IAEC meeting every three months. The IAEC has been empowered to approve project proposals on rodents and rabbits. The project proposals using larger animals such as cats, dogs, primates, sheep, and other larger animals need to be submitted with the recommendations of the IAEC to the CPCSEA National Committee for approval. The process may take 6 months to 2 years to obtain a single approval.

The IAEC's oversight emphasizes mostly the ethical aspects of animal usage and the animal welfare measures adopted. The 3Rs is normally the basis of most of the discussions. The CPCSEA rules insist that researchers submit their requisition for animal experimentation in the prescribed "Form B" to the IAEC.[7] Form B contains a clause instructing the researcher/experimenter to review the literature and declare whether the experiment has already been done in vitro/in vivo elsewhere. It requires the researcher to explain why the proposed experiment cannot be performed in vitro. In addition, the CPCSEA has introduced the 4th R that denotes rehabilitation of the animals after use in experiments. The CPCSEA insists that the 4th R be applied to all animals. Implementation of the 4th R principle in the case of rodents and rabbits which are probably the most used animals could impact future research, and experience has already shown that application of the 4th R in the case of primates in a couple of instances did not give the desired results. The present approach may be untenable and deserves further discussion and strategic planning.

Besides the above mentioned Act, there are other Acts of the Indian Parliament that govern animal experimentation and research in India. The Indian Veterinary Council Act 1984 (Act 52 of 1984)[8] read along with its Amendment of 1992 [GSR 395 (E) Section 2(f)][9] includes laboratory animal medicine, animal experimentation, and vaccine production within the definition of the Veterinary Profession. Section 30 of the IVC Act 1984 legally authorizes only veterinarians registered with the Veterinary Council of India to practice laboratory animal medicine and animal experimentation. It is mandatory for all graduates in veterinary sciences in India to register with the respective State Veterinary Councils to practice veterinary medicine and surgery and the rights of the registered practitioner are protected by the Council. This is irrespective of whether the individual is working in the government or in a private institution.

The animal facilities of small educational institutions typically consist of 2–4 rooms (mice, rat, guinea pig, and rabbit rooms) with a separate area for washing cages and, if available, a separate room for storing feed and cleaned cages. Such facilities have a working window/split air conditioner and the facilities are used only once or twice a year for short durations for experiments ranging between 14 days and 90 days. The provisions for animal housing are the bare minimum and the personnel working in the facility are mostly the research scholars. A majority of the animal facilities do not have personnel trained in laboratory animal science. The animal facility-in-charge individual in most institutions is the Head of the Department of Pharmacology. The veterinarians in such facilities generally work as consultants on an ad hoc basis. There is an urgent need for training personnel from such institutions.

The government sector, especially India's premier national research institutions have good animal facilities and sound programs of animal care. However, challenges remain to improve the quality of animals from such institutions. A majority of the animal facilities in such institutions are headed by veterinarians who have had their training in laboratory animal science/medicine either in Europe or the United States, but the animal technicians have not had the benefit of formal training. On the research front, scientists and researchers working in these government institutions are not required to have formal training in laboratory animal sciences or the application of in vivo animal research methodologies, and this may impact the quality of animal experimentation and research.

More than a dozen pharmaceutical companies in the Indian private sector have attained Good Laboratory Practices certification, and just over a dozen of the institutions have also achieved AAALAC international accreditation reflecting the strength of their programs of animal care and quality of their facilities. Many of the institutions in the pharmaceutical industry as well as in the government sector have now switched over to individually ventilated cage systems in an effort to upgrade laboratory animal management practices. However, the quality of the animals (both microbiologically and genetically defined) needs improvement.

The pharmacy, medical, veterinary and basic science colleges, which are the largest number of registered institutions with the CPCSEA, use the commonly available random bred laboratory rodents and rabbits for experimentation.

In the government sector several institutions have been using outbred and inbred mice and rats, Syrian Golden hamsters and New Zealand White Rabbits. Almost all the facilities with a couple of exceptions predominantly use random bred animals. However, a few government institutions have imported unique or specialized stocks, inbred strains, and other transgenic and knockout mice for their particular research purposes. The initial stocks of the above mentioned strains of animals were procured from the National Institutes of Health, Bethesda, USA, and high quality commercial vendors in the USA, Europe or Asia. However, no database is available about the different strains and numbers of each animals imported from various sources. Similarly, information about the availability, or the practices used for the maintenance of these strains in the various institutions in India is not available.

In the pharmaceutical industry, the spectrum of research animals used is similar to that of the government sector. However, in facilities where research and development in drug discovery is in progress, individual companies may import specialized strains, transgenic and knockout animals from suppliers and laboratories globally. The procedure to import animals in such cases requires submission of a proposal to the IAEC of the institution. Once the IAEC approves the proposal, a No Objection Certificate (NOC) issued by the CPCSEA must be obtained. The proposal with the NOC goes to the Director General of Foreign Trade (DGFT). However, import of larger animals such as beagle dogs and others requires permission from the CPCSEA National Committee. Until recently, beagles were imported from a commercial vendor with breeding facilities in Europe or China. Recently, a European beagle supplier started a beagle dog breeding facility in India and the animals are available within India.

Looking at the distribution of animal facilities in the Indian Subcontinent, the majority of the registered animal facilities are available in the states of Maharashtra (272), Andhra Pradesh (230), Karnataka (190), Tamil Nadu (184), and Gujarat (138). The establishment of animal facilities in the different states of India also reflects the developments in the pharmaceutical, biotechnology and contract research organizations.

LEGAL AND REGULATORY SCOPE/APPLICABILITY

The CPCSEA utilizes the very inclusive definition of "animal" provided in the Indian Penal Code[2] and the PCA Act[3] as noted above which means "any living creature other than a human being." Chapter I of the PCA Act also provides definitions for "captive animal" and "domestic animal," but these subcategories have minimal representation in or bearing on the Indian regulations pertaining to experimentation on animals. The privilege to conduct animal experimentation is protected in the language in Chapter IV of the PCA Act which states, "Nothing contained in this Act shall render unlawful the performance of experiments (including experiments involving operations) on animals for the purpose of the advancement by new discovery of physiological knowledge or knowledge which will be useful for saving or prolonging life or for combating disease, whether of human beings, animals or plants."

It is estimated that there could be between 4000 and 5000 laboratory animal facilities in India; of these about 1700 animal facilities have been registered with the CPCSEA. A majority of the animal facilities are with the pharmacy, medical, veterinary, and basic science colleges numbering about 1050, followed by government research, training, and testing institutions at 350 and another 400 organizations in the private pharmaceutical and biotech industry.

A majority of the animal facilities in the educational institutions are for the purpose of research only, as use of animals for education and training has been banned.

REGULATORY AUTHORIZATION AND ENFORCEMENT MECHANISMS INCLUDING PENALTIES

The CPCSEA under the Ministry of Environment and Forests, Government of India is the only organization that has the power under the law to authorize and register institutions that breed or use laboratory animals for experimentation.

Institutions or organizations intending to breed and trade in laboratory animals or proposing to use laboratory animals for basic research, regulatory testing or for the purpose of education need to apply to the CPCSEA to register their facility. The CPCSEA is constituted of 19 members, 10 of whom are government officials, four are animal activists, three are veterinarians, and another two represent the basic sciences. Stakeholders from professional organizations such as the Laboratory Animal Scientist's Association of India, representatives of the pharmaceutical industry, the contract research organizations and the biotechnology industry who are the actual breeders and users of laboratory animals are not included presently, and would offer valuable perspectives as members of the CPCSEA national committee.

Facility registration entails submission of the basic floor plan/design of the facility along with the application for approval of an animal production facility. Once the facility is ready, the CPCSEA authorizes one of its nominees (most nominees are personnel already working in animal facilities, have basic qualifications and background in lab animal science and are appointed on an honorary basis) to inspect the facility and submit a report to the CPCSEA. There are instances where the CPCSEA nominates personnel who do not have a basic degree or diploma in laboratory animal sciences to inspect a facility or as a main nominee to the IAEC. As a consequence of the nominee's lack of knowledge and preparation, facility registration may be delayed due to incomplete or inconclusive inspection reports. Moreover, subsequent inspections may reveal the lacunae in the previous inspections. The need to bring consistency and credibility to the inspection process through the appropriate selection of members on the regional teams becomes evident in these examples.

The CPCSEA nominee is provided with a prepared list that identifies items to be inspected and recorded during the inspection. If the inspection report submitted is satisfactory, the CPCSEA allots a registration number to the institution and appoints four members of the IAEC and approves the other members proposed by the institution. The four members appointed by the CPCSEA include the nominee, the link nominee, the scientist from outside the institution, and the nonscientific socially aware member. The IAEC is expected to make site visits to the animal facilities before every meeting. The frequency of meeting depends on the number of experiments conducted in an institution. However, a minimum of two meetings annually is mandatory and the Chairman of the IAEC and the nominee are duty bound to submit a copy of the minutes to the CPCSEA for record. Failure to submit the IAEC minutes for 3–4 consecutive meetings would attract disqualification and de-recognition requiring the institution to reapply for registration.

Besides the nominee, the CPCSEA may authorize any of its representatives to inspect a facility at any given time. A person so authorized would have the power to make entry at the facility at any time convenient to him and after inspection would have to submit a report to the CPCSEA. However, the authorized person has no power to seize the animal. Based on the inspection report submitted by the person so authorized, the CPCSEA national committee has the power to cancel registration or suspend registration of an organization permanently or for a specified time, after giving a chance to be heard to the organization. Normally, the head of the institution is held responsible and would be the person subject to penalty.

It is worth mentioning that the CPCSEA has submitted a revised PCA Act 2011 which in a draft state was circulated for public opinion.[10] The Revised PCA Act would be called the PCA Act 2011. In this particular revised Act, the penalties have been revised and are very harsh. The scientific community has objected and demanded revision. In the revised Act; Section V, page 17, Section 28: Penalties, it states that any person contravening the provisions under the Act would be liable for a fine of Rs 50,000 or imprisonment of not less than 1 year extended up to 3 years. On second or subsequent offence a fine of Rs 75,000–100,000 with imprisonment of not less than 2 years and extended up to 5 years has been proposed. As per the Act any scientist who contravenes the Act would be liable for prosecution. This would be true even in instances where the person conducting an experiment was a public servant performing his duty in good faith.

The scientific community has taken serious objection to the revised Act and has submitted its objections to the ministry. Such provisions would not only impede biomedical research but would also dissuade the younger generation from pursuing careers in the biology discipline. The final Act is pending and scientists are hoping the objections will be addressed.

NONHUMAN PRIMATES: SPECIAL CONSIDERATIONS

Over half a dozen institutions in India have primate facilities. These institutions would obtain a license from the respective State Wildlife Warden to capture wild rhesus or bonnet monkeys for research whenever required. The license would be given to a private wild animal trapper who would provide the institutions with the required number of animals. In fact, due to massive deforestation, wild animals, especially monkeys in search of food, have been moving over to the urban areas where they have become a nuisance. The population of these animals both in the wild and urban areas has grown exponentially as all wildlife is protected under the law. The availability of the animals in large numbers in the wild was partly a deterrent to the establishment of primate breeding facilities. The wildlife authorities have now stopped issuing licenses to trap and use monkeys for research, insisting that scientists procure the animals from breeding facilities. The lack of foresight and improper planning on the part of the scientific community in establishing monkey breeding facilities has now led to a situation where the animals are not available for research. An ambitious program of setting up a primate breeding facility was started at the National Institute for Research in Reproductive Health with a substantial financial grant from the National Institute of Health, USA. Unfortunately, the project has not progressed.

As such, primates are revered in India and the Monkey God "Hanuman" is said to be the reincarnation of God. Therefore, the people have serious issues regarding use of monkeys in research. Moreover, permission from the CPCSEA to use monkeys in research is not easily acquired. Presently, a few institutions, namely the Indian Institute of Science, Bangalore, the National Institute of Research in Reproductive Health, Mumbai, The National Institute of Virology, Pune, the All India Institute of Medical Sciences, New Delhi, Central Drug Research Institute, Lucknow, the National Institute of Nutrition, Hyderabad, and the Zydus Research Center, Ahmedabad, have small monkey colonies which may total up to a 1000 primates. Except the Zydus facility none of the other facilities have a primate breeding program. The Indian macaque, especially the rhesus, are in great demand the world over, however, there has been a ban on export of these monkeys since 1979. As such, research and testing on primates is not banned in India, however, the use of monkeys as trained performing animals for exhibition purposes is prohibited.[11]

GENETICALLY ALTERED ANIMALS: SPECIAL CONSIDERATIONS

As such there are no laws restricting the development of genetically altered animals in India. The CPCSEA rules permit the researcher to import transgenic and knockout mice. However, all genetically modified animals are on the restrictive import list and need to go through an elaborate import procedure.

An end user can only import an animal species/strain that is not available in India. The user needs to submit a research proposal to the IAEC. The IAEC issues a NOC. The user needs to obtain a NOC/clearance certificate from the State Forest Officer. An application form along with all the necessary documents (viz., registration certificate, invoice of animals/free gift letter, gazette notification, and application fee) should be submitted to the Joint Director General Foreign Trade in the respective states of India. One copy of the application should be sent to the Director General of Foreign Trade at New Delhi. The DGFT office forwards the application to the Animal Husbandry Commissioner, Ministry of Agriculture, Government of India, New Delhi for providing a license. The approved application comes back from the Animal Husbandry Commissioner's office to the Joint DGFT of the respective state. The animals can be imported through any of the airports in India where an approved quarantine facility exists. Presently, the quarantine facilities exist at Mumbai, Chennai, New Delhi, Hyderabad, and Bangalore airports.

The Biological Diversity Act, 2002 (Act No. 18 of 2003) governs the use and or trade of any biological material in the country.[12] Under the provisions of the act, biological samples cannot be imported or exported out of the country without proper approvals from the competent authority.

INSTITUTIONAL AND DESIGNATED PERSONNEL

The CPCSEA holds the head of the institution/organization responsible for the maintenance, upkeep and smooth functioning of the animal facility. The head of the institution is also the individual who would be held responsible for any eventuality or accident in the animal facility. The overall responsibility of care, breeding, management, and experimentation in an organization lies directly with the veterinarian and staff working in the animal facility. However, the IAEC has the overall responsibility to oversee the functioning of the facility and the conduct of animal experiments.

The three most important persons representing the institution in an IAEC include the Director/Officer-in-charge of the Institute, who typically also is the Chair of the IAEC, and the Member Secretary who arranges all the IAEC meetings and ensures that all animal experimental protocols that need to be discussed and approved in the IAEC are in the prescribed format. The third most important person is the veterinarian (the attending veterinarian) who has the responsibility to procure the animals and ensure that the experiments are conducted as per the prescribed laws and also ensure animal welfare. The attending veterinarian is also authorized to suspend any animal experiment, if he/she finds that the animals are under extreme stress or animal welfare laws are being flouted.

OVERSIGHT AND ETHICAL REVIEW

India utilizes IAECs for the purpose of control and supervision of experiments on animals performed in an establishment which is constituted and operated in accordance with procedures specified for that purpose by the CPCSEA.[13] The primary duty of IAEC is to work for achievement of the objectives as mentioned above. The IAEC will review and approve all types of research proposals involving small animal experimentation (i.e. species smaller than the laboratory rabbit and not to include any nonhuman primate) before the start of the study. For experimentation on large animals, the case is required to be forwarded to CPCSEA in the prescribed manner with recommendation of IAEC. IAEC is required to monitor the research throughout the study and after completion of study through periodic reports and visits to the animal house and laboratory where the experiments are conducted. The committee has to ensure compliance with all regulatory requirements, applicable rules, guidelines and laws.

The members are appointed to the IAEC for a period of 3 years coincident with the CPCSEA registration of the IAEC. When the IAEC is reconstituted at least half of the members must be replaced. Members are expected to observe absolute confidentiality and declare conflicts of interest pertaining to IAEC discussions and actions. The IAEC also is expected to formulate a Standard Operating Procedure (SOP) for its working requirements and follow it in all the meetings.

A minimum of six members is required to compose a quorum and all decisions should be taken in meetings and not by circulation of project proposals. The presence of the CPCSEA nominee is required unless excused by written notification of unavailability to the Chairman of the IAEC, in which case the link nominee may substitute. The socially aware member's presence is compulsory in large animal cases referred to CPCSEA and in at least one meeting in a calendar year. Minutes of IAEC meetings must be prepared by the Chair or his/her alternate and circulated to the CPCSEA Member Secretary within 15 days for the meeting to be valid.

Meetings of the IAEC are held at scheduled intervals as prescribed in the IAEC's SOP and additional meetings may be held if there are reasons to do expedited review. The proposals are sent to members at least 15 days in advance of the meeting. Decisions are derived through consensus after discussions, and dissenting view points are recorded in the minutes. If consensus is not reached, the case should be referred to CPCSEA. Researchers will be invited to offer clarifications if necessary, and independent consultants/experts will be invited to offer their opinion on specific research proposals if needed. All decisions are recorded in the IAEC meeting minutes and the Chairperson's approval is confirmed in writing with signatures of all the IAEC members present.

The IAEC must maintain appropriate records and archiving standards including: Curriculum Vitae of all members of IAEC including training programs in animal ethics attended; copies of all study protocols with enclosed documents, progress reports; minutes of all meetings duly signed by the Chairperson and the members; copies of all existing relevant national and international guidelines on animal ethics and laws along with amendments; copies of all correspondence with members, researchers and other regulatory bodies; final reports of the approved projects; records of breeding of animals, supply, etc., if breeding of animals is undertaken; records of import of animals with species, source, quantity, usage, etc.; records of all contract research, if conducted at the institute; and, records of rehabilitation of large animals if done. All documents should be archived for a period as prescribed in the IAECs; however, this should not be less than one year.

IAEC members should be made aware of all relevant new guidelines and amendments to the rules and PCA Act. Also, members should be encouraged to attend national and international training programs/workshops/conferences in research ethics for maintaining quality in ethical review and be aware of the latest developments in this area.

In addition to the submission of minutes of the IAEC meeting to CPCSEA within 15 days, the IAEC must also submit to the CPCSEA an inspection report of the animal house with photographs once in a calendar year. If action is required, the facility must provide a report and take action within 30 days.

Initially, when the IAECs started functioning, the emphasis was on reducing the number of animals used in each individual experiment. In the IAECs of institutions where the representatives of the CPCSEA (called CPCSEA nominee) was a qualified veterinarian or scientist, questions and/or discussion on the scientific content of the research protocols were vehemently opposed by the scientific community. However, in some institutions, internal review committees are in place to scrutinize and review the animal experimental protocols before their submission to the IAEC. In many of the institutions with IAECs, members are struggling with their role due to the lack of specific guidance available in "Form B" which is the form specified by the CPCSEA for its approval of studies.[7] In broad terms, Form B solicits the following information: particulars about the kind of animal used for conducting the experimentation; the health of the animal; the nature of the experiments performed; and the reasons for conducting such an experiment in the particular species. The IAEC may put conditions as it deems fit to ensure that animals are not subjected to undue pain and suffering.[5] Some laboratory animal scientists in India are of the opinion that during subsequent CPCSEA review, protocols may be approved without rigorous discussion or scrutiny by the representatives of the CPCSEA due to their lack of relevant laboratory animal science background. The total number of organizations registered with the CPCSEA exceeds 1700 and the CPCSEA finds it difficult to appoint nominees and socially aware nonscientific members on the IAECs who have the relevant qualification and training for service. The CPCSEA has been trying to conduct training programs for the members of the IAEC using resource personnel who themselves have limited backgrounds in laboratory animal science and experimentation in some instances. The CPCSEA itself is constituted of 19 members and none have academic training or experience in laboratory animal science. The lack of involvement of trained individuals has resulted in occasional circumstances where the IAECs approve animal experimental protocols with an arbitrary reduction in the number of animals used in an experiment that has impacted the statistical validity of the study. Some institutions do invite a biomedical statistician during the IAEC as a special invitee. However, efforts to make it mandatory to have a statistician on the IAEC have not succeeded.

The IAEC has a significant representation from the institution. The chairman, member secretary, named/attending veterinarian, two scientists from different biological disciplines constitute five out of the eight members of the IAEC that represents the institution. It has been observed that in the majority of the institutions, these IAEC members consider the other three members appointed by the CPCSEA as activists who intend to stop all animal experimentation. Therefore, the readiness to discuss science and improve the design and/ or techniques used in animal experiments is lacking. Also, the poor preparation of the nominee, the outside scientist, and the socially aware nonscientific person compounds this problem.

Institutions where internal review committees exist also encounter problems. The members of the internal review committees are senior scientists or researchers who may lack the relevant educational qualification in laboratory animal science and in other pertinent subjects. Though the Government of India has been striving to usher in the Good Laboratory Practices in research institutions, the concept languished in government institutions due to various reasons. However, since the implementation of the Breeding off and Experimentation on Animals Rules in 1998, there has been a significant improvement in the physical structure of the animal facilities, improvement in the care of laboratory animals and a very significant improvement in the awareness among people about animal welfare issues. The CPCSEA has embarked on an ambitious project of establishing an Animal Welfare Institute called the National Institute of Animal Welfare near New Delhi, where it has been conducting regular training programs in Animal Welfare. However, as the training programs are not job oriented, the response of the people has been somewhat lackluster.

Associations such as the Laboratory Animal Scientist's Association of India have been striving to update professionals by conducting regular conferences, workshops and training programs. Collaborations and global partnering with international organizations and associations has given a tremendous exposure to the professionals. Efforts which have succeeded in starting the publication of a journal in laboratory animal sciences, and in introducing the FELASA training programs by these associations could go a long way in improving the quality of animals, experiments and research.

ANIMAL REUSE

The CPCSEA regulations do permit the reuse of animals (rodents, lagomorphs, dogs, and primates) after a specified washout period or after a considerable period of rest as indicated in Annexure I, Experimental Animal Health Records if such reuse is scientifically sound.[14] Nevertheless, the reuse of rodents is not practiced as most of the experiments conducted are terminal experiments and in cases where the experiment is not terminal, the advanced age of the animal after use in an experiment does not permit a researcher to reuse them. However, used animals, especially rodents, had been used again in teaching by supplying these animals to pharmacy colleges. But now, use of animals in teaching has been banned and researchers are forced to conduct terminal experiments. There are also instances where some profit making animal breeders locally have been purchasing used animals and are supplying them along with fresh animals to some organizations at a cheaper cost. This trend is gaining momentum as the researcher has no say in the purchase of the animals. Most purchases in the institutions from the government sector as well as in the industry are made by the purchase sections of the institution where the decision lies with a purchase officer who would in most cases be an economics graduate. The problem is compounded by the Indian tradition of selecting the lowest quotations for purchase of any commodity.

In the case of rabbits, several organizations use these animals for routine pyrogen testing as mandated by regulatory requirements. Though an alternative is available, the regulatory authorities have been accepting data based on animal experiments and the trend continues.

In the case of larger animals, such as dogs (beagles), reuse of animals especially in pharmacokinetic and pharmacodynamic studies is permitted. However, the number of experiments per dog varies and the decision would lie with the IAEC to decide. The case with research on primates is similar though the number of animals reused would be much less than that of dogs. Although the CPCSEA guidelines approve the reuse and rehabilitation of animals, the underlying law is silent on this matter.

SETTING FREE/REHABILITATION OF EXPERIMENTAL ANIMALS

The CPCSEA rules do not permit the setting free of experimental animals. The CPCSEA Amendment of 2006 insists that the person performing animal experiments and the organization where animal experiments are performed are responsible for the aftercare and rehabilitation of animals after experimentation, and that they shall not euthanize the animals except in situations as defined in the law.[14] Moreover, the law says that the cost of aftercare and rehabilitation of animals after experimentation shall be made part of research costs and shall be scaled in positive correlation with the level of costs involved in such aftercare and rehabilitation of the animals. The law also states that the rehabilitation shall extend till the point the animal is able to resume a normal existence by providing a lump-sum amount, as costs for rehabilitation and care of such animal should cover its entire statistical expected life span.

However, laws such as the Organized Crime and Police Bill of Britain and the Animal Enterprise Terrorism Act of the USA for persecution of individuals involved in animal rights extremism do not exist in India. It may be mentioned here that to date not a single incident of animal rights extremism has been reported in any part of India. A few cases where animals were seized and rehabilitated were done under the laws following all legal procedures by authorized personnel. The need for laws such as the ones in Britain and the USA may not arise even in the distant future in India for several reasons. Only a small percentage (maybe 10–15%) of the urban population is aware of animal welfare and the rules that govern them. From among this population only 10% would have any information about laboratory animals and experimentation. Among the small percentage of individuals likely to be aware of animal experimentation only a handful would likely have extremist views. Moreover, the police and the security apparatus in India is feared by the general population and the legal persecution and after effects deter most people from committing arson. Also, the easy availability of the judicial system to the people to have their grievances redressed could be another factor dissuading them from acts of extremism.

OCCUPATIONAL HEALTH AND SAFETY

The Article 21 of the Indian Constitution guarantees the protection of life and personal liberty of a person. The right to health, medical aid to protect the health and rigor to a worker while in service or post retirement is a fundamental right under article 21 of the Indian Constitution read with Article 39(e), 41,43,48 A. The Factories Act 1948, the Mines Act 1952, the Dock workers (Safety, Health and Welfare) Act 1986 are some of the laws which regulate health of workers in an establishment.[15] The employees State Insurance Act 1948 and the workers Compensation Act 1923 are compensatory in nature. However, the implementation of the Occupational Health Laws in India is reportedly inconsistent.

In a majority of animal facilities in India, neither the technical staff nor the scientific staff is aware of the diseases of laboratory animals. The veterinarian is the only one who is aware of the diseases of zoonotic importance but, in most instances due to inadequate preparation, the veterinarian also has no experience or training in laboratory animal diseases.

In the laboratory animal facilities, most of the acts mentioned above are irrelevant; hence the application of these acts is limited. The CPCSEA rules are silent on the issue of occupational health of people working in animals facilities. In a majority of animal facilities in both the government and private sector, the animals have a wide variety of endemic infections and external and internal parasitic infestations and zoonotic agents have been recorded. The exposure of personnel working in these animal facilities and their immediate families is not known and it is doubtful that few medical practitioners would be aware of the zoonotic significance of laboratory animal infections/infestations.

CONDUCT OF EXPERIMENTAL PROCEDURES

According to the Indian Veterinary Council Act 1984, Section 30 and its amendment of 1992 G.S.R 395 (E), laboratory animal medicine, animal experimentation, and vaccine production come under the veterinary profession.[8] As per the Act, registered veterinary practitioners alone are permitted as per the law to practice laboratory animal medicine and perform animal experiments. The Ministry of Environment & Forests, Government of India, passed the S.O. 134 (E) [15/2/2001]—The Breeding of and Experiments on Animals (Control and Supervision) Amendment Rules 2001.[6] According to the regulations, the IAEC is expected to ensure that experiments are conducted with care and humanity, and the IAEC may put conditions as it may deem fit to ensure that animals are not subjected to unnecessary pain or suffering before, during or after the performance of experiments on them.[5] While constituting the IAEC, it has made the named/attending veterinarian a mandatory member of the committee. Further, the rules have given permission to degree holders in medicine or veterinary sciences, post graduate and above in life sciences/pharmaceutical sciences or any other natural sciences, degree or diploma holders in pharmacy, diploma or certificate holders in laboratory animal techniques sciences from a recognized institution as identified by the CPCSEA the liberty to perform animal experiments. All the professional degree courses mentioned above, with the exception of Veterinary Sciences where laboratory animal sciences are included in the syllabus, do not have laboratory animal sciences or laboratory animal medicine in their curriculum. The permission to conduct animal experiments without significant education and training in laboratory animal sciences has led to a situation where some research has progressed unfettered under conditions that may compromise the quality and conduct of the research.

The regulations and Form B do not specifically draw attention to the application of the 3Rs and offer no schema for severity classification of experiments. Only some researchers in some of the recognized government institutions and multinational companies in the private sector are aware of the 3Rs principles. In a majority of the institutions the IAEC members arbitrarily insist on the reduction of the number of animals in each experiment (e.g. six animals per group in the case of rodents) and quite often the other common recommendation is to use anesthetics in case the animal is under pain or stress. Most institutions lack in vitro facilities and the IAECs lack awareness and resources to determine if or where alternatives are available. In the case of regulatory data generation, most institutions follow the OECD (Organization for Economic Cooperation and Development) guidelines or the DGCI (Director General of Commercial Intelligence and Statistics) rules which are almost similar to the OECD guidelines. The LD50 test has been replaced with the fixed dose method or the up and down method. The Draize test has been replaced by the vaginal mucous membrane test in rabbits and the pyrogen test has been replaced by the Limulus Amoebocyte Lysate assay (LAL) test using the blood from horse-shoe crabs. However, the Indian regulatory authorities continue to accept the in vivo pyrogen test results from rabbits, perpetuating the indiscriminate use of this model.

The CPCSEA rules insist that the researcher use animals that are lower on the phylogenetic scale. However, the use of earthworms and frogs in education and teaching at high school (10+2) level and at undergraduate level has been banned. Moreover, several species of frogs are now listed on the endangered list and hence their use is banned.

EUTHANASIA

The Indian Veterinary Council Act 1984 (Act 52 of 1984) authorizes only registered veterinary practitioners to practice veterinary medicine and surgery. As per the law only registered veterinarians are supposed to anesthetize or euthanize animals, conduct necropsies, and issue health or death certificates in animals. In addition the law authorizes only registered veterinarians to give evidence at any inquest or in any court of law as an expert on any matter relating to veterinary medicine.

As per the CPCSEA rules the attending veterinarian has been entrusted with the authority as well as duty to administer anesthesia or euthanasia to animals.[5] Other professionals have also been permitted to use anesthetics and euthanasia agents but, under the supervision or advice of the attending veterinarian. The guidelines also emphasize that the animal should remain under veterinary care till it completely recovers from anesthesia and postoperative stress.

The CPCSEA Rules SO 42 (E) 14/10/2001 Breeding of and Experiments on Animals (Control and Supervision) Amendment Rules 2005 have laid down the parameters for application of euthanasia in experimental animals. An animal is to be euthanized if:

1. it is paralyzed and is not able to perform its natural functions;
2. it becomes incapable of independent locomotion;
3. it can no longer perceive the environment in an intelligible manner;

4. it has been left with a recurring pain during the course of an experimental procedure wherein the animal exhibits obvious signs of pain or suffering; or
5. if the continuation of life of the experimental animal will be life threatening to human beings or otherwise.

The CPCSEA Guidelines give details of the recommended methods to be followed for euthanasia. Under physical methods the guidelines recommend exsanguination in rats, hamsters, guinea pigs, rabbits, and cats but not in mice, dogs, and monkeys. Decapitation and cervical dislocation are recommended in mice, rats, and hamsters. Inhalation euthanasia agents that have been approved are: carbon monoxide in all animals, carbon dioxide in all animals except dogs and monkeys; a combination of carbon dioxide and chloroform in all animals except dogs and monkeys; and halothane in all animals. The drugs that have been approved are: an overdose of barbiturates in all animals; chloral hydrate is recommended only in rabbits, cats, dogs, and monkeys and not in mice, rats, hamsters, and guinea pigs. Ketamine has been approved for use in all animals, similarly sodium pentathol (sodium pentobarbitone) has been approved for use in all animals. The table of acceptable euthanasia methods from the CPCSEA Guidelines is reproduced below (Table 8.1)[16]:

The methods that have been disapproved are electrocution, decompression, and stunning under physical methods. Nitrogen flushing and Argon flushing under the inhalation drugs category are not recommended. Drugs such as Curare, Nicotine Sulfate, Magnesium Sulfate, Potassium Chloride, Strychnine, Paraquat (N,N′-dimethyl-4,4′-bipyridinium dichloride), Dichlorvos (2,2-dichlorovinyl dimethyl phosphate), and air embolism are not acceptable.

EQUIPMENT AND FACILITIES

The SOP for IAECs sets very general expectations and recommendations for the design of facilities and types and characteristics suitable for the programs undertaking research in animal subjects.[17] These guidelines suggest that animal housing areas should be separated from areas used by personnel for nonanimal purposes while maintaining proximity to laboratory areas dependent upon animal studies. Facilities should provide for the macro- and micro-environmental needs of research animal subjects and accommodate separation by species and project, allow isolation and quarantine for the program of veterinary care, and provide for a full array of specialized animal support and research functions as necessary for the program.

Facilities should be built with durable materials, allow for effective vermin control and sanitation, provide adequate power and lighting, ventilation, and noise control. Construction topics include: building materials, corridors, utilities, animal room doors, exterior windows, floors, drains, and walls and ceilings. The guidelines also address the need for areas to provide for basic animal care and use support functions (e.g. cage wash, food, and bedding storage, etc.). While this discussion of these topics serves to bring them to the attention of

TABLE 8.1 Acceptable Euthanasia Methods in the CPCSEA Guidelines

ANNEXURE – 6
EUTHANASIA OF LABORATORY ANIMALS
(A – Methods Acceptable   NR – Not Recommended)

Species	Mouse	Rat	Hamster	Guinea pig	Rabbit	Cat	Dog	Primate
a) PHYSICAL METHODS								
Electrocution	NR	NR	NR	NR	NR	NR	NR	NR
Exsanguination	NR	A	A	A	A	A	NR	NR
Decapitation (for analysis of stress)	A	A	A	NR	NR	NR	NR	NR
Cervical dislocation	A	A	A	NR	NR	NR	NR	NR
b) INHALATION OF GASES								
Carbon Monoxide	A	A	A	A	A	A	A	A
Carbon Dioxide	A	A	A	A	A	A	NR	NR
Carbon Dioxide plus Chloroform	A	A	A	A	A	A	NR	NR
Halothane	A	A	A	A	A	A	A	A
c) DRUG ADMINISTRATION								
Barbiturate Overdose (route)	A(IP)	A(IP)	A(IP)	A(IP)	A(IV,IP)	A(IV,IP)	A(IV,IP)	A(IV,IP)
Chloral hydrate Overdose (route)	NR	NR	NR	NR	A(IV)	A(IV)	A(IV)	A(IV)
Ketamine Overdose (route)	A(IM/P)	A(IM/P)	A(IM/IP)	A(IM/IP)	A(IM/IV)	A(IM/IV)	A(IM/IV)	A(IM/IV)
Sodium Pentothol [Overdose (route)]	IP	IP	IP	IP	IV	IV	IV	IV

Methods Not Acceptable for any species of animals

a) PHYSICAL METHODS:
(i) Decompression   (ii) Stunning
b) INHALATION OF GASES
(i) Nitrogen Flushing (ii) Argon Flushing
c) DRUG ADMINISTRATION
(i) Curariform drugs (ii) Nicotine Sulphate (iii) Magnesium Sulphate (iv) Potassium Chloride (v) Strychnine (vi) Paraquat (vii) Dichlorvos (viii) Air Embosium

IV = Intravenous
IM = Intramuscular
IP = intraperitoneal

facility planners, the guidelines are not sufficiently detailed with specific information to facilitate critical design or construction decision-making.

There have been serious efforts to develop high quality animal facilities to support contemporary research needs in India and a few examples follow. The first specific pathogen free colony of animals was established in 1988 at the National Center for Laboratory Animal Sciences (NCLAS) (formerly Laboratory Animals Information Service Centre), National Institute of Nutrition, Hyderabad. Isolators were installed and technicians with specific training were responsible for maintaining the specific pathogen free (SPF) animals. However, the facility encountered problems and was compromised due to power outages, inconsistent professional staffing, inadequate demand for the animals, and other impediments. A few other institutions like the National Institute of Immunology, New Delhi, the National Institute of Virology, Pune, and others embarked on the ambitious plan but have not yet consistently produced and maintained high quality animals compared to international standards.

Efforts also were made at the NCLAS beginning in 1988–1990 to start an embryo freezing unit to preserve the fetuses of specialized strains of animals as a back up and also to have a repository of different strains of experimental animals. The project did not come to fruition due to the pitfalls noted above concerning the pathogen free effort. The Advanced Center for Treatment, Research and Education in Cancer, at Navi Mumbai is the only center presently in India that is in the process of establishing a fully functional embryo freezing facility.

A majority of the animal care and use programs in India (about 95%) are conducted in conventional facilities. Animal facilities commonly are comprised of a couple of rooms with or without air conditioners. Such facilities exist in pharmacy, medical and other basic science colleges. Only a handful of veterinary schools have facilities designed and functioning to support larger scale rodent housing activities.

A few animal facilities in the private as well as in the government sector have installed heating, ventilation and air conditioning (HVAC) systems equipped with high-efficacy particulate air (HEPA) filtration of 3 microns to maintain barrier conditions. Nevertheless, the costs of running these barrier facilities are high and have been found to be economically challenging. In the government sector, facilities have deteriorated significantly as a consequence of lack of funding or lack of commitment to run such facilities. In the private sector, the economic costs also have negatively impacted maintenance of facilities, and some facilities have been re-programmed for other purposes due to the inadequate market for barrier maintained animals.

The introduction of individual ventilated cage (IVC) systems was touted to revolutionize the laboratory animal industry by reducing the cost and improve the ease of maintaining high quality rodents. However, IVC housing has not yet catalyzed widespread improvements in rodent health. Presently, the availability of specific pathogen free animals is restricted to only a small number of organizations in the government and private sector. Usually the particular research group or the pharmaceutical company undertakes importation of these animals (mostly transgenic, knockout and other specialized strains) for their specific research purpose. Such organizations normally execute restrictive "No breeding and supply" agreements with the foreign suppliers. Therefore these animals are not available to other research groups or institutions.

Biosafety level facilities have been established in some institutions. However, the majority of these facilities are not for laboratory animal use. It is a matter of concern that except in a couple of institutions containment facilities do not exist. Nevertheless, this has not deterred some research groups from conducting animal research involving pathogens and hazardous material. Annexure 8 of the CPCSEA Guidelines indicates that it is the responsibility of the Institutional Biosafety Committee and the IAEC to review studies involving biohazardous agents, but the guidelines offer no elaboration on specialized facility conditions deemed relevant or necessary for the conduct of studies.[18] There is an urgent need to improve the facilities and also educate the researchers. In contrast, the use of radioactive nucleotides and materials is regulated by the government and there are strict conditions prevalent in most institutions which include the availability of trained radiation safety officers in most institutions.

One of the essential components of the IAEC is to ascertain from the researcher whether any use of pathogens, radionuclides, chemical carcinogens or other hazardous chemical agents is proposed. In cases where the researcher intends to use these chemicals, it is mandatory for the principal investigator to obtain a clearance from the Institutional Biosafety Committee which is constituted by the institutions with representatives from the Department of Biotechnology, Government of India.

There are more than a dozen institutions in India where imaging systems and equipment are being used especially in drug discovery research and toxicology. However, the maintenance of such equipment and availability of trained personnel to use such equipment hampers development.

EDUCATION, TRAINING, AND COMPETENCE OF PERSONNEL

Institutions recognized by the Indian government as eligible to conduct scientific experimentation in animal subjects have the requirement to ensure the education, training and competence of personnel in several areas according to the CPCSEA Guidelines.[19] The guidelines specify education and training expectations at every level of involvement in the program of care and use of research animals including the IAEC oversight body, the supervisory veterinarian or scientific personnel, investigators, and technical staff involved in animal experimentation, and personnel involved in the basic care of animals.

The IAEC members are expected to stay abreast of all new guidelines and amendments to the rules and the Prevention of Cruelty to Animals (PCA) Act 1960 as amended in 1998. The CPCSEA also stipulates that IAEC members should be encouraged to attend national and international training programs/ workshops/conferences in research ethics for maintaining quality in ethical review and be aware of the latest developments in this area. Such knowledge is essential to the core mission of the IAEC to control and supervise experiments on animals and ensure that animal studies are conducted under satisfactory conditions using appropriate techniques in compliance with the PCA Act and CPCSEA guidelines. The Curriculum Vitae maintained for all IAEC members in the IAEC records serves to document attendance in training programs in animal ethics and other pertinent qualifications.

The IAEC must also ensure that leadership for the programs of animal experimentation are performed in every case by or under the supervision of a person duly qualified by Degree or Diploma in Veterinary Science or Medicine or Laboratory Animal Science awarded by a university or an institution recognized by the government for the purpose and under the responsibility of the person performing the experiment. A similar requirement pertains to the oversight of the program of laboratory animal management encompassing animal living conditions, housing, feeding, care, health, and comfort. Also following this precedent, although adequate veterinary care must be provided, the responsibility for and the provision of the care by a veterinarian is not an absolute requirement. Others with training or experience in laboratory animal sciences and medicine may do so under veterinary direction.

The PCA Act and CPCSEA have not provided a specific set of task-oriented expectations for the IAEC to incorporate into its oversight of training and education of research investigators and research technical personnel. Therefore, although the IAECs have the option of improving the training oversight mechanism by moving from a cursory credentialing process to a performance-based, skills analysis process, the regulations offer no assistance in this direction. The IAEC checklist for investigators submitting proposals for review that involve survival surgery illustrate this approach; investigators are expected to provide the names, qualifications, and experience levels of operators. The degree to which individual IAECs have moved beyond the good faith principle to embrace a system of oversight involving the direct affirmation of skills in sensitive procedures is unknown.

The CPCSEA guidelines note that animal care programs require technical and husbandry support and should employ people trained in laboratory animal science or provide for both formal and on-the-job training to ensure effective implementation of the program. The CPCSEA guidelines specify the qualifications and knowledge necessary for laboratory animal attendants. The basic educational requirement indicates that candidates should have completed 10 years of primary and secondary school education. During subsequent training, a few weeks must be dedicated to teaching newly recruited staff about the animal handling techniques, cleaning of cages and importance of hygiene, disinfection and sterilization. Staff should also become familiar with the activities of normal healthy and sick animals in order to detect sick animals during their daily husbandry routines and animal observations. The CPCSEA Annexure 7 also defines many other areas of specific knowledge helpful to the attendant's understanding of the workplace and completion of particular job tasks. These include: an overview of the types and diversity of animals used in research; animal cage selection and housing densities for the various species; cleaning of the animal room environment and cages; methods of safe animal handling and associated precautions and protections; common injuries and ailments in animals; personal hygiene practices to prevent exposure to chemicals and animal related hazards (e.g. zoonoses, venoms); first aid measures; and dealing with emergency situations. Records should be kept by the animal facility on all staff, both technical and nontechnical, and should include the training records of all staff involved in animal activities.

VETERINARY CARE

The discussion of veterinary care in the CPCSEA guidelines is closely aligned with and appears to be derived from the recommendations of the Guide for the Care and Use of Laboratory Animals, 7th Edition. (Guide, 7th Edition).[20] As noted above, adequate veterinary care must be provided although the direct involvement of a veterinarian is not required. Daily observation of animals can be accomplished by someone other than a veterinarian provided that a mechanism of direct and frequent communication allows timely and accurate information on problems in animal health, behavior, and well-being to be conveyed to the attending veterinarian. Other areas of potential veterinary contribution are also recognized including the establishment of appropriate policies and procedures for ancillary aspects of veterinary care, such as reviewing protocols and proposals, animal husbandry and animal welfare; monitoring occupational health hazards containment, and zoonosis control programs; and supervising animal nutrition and sanitation. Institutional requirements will determine the need for full-time or part-time or consultative veterinary services. The CPCSEA guidelines highlight other interrelated and integrated programmatic considerations typically involving veterinary direction and oversight. These include animal procurement, source assessment, and receiving practices; quarantine, stabilization, and separation practices; and disease surveillance, diagnosis, treatment, and control provisions. The CPCSEA guidelines introduce the topics and discuss the rationale for including each subject area and a brief and basic synopsis of considerations in program development. However, substantial professional input and knowledge of the current literature and practices would be needed to fulfill the scientific objectives of many programs.

HOUSING AND ENRICHMENT

The CPCSEA guidelines recognize the importance of the cage environment noting specifically that it affects the quality of the animal. The guidelines also suggest that cages should be constructed of sturdy, durable materials and designed to minimize cross-infection between adjoining units, and additional information is provided on the type of materials that are suitable for cage construction for small and large laboratory animal species. The space recommendations for laboratory housing in Annexure 3 closely parallel the recommendations of the Guide, 7th Edition.[20]

The cage sizes recommended in the table provided in Annexure 3 of the CPCSEA guidelines are expected to accommodate the normal behavioral activities of laboratory animals adequately. The importance of monitoring animals for behavioral aberrations is mentioned in the CPCSEA guidelines in connection with veterinary care, species separation in housing to avoid interspecies conflicts, the acclimatization to restraint devices, and population density. The PCA Act and CPCSEA guidelines do address the importance of the social environment to the expression of the basic natural behaviors of the species in question and the ability of animals, especially those housed chronically, to exhibit natural locomotor patterns. Environmental enrichment is not given any discussion as a behavioral management tool or as a means to expand the repertoire of the natural behaviors that might be expressed by laboratory animals.

HUSBANDRY AND ENVIRONMENT

Information on all dimensions of the husbandry of laboratory animals are provided in the CPCSEA guidelines in a seven page narrative closely aligned, albeit much more general and abbreviated, with other national and international guidance documents in this area, particularly the Guide, 7th Edition.[20] Topics summarized encompass housing systems; sheltered or outdoor housing; social environment; activity; food; bedding; water; sanitation practices; assessing the effectiveness of sanitation; waste control; vermin control; emergency, weekend and holiday care; record-keeping; and guidelines for laboratory animal management standard operating procedures.

Facility systems important to the control of environmental factors also receive mention, citing widely accepted general standards for temperature and humidity control, ventilation, power, and lighting and noise control. The parameters given in these aspects of environmental control closely mimic those provided in other national and international guidance documents.

TRANSPORTATION

Regulations and guidance for the transportation of laboratory animals is contained in the PCA Act,[1] the Transport of Animals Rules,[21] and the Transport of Animals (Amended) Rules[22] as well as in the CPCSEA Guidelines and Annexure 4.[23] The CPCSEA guidelines note the importance of careful attention to transportation of laboratory animals citing the mode of transport, container type, population density in cages, food and water during transit, protection from transit infections, injuries, and stress as relevant considerations. Animals can be transported by many acceptable methods provided that transport stress is avoided and the containers used are of an appropriate size so as to enable these animals to have a comfortable, free movement and protection from possible injuries. The food and water should be provided in suitable containers or in a suitable form to ensure adequate provisioning during transit. The transport containers (cages or crates) should be of appropriate size and each container should be populated to avoid overcrowding and infighting. Annexure 4 of the 2005 CPCSEA regulations provides specifications for containers and population densities for animals during transit.

REFERENCES

1. The Constitution of India, Part IV, Directive Principles of State Policy, 48. Organization of Agriculture and Animal husbandry.
2. Indian Penal Code 1860. Act No. 45 of 1860.
3. The Prevention of Cruelty to Animals Act, 1960. No. 59 of 1960.
4. S.O.732 (E), [26/8/1998]—The Experiments on Animals (Controls and Supervision) (Amendment) Rules, 1998. Available at: http://envfor.nic.in/legis/animal_welfare.htm. Verified on April 16, 2013.
5. S.O.1074, [15/12/1998]—The Breeding of and Experiments on Animals (Control and Supervision) Rules, 1998. Available at: http://envfor.nic.in/legis/animal_welfare.htm. Verified on April 16, 2013.
6. S.O.134 (E), [15/2/2001]—The Breeding of and Experiments on animals (Control and Supervision) Amendment Rules, 2001. Available at: http://envfor.nic.in/legis/animal_welfare.htm. Verified on April 16, 2013.
7. Ministry of Forestry and Environment, Animal Welfare Division, Standard Operating Procedure for Institutional Animal Ethics Committee, Guidelines on the Regulations on Scientific Experiments on Animals, Form B, p. 103-8-31, 2007. Available at: http://www.aaalac.org/resources/SOP_CPCSEA_inner_page.pdf. Link verified April 12, 2013.
8. The Indian Veterinary Council Act 1984 (Act 52 of 1984).
9. The Veterinary Council of India [Standards of Professional Conduct, Etiquette and Code of Ethics for Veterinary Practitioners] Regulations 1992, GSR 395. Available at: http://www.dahd.nic.in/dahd/upload/codeethics.pdf. Verified on April 16, 2013.
10. The Animal Welfare Act, 2011 (Draft). Available at: http://moef.nic.in/downloads/public-information/draft-animal-welfare-act-2011.pdf. Link verified on April 12, 2013.
11. G.S.R. 528 (E), Ministry of Environment and Forests Notification, 2011. Available at: http://envfor.nic.in/downloads/public-information/gsr-528-e.pdf. Link verified on April 16, 2013.
12. Biological Diversity Act, 2002 (18 of 2003). Available at: http://nbaindia.org/content/25/19/2/act.html. Link verified on April 16, 2013.
13. Ministry of Forestry and Environment, Animal Welfare Division, Standard Operating Procedure for Institutional Animal Ethics Committee, Guidelines on the Regulations on Scientific Experiments on Animals, p. 1–6, 2007. Available at: http://www.aaalac.org/resources/SOP_CPCSEA_inner_page.pdf. Link verified April 12, 2013.
14. Ministry of Forestry and Environment, Animal Welfare Division, Standard Operating Procedure for Institutional Animal Ethics Committee, Guidelines on the Regulations on Scientific Experiments on Animals, Annexure I, p. 120–126, 2007. Available at: http://www.aaalac.org/resources/SOP_CPCSEA_inner_page.pdf. Link verified April 12, 2013.
15. Pingle S. Occupational safety and health in India: now and the future. *Ind Health* 2012;**50**: 167–71.

16. Ministry of Forestry and Environment, Animal Welfare Division, Standard Operating Procedure for Institutional Animal Ethics Committee, Guidelines on the Regulations on Scientific Experiments on Animals, Annexure 6, p. 56, 2007. Available at: http://www.aaalac.org/resources/SOP_CPCSEA_inner_page.pdf. Link verified April 12, 2013.

17. Ministry of Forestry and Environment, Animal Welfare Division, Standard Operating Procedure for Institutional Animal Ethics Committee, Guidelines on the Regulations on Scientific Experiments on Animals, p. 25–31, 2007. Available at: http://www.aaalac.org/resources/SOP_CPCSEA_inner_page.pdf. Link verified April 12, 2013.

18. Ministry of Forestry and Environment, Animal Welfare Division, Standard Operating Procedure for Institutional Animal Ethics Committee, Guidelines on the Regulations on Scientific Experiments on Animals, p. 27 and 59 (Annexure 8), 2007. Available at: http://www.aaalac.org/resources/SOP_CPCSEA_inner_page.pdf. Link verified on April 17, 2013.

19. Ministry of Forestry and Environment, Animal Welfare Division, Standard Operating Procedure for Institutional Animal Ethics Committee, Guidelines on the Regulations on Scientific Experiments on Animals, p. 40 and 58 (Annexure 7), 2007. Available at: http://www.aaalac.org/resources/SOP_CPCSEA_inner_page.pdf. Link verified on April 17, 2013.

20. Committee to Revise the Guide for the Care and Use of Laboratory Animals, Institute for Laboratory Animal Research, Division of Earth and Life Sciences, National Research Council. *Guide for the care and use of laboratory animals.* 7th ed. Washington, D.C: National Academy Press; 1996.

21. No. 18-6/70-LDI, [23/3/1978]—The Transport of Animals Rules, 1978. Available at: http://envfor.nic.in/legis/animal_welfare.htm. Link verified on April 17, 2013.

22. S.O.269 (E), [26/3/2001]—The Transport of Animals (Amendment) Rules, 2001. Available at: http://envfor.nic.in/legis/animal_welfare.htm. Link verified on April 17, 2013.

23. Ministry of Forestry and Environment, Animal Welfare Division, Standard Operating Procedure for Institutional Animal Ethics Committee, Guidelines on the Regulations on Scientific Experiments on Animals, p. 25, 26, 40-1 and 54 (Annexure 4), 2007. Available at: http://www.aaalac.org/resources/SOP_CPCSEA_inner_page.pdf. Link verified April 12, 2013.

Chapter 9

Oversight of Animal Research in China

Kathryn Bayne[1] and Jianfei Wang[2]

[1]AAALAC International, Frederick, MD, USA, [2]GlaxoSmithKline R&D China, Pudong, Shanghai, P.R. China

Chapter Outline

China has developed into a significant world economy, evidenced in part by its strong commitment to supporting biomedical research as a key economic engine and an increasingly dominant role in the global life sciences landscape. Numerous institutions in diverse countries are partnering with China through outsourcing and collaborations for animal-based research and testing objectives. The research and laboratory animal veterinary communities share an interest in the standards of animal care and use applied in institutions throughout China. Yet, little is known about the complex set of regulations governing the use of animals in research in Western countries, in part due to language obstacles—with English translations of

Laboratory Animals. http://dx.doi.org/10.1016/B978-0-12-397856-1.00009-X

the information scarce—or because of challenges to the non-Chinese scientist or veterinarian in finding the information. In addition, the regulatory environment in China is evolving and thus it is challenging to stay current on the regulations. This chapter aims to provide a comprehensive review of the regulatory environment in China as a single source to facilitate understanding of the country's regulations today with indications of how the regulations are changing.

INTRODUCTION

The most recent statistics from the World Health Organization (WHO) note that the global population growth and a concomitant increase in the longevity of the human population are leading to a rapid increase in the total number of middle-aged and older adults, with a corresponding increase in the number of deaths caused by noncommunicable diseases.[1] The WHO projects that the annual number of deaths due to cardiovascular disease will increase from 17 million in 2008 to 25 million in 2030, with annual cancer deaths increasing from 7.6 million to 13 million. These health statistics also predict that the total number of annual deaths from noncommunicable disease will reach 55 million by 2030, while deaths from infectious disease are projected to decrease over the same period. Countries with rapidly expanding populations and resources to invest in health research, such as China, will clearly be among the leaders in advances in health research. Because health-based research generally involves animals at some stage of the research, animal-based research will continue to have a pivotal role in the health research portfolio of China for the foreseeable future. The Pan American Health Organization report on "Health in the Americas"[2] states: "The world's increasing connectivity, integration, and interdependence in the economic, social, technological, cultural, political, and ecological spheres... is one of the greatest challenges confronting the health sector." Thus, the linkage between China's approach to the conduct and oversight of animal research has significant implications for the rest of the world.

As evidence of this linkage, in recent years China has positioned itself to be an attractive provider of outsourcing capabilities in preclinical safety assessment studies, recognizing that biomedical research conducted in accordance with global standards can be a strong economic engine. China has also dramatically ramped up its animal research program to benefit the health of the domestic Chinese population. In addition, the cosmetic industry has recently been increasing its presence in China, both in terms of sales to Chinese women, but also for safety testing purposes. These initiatives have prompted increased attention to global standards of laboratory animal care and use resulting in the construction of new research facilities, additional government-approved producers of laboratory animals, increased government support for biotech and pharmaceutical companies, training opportunities for researchers and research support personnel, and the adoption of regulations that incorporate key globally accepted principles in laboratory animal care and use.

China's emergence onto the global stage of biomedical research and testing has not gone unnoticed. China is not immune to attention from animal rights/protection organizations. Animal activists have criticized China for requiring animal testing of numerous chemicals and products. For example, PETA (People for the Ethical Treatment of Animals) Asia-Pacific advocates ceasing animal experimentation on its website. The Chinese Animal Protection Network addresses many animal issues including what they refer to as "academic research of animal ethics" and an "awareness campaign of lab animal protection" and in 2008 organized the first World Lab Animal Day in China. Multinational corporations and academic institutions see opportunities for collaboration and expansion of scientific inquiry. The general public views this development with caution due to the number of recalls of products manufactured in China, ranging from pet food to toys, clothing, tainted candy sold in Asia, and other goods. Yet, China's rapid pace in maturing its biomedical research enterprise is a strong indicator that the culture readily embraces new knowledge and positive change.

The National Science Foundation (NSF) reports[3] that the pace of growth over the past 10 years in China's overall research and development (R&D) sector is approximately 20% annually, placing it as the second largest R&D country in the world. Indeed, since 1999, the percentage of gross domestic product invested in R&D has almost tripled to 1.7%. Multinational companies are continuing to invest in Asia, and especially China. As the NSF reports, "the share of R&D performed by Asia-located affiliates (other than in Japan) increased from 5.3% to 14.4% from 1997 to 2008. In particular, the share of U.S.-owned affiliates R&D performed in China, South Korea, Singapore, and India rose from a half percentage point or less in 1997 to 4% in China...." To support this burgeoning field of research in China is a similar increase in the number of scientists, with the number tripling between 1998 and 2008, in sharp contrast to the reduced proportion of researchers in the United States and Europe that comprise the global researcher pool over the same period. From 1995 to 2002, the United States and Europe experienced an annual growth rate in number of researchers of 3–4%, while the annual growth rate for China was 12%. If one views the number of research publications as an indicator of the contribution a country makes to the body of scientific knowledge, then the trend exemplified by China is noteworthy. The NSF report notes that the combined share of published articles by Europe and the United States declined from 69% to 58% in the period from 1995 to 2009, while Asia's share increased from 14% to 24%, largely due to China's 16% average annual growth in publications, with China displacing Japan as having the second highest publication rate in science and engineering. A *Newsweek Magazine* article entitled "It's China's World; we're just living in it"[4] highlighted the pivotal role China has assumed in major global issues, including trade, climate change, currency, and technology. As the NSF data demonstrate, China has assumed a similarly prominent role in life sciences research.

Regardless of the countries involved, when considering outsourcing R&D or engaging in an international collaboration that involves the use of animals, any potential differences in key animal care and use program elements should be identified and the impact of those differences on the animals' welfare and the quality of the science should be assessed. Critical areas that could directly impact laboratory animal welfare include animal procurement, transportation of the animal from the vendor to the place of study, the provision of adequate veterinary care, the training and competency of the personnel associated with the animal program, the animal's environment (both in the primary enclosure and in procedure areas), and the method of review and approval of the proposed work (i.e. ethical review).[5] Each of these program elements may be influenced by the economic, religious, and cultural experience of the personnel at the institution, as well as the degree of their training and level of authority within the institution. Exposure to animal facilities outside of one's own country is extremely informative; however, the cost and logistics of setting up such travel can be a factor that impedes this form of training. But, the value of such interactions cannot be overstated. Indeed, China's "sea turtles" (individuals who train outside of China and then return to the country with specific expertise) return with an expanded concept of animal use and often retain contact with their overseas colleagues, thereby keeping open channels of discourse, learning, and awareness of changing global societal mores.[6,7] Also, increased international collaborations and participation in international conferences further inform Chinese scientists and regulators regarding global standards in the care and use of laboratory animals.[7]

OVERSIGHT OF RESEARCH ANIMALS

The use of animals for research in China dates back to 1918, but the first formal national laboratory animal science meeting, organized by the Ministry of Science and Technology (MOST), was held in 1982.[8] In China, MOST is the responsible government agency for establishing regulations pertaining to the conduct of research. In October 1988 the State Council approved the "Statute on the Administration of Laboratory Animal Use" which was promulgated by Decree No. 2 of the State Council in November 1988. Thereafter, MOST provided oversight of laboratory animal science development in China. Implementation of the Statute was achieved through the issuing of provincial laws, beginning with Beijing in 1996.[8] The regulations address quarantine, import/export, animal quality, personnel training, licensing for production and use of laboratory animals, biosafety, required documentation, and other relevant subjects. Specific topics within these regulations were subsequently expanded and updated, and new overarching regulations were published (most recently in 2006, "Guidelines on the Humane Treatment of Laboratory Animals").[9] Over the years, the State regulations and standards regarding the care and use of laboratory animals have been supplemented with guidelines and policies developed in the provinces, autonomous regions, and municipalities. A Provincial Department of Science and Technology (PDST) oversees laboratory animal use at the provincial government level, with a local Administration Office of Laboratory

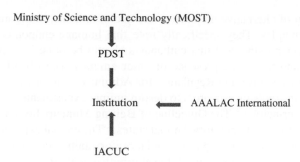

FIGURE 9.1   Oversight of laboratory animal use.

Animal Use (e.g. the Beijing Administration Office of Laboratory Animal, BAOLA) being established to enforce regulations, issue use licenses, as well as conduct inspections and related management activities.[10] Both individual and institutional licenses are issued. Institutional self-monitoring is achieved through Institutional Animal Care and Use Committees (IACUC). In addition, the voluntary accrediting organization, the Association for Assessment and Accreditation of Laboratory Animal Care (AAALAC International) provides a third party evaluation of animal research, testing, and teaching programs around the world and accredits approximately 40 institutions in China at the time of this writing. Figure 5.1 shows the oversight framework for laboratory animals in China.

ANIMAL WELFARE PRINCIPLES

Kong and Qin have summarized the progression of inclusion of animal welfare in Chinese regulations.[11] Reference to research animal welfare dates back to 1988 in the Statute on the Administration of Laboratory Animals, which essentially addressed animal welfare through the provision of nutritious food, potable water, and qualified personnel working with the animals. Several years later, in 1997, mention is made of the Three Rs in MOST documents. Since 2005, the guidelines published by MOST have increasingly strengthened recommendations related directly to laboratory animal welfare, to include a requirement for provincial ethical review of proposed work with animals and regulations addressing husbandry, use, and transportation issues. Indeed, MOST has funded projects to study and translate animal welfare laws, regulations, and policies from other countries, such as the United States and Europe. The 2006 regulations (Guidelines on the Humane Treatment of Laboratory Animals) cite noncompliance with laboratory animal policies as one of six "dishonorable behaviors" which can result in disciplinary action.[9] The MOST guidelines set the general requirement that each institution must establish a local committee to oversee all aspects of the care and use of laboratory animals. The scope of these guidelines includes breeding animals, transportation of laboratory animals, and the humane treatment of animals during the experiment. The guidelines encourage consideration of alternative methods to the use of animals, in accordance with scientific principles. They specifically note that humane endpoints should be used whenever possible and that euthanasia should be done in a manner that avoids pain and not in the presence of other animals. The MOST guidelines are extended by the Beijing Regulation for Administration of Laboratory Animals, which instructs investigators to design animal experiments to conform to the Three Rs principles.[8] The Guideline of Beijing Municipality for Review of Welfare and Ethics of Laboratory Animal states: "The animal experiment procedure should be refined for the purpose of laboratory animal protection; reducing unnecessary number of animal used; the alternative methods should be used for replacement of animal experiment after evaluation."[10]

Examples of steps taken to implement the Three Rs include the acceptance of the Swordtail fish (*Xiphophorus helleri*) as a laboratory animal by the State Evaluation Committee of Fisheries Stock, for use in quality monitoring of water, testing of chemical toxicity, and as an animal disease model. In addition, the chicken egg has been established as an alternative to the Draize test and for evaluation of cosmetics ingredients. Also, the Limulus Amoebocyte Lysate test has been embraced as an alternative to rabbit pyrogen testing. Notably, the proposed Animal Protection Law of the People's Republic of China, Article 89[12], affirms "The State encourages domestic and international sharing of experimental data and information to reduce the number of experimental animals; promote alternative experimental methods, to reduce unnecessary animal experiments; optimization of the experimental method, technology, content and procedures; and avoid unnecessary suffering to the animals and injury." Reinforcing this approach, Article 95 of the proposed Animal Protection Law[12] states that procedures should be selected that use the least number of animals and that produce the minimum pain or distress that can produce satisfactory results.

AUTHORIZATION FOR THE USE OF LABORATORY ANIMALS

National Level

A Regulation on the Management of Laboratory Animal License System was issued by MOST in 2002,[13] which created a system of licensing of institutions that breed or use animals for research, is administered at the level of the province, municipality or autonomous region. Kong and Qin report that 29 of 31 provinces have established administrative offices of laboratory animals (AOLAs).[11] In Beijing, for example, animals cannot be bred for research or used in research without the appropriate license. Institutions using animals for research are obligated to use "qualified" animals, in other words, animals produced at licensed breeding facilities which meet specific genetic and microbiological criteria. The institution applying for a license must have standard operating procedures in place; have animal welfare and biosafety management systems; have provisions in place for training of staff and for the health protection of staff; meet the national standards for facility environment; meet the national standards for quality of food and water, bedding, cages, and support equipment; as well as employ staff who must receive the training organized by local administrative office for laboratory animals, have knowledge of laboratory animals, and have passed the qualifying examinations. There are additional requirements for animal research institutions that conduct studies using infectious agents, radiation, or chemicals (i.e. where public safety may be compromised) or that conduct research on wildlife. The application submitted to BAOLA must be accompanied by several supporting documents, such as testing reports for the facility environment, proof of medical examination of the personnel, a description of the ethical review system, animal facility plans, etc. Licenses are valid for a period of 5 years. Upon expiration, a reinspection process is initiated for license renewal. An individual has to attend the training organized by the local administrative office for laboratory animals again should he or she wish to maintain the license. Continuing education for the license holder is encouraged. Should the institution wish to change the scope of work conducted or make facility modifications, it must apply to the Science and Technology Commission of the Municipality for authorization of the change(s).

Annual inspections of animal facilities are also under the authority of the Science and Technology Commission of the Municipality. The inspection process relies on a combination of random sampling and a "self-check." The inspection includes a review of the work records; records of training of personnel; records of physical examinations of personnel; sources of the animals and methods of transportation to the facility; the quality of food, bedding, cages, and ancillary equipment; the management system and operating rules; the operating conditions of the animal facility; implementation of biosafety practices; animal welfare and ethical review systems; and whether the number and types of animals used conforms with the license. Violations or forging of the license is punishable by law. Work conducted outside the scope of the license renders the animal experiments invalid. The results of the inspection are submitted to the State Science and Technology Commission of the People's Republic of China. Institutions that fail the annual inspection have a 3 month period to effect corrections and are then subject to reinspection. Should the institution again fail the inspection, the license is revoked. However, the institution can reapply for a license once all corrections have been made and the institution complies with the standards.

Local Level

A local authorization system serves as an adjunct to the national licensing system. The Regulations of Beijing Municipality for the Administration of Affairs Concerning Laboratory Animals[14] defines a laboratory animal as "those animals that are artificially fed and bred, the microorganisms and parasites on or in whose bodies are kept under control, whose genetic backgrounds are definite or whose sources are clear and that are to be used in scientific research, teaching, production, examination and verification and other scientific experiments. In accordance with the controls on microorganisms and parasites, laboratory animals are classified into conventional animals, clean animals, animals carrying no specific pathogens and animals carrying no bacteria."

In Beijing, the Welfare and Ethics Committee (referred to as the Ethics Committee) is responsible for the inspection of the welfare of the animals and for implementing an ethical review system. The Committee must be comprised of at least five people. The Chair of the Ethics Committee should be a specialist in laboratory animals, and there is a preference for a veterinary surgeon to fulfill this role. By regulation, the institution is responsible for providing training of the Ethics Committee members and to ensure Committee membership is maintained in accordance with the 3–4 year appointments. Also, in accordance with the regulations, the Ethics Committee members "shall commit to preserve the welfare and ethics of laboratory animals."

LABORATORY ANIMAL QUALITY CONTROL

To guarantee the quality of laboratory animals, various national standards (referred to as "GB") have been established in China. In 1994, the China State Bureau of Technical Supervision issued a series of national standards, 47 of which applied to laboratory animals. The 2001 revision of these standards increased the number to 83.[15] In 2005, the National Technical Committee on Laboratory Animal Science of the Standardization Administration of China was established and set 93 standards for laboratory animal quality control covering five areas: microbiology, genetics, feedstuff, environment, and specific pathogen free (SPF) chickens. In addition, other government agencies, such as the ministries of medicine and agriculture, have set laboratory animal quality standards. Also, some of the more developed provinces and cities (e.g. Beijing, Shanghai, Guangdong, and Jiangsu) impose their own standards, as do some larger companies or enterprises for their products.

The Laboratory Animal Development Program for the Ninth Five-Year Plan established a national lab animal quality control network, which determines requirements for laboratory animal quality monitoring centers overseen by MOST and the PDSTs. MOST subsequently issued the "Guideline and Detailed Criteria for the Review of Provincial Laboratory Animal Monitoring Centers."[16] The task of these centers is to explore research on testing techniques, train technicians, carry out site inspections and annual review of provincial laboratory animal monitoring centers, and arbitrate disputed monitoring results. The quality monitoring network comprises six national centers covering microbiology, parasitology, genetics, pathology, diets, environment, and facilities, as well as 26 provincial centers in 23 provinces.

OVERSIGHT AND ETHICAL REVIEW PROCESS

In accordance with the "Examination Guideline for the Welfare and Ethics of Laboratory Animals," Article 4, the institution must establish a Welfare and Ethics Committee comprised of staff from management, scientific and technical personnel, professional veterinary staff, and persons who are not members of the institution. The Committee is responsible for the welfare and ethical review of laboratory animals and supervision and administration of this activity. The Committee's role is increasing in importance as MOST implements reforms to discourage scientific misconduct.[17]

The Beijing Experimental Animal Welfare Ethics Review Guidelines specify several activities of the Ethics Committee.[10] These duties include review and supervision of breeding, transportation, feeding, experimental design, and procedures. The Ethics Committee is charged with ensuring implementation of animal welfare and ethical principles. The Committee is also instructed to take into account the interests of animal welfare and the value of the research in its assessment of the proposed work. The research application (protocol) should include information regarding the significance of the project, the need for animals, the expected harm to animals, and a detailed description of the animal welfare and ethical issues involved in the project. The animal welfare issues that must be addressed include overall general requirements of the animal, husbandry and care, experimental objective, transportation, and procedural methods. Prevention and mitigation of pain and distress are a particular focus of the Guidelines, which call for adherence to the Three Rs during each stage of the animal experiment. In addition, the final disposition of the animals must be described, and the method of euthanasia must be the least painful. The Guidelines state that the review by the Ethics Committee should be independent, impartial, scientific, democratic, transparent, confidential, not influenced by politics, commercial interests, or self-interest. Finally, the Guidelines outline several reasons that would result in a protocol being rejected or suspended. Activities of the local Ethics Committees that are found to be noncompliant with the Guidelines are subject to a "rectification decision" by the Beijing Experimental Animal Management Office.

The Ethics Committee is responsible for establishing institutional regulations, review procedures, supervision systems, conducting regular meetings, establishing a reporting system, establishing training programs, and reporting the list of members to the overarching Welfare and Ethics Committee of Laboratory Animals of Beijing. Of note, the standard requires that the Ethics Committee work independently in its inspections and supervision of the research, breeding, feeding, production, management, and transportation of the laboratory animals and in confirming that the experiments conform with the animal welfare and ethical principles, to include a harm–benefit analysis. Specifically, the Ethics Committee is charged with taking into account the benefits of the animal research while "comprehensively evaluating the injuries that animals suffer from the necessity of using animals." The Committee must provide a report of this ethical review. To assist the Committee in this determination, the study application must describe the significance of the project, its purpose and the expected harm to the animal.

A draft "Animal Protection Law of the People's Republic of China" was developed in 2009 by a committee chaired by Chang Jiwen, Director of Social Law Research at the Chinese Academy of Social Sciences.[12] The draft law is broadly encompassing, addressing wildlife, companion animals, food animals, laboratory animals, and other animal uses. Chapter 6 of the draft law focuses on "Legal Protection for Laboratory Animals." Of note is a recommendation that a national laboratory animal ethics committee should be established to supervise and manage the scientific use of laboratory animals, which is comprised of at least one veterinarian and one representative of an animal protection organization. There is a further recommendation that each province also establish an animal ethics committee with similar composition to the national committee. The draft law proposes that the national and provisional ethics committees publish basic principles or guidelines on animal experimentation.

REUSE

The Guideline on Humane Treatment of Laboratory Animals addresses, in general terms, the welfare of animals used in research, but does not specifically touch on the topic of reuse.[9] However, it should be noted that the subject is under consideration. Specifically, the draft "Animal Protection Law of the People's Republic of China" (Article 95) proposes, "The recording of the number of times laboratory animals are used and the aims of the experiments should be the responsibility of a specified person."[12] The draft law also states (Article 96) that "The same animal may not undergo the same experiment repeated in the same test cycle. If such an experiment is necessary it should be examined and approved by the laboratory animal ethics committee."

SETTING FREE/REHOMING

The draft "Animal Protection Law of the People's Republic of China" does not support the release of research animals.[12] Specifically, it states in Article 99, "Work units, individuals and organizations engaged in animal experimentation should safeguard biosecurity and public health and safety, and the abandonment or release into the wild of laboratory animals is prohibited."

OCCUPATIONAL HEALTH AND SAFETY

China has established a classification system for hazardous agents that is based on pathogenicity to people or animals, as follows[18–20]:

- "Hazard Level IV (High Individual Hazard, High Group Hazard)": pathogens that can cause serious disease; they cannot be cured in general cases and can be transmitted between people or between animals and people, or between animals through direct contact, indirect contact, or "casual" contact.
- "Hazard Level III (High Individual Hazard, Low Group Hazard)": pathogens that can cause serious disease or cause serious economic loss, but cannot be transmitted through casual contact in general cases, or be treated with antibiotics or antiparasitic drugs.
- "Hazard Level II (Medium Individual Hazard, Limited Group Hazard)": pathogens that can cause people or animals to become infectious, but do not pose a serious hazard to healthy operators, groups, domestic animals, or the environment in general cases. Laboratory infections do not cause serious disease; there are effective therapeutic and preventive measures and a limited transmission risk.
- "Hazard Level I (Low Individual Hazard, Low Group Hazard)": bacteria, epiphytes, viruses, parasites, and other biological agents that do not harm healthy operators or make animals infectious.

Microorganisms in Hazard Levels III and IV are collectively referred to as "highly infectious" agents and can only be studied in laboratories that are authorized by the government as Class Three or Class Four. Protective measures for the use and transportation of these agents depend on the biosafety level of the agent. Both engineered safety features in the facility (e.g. airlocks, HEPA (high-efficiency particulate air) filtration, directional airflow, biological safety cabinets, and safety hoods) and use of protective equipment at the individual level are required to achieve personnel protection (e.g. goggles, safety shield, mask or respirator, gloves, shoes). In addition, procedures such as hand washing, minimizing aerosols in the workplace, safe use and disposal of sharps, etc. are described for each biosafety level. Biohazard signage is required for Level 2 and higher areas and must be prominently displayed. Biosafety equipment must be inspected annually to ensure appropriate function and personnel protection.[19]

Chemical and radiological safety standards address storage, handling, use, and disposal practices. Documentation of control measures is required and a safety supervisor is designated at the institution.[21]

A plan for emergency evacuation of personnel must be in place at the institution.[18] Personnel must participate in at least one fire drill each year and must be familiar with the evacuation plan.

In addition to the national standards addressing occupational health and safety, Beijing Municipality regulations (Article 12)[10] require that "Institutions dealing with laboratory animals shall take protective measures to ensure the health and safety of employed persons... [and] employed persons shall be physically examined each year. Persons who are unsuitable for undertaking affairs relating to laboratory animals due to health conditions shall change their positions."

ALTERNATIVES

Consideration of the use of alternatives in animal testing is gaining significant momentum in China. The MOST Guidelines[10] encourage consideration of the Three Rs when designing an animal study, and laboratories in government agencies have been funded to evaluate Replacement strategies (e.g. swordtail fish, silkworm, horseshoe crab), especially in toxicology studies. Factors contributing to this include increasing public attention within China to animal welfare issues, such as harvesting bear bile and the consumption of shark fin soup, as well as China's entry into the World Trade Organization. China could encounter trade restrictions if animal testing procedures are not conducted at an international level or, in some cases, if nonanimal alternatives are not used. Much of this pressure has been felt in the cosmetics industry. A 2003 amendment to the EU's Cosmetics Directive called for a timetable to implement a progressive ban on the use of laboratory animals in testing cosmetic products. Since 2004, animal testing of final cosmetics products was banned, and in 2009 a timetable to prohibit animal testing of the ingredients of cosmetics was established. A result of this action is the pressure on countries, such as China, that manufacture cosmetics ingredients.

The First International Symposium on Cosmetics—Alternatives to Animal Experimentation for Cosmetics was held in Beijing in 2011 and participants called for more focused action to reduce the number of animal toxicity tests and to increase the use of alternatives, with a specific request for assistance from the West in developing nonanimal alternatives.[22] The State Food and Drug Administration (sFDA) requires animal toxicology test reports prior to licensing a new cosmetic ingredient; however, this requirement is undergoing review due to concern expressed by countries where nonanimal testing methods have been demonstrated to be valid and due to the economic pressure on Chinese companies that conduct the nonanimal tests for the export of ingredients, but still must conduct animal tests for the sFDA and domestic Chinese market. In February 2012, the sFDA proposed the first in vitro method for cosmetic phototoxicity.[23] The test uses a BALB/c 3T3 fibroblast cell line to test acute phototoxic effects of cosmetic ingredients on the skin. The test conforms to "OECD (Organisation for Economic Co-operation and Development) guidelines for the testing of chemicals 3T3 NRU (Neutral Red Uptake) phototoxicity test (no. 432,2004)."

Despite this significant step toward increasing reliance on nonanimal testing methods, recent actions taken by the sFDA in recategorizing some cosmetics will likely increase the use of animal testing.[24,25] In China, cosmetics are categorized as either ordinary use (e.g. hair care, nail care, skin care, perfumes, etc.) or special use (e.g. hair growth, hair color, hair removal, spot removal, sun block, etc.).[26] Ordinary cosmetics are reviewed at the provincial level and often are sold without animal testing. Special use products require a State registration which necessitates animal testing for eye and skin irritation. In recent years there have been reports of safety issues associated with nonspecial use cosmetics. The proposed recategorization is intended to strengthen consumer safety by moving some cosmetics into the special use category, which requires more intense testing, including animal testing to produce safety assessment data. Nevertheless, in the long run, the cosmetics industry may become the first sector to eliminate animal testing as a result of legislation. But, for the pharmaceutical, agrochemical, and the chemical sectors in general, there is no clear timetable for replacing animal testing.

EDUCATION AND TRAINING

The training provided to personnel working in the animal research environment is generally determined by the provincial government.[27] For example, in Beijing the BAOLA is responsible for the professional training of supervisory staff and assessing their training. The supervisory staff at an institution are provided with a training syllabus for the personnel in their workplace. Staff are required to take an examination established by BAOLA, using a randomized set of questions based on the syllabus, and administered at the institutional level. In Beijing, for personnel to work with animals, they must obtain a "Position Qualification Certificate of the Laboratory Animal Practitioners in Beijing." Continuing education and a training plan for personnel must be in place. Records of training must be maintained. Supervisory staff must be university graduates with a degree in a relevant field (e.g. medical science, biology, zootechnical science, veterinary medicine).

The Chinese Association for Laboratory Animal Science (CALAS) was established in 1987, and although it is independent from MOST, it has a nationwide significant role in the self-regulation of laboratory animal science through supervising provincial associations and in providing continuing education and training. Several colleges of medicine, veterinary medicine, pharmacy, or biotechnology throughout the country offer undergraduate and graduate programs in laboratory animal science.

China has a long-established open attitude to inviting non-nationals to provide training. As early as 1992 China initiated training of laboratory animal scientists and technicians through collaboration with the Japanese International Cooperation Agency. The formal collaboration continued through 1997. The goal was to organize a system of study and to elevate training to an international level, with an emphasis on breeding, managing and using laboratory animals. A significant output of this initiative was the development of knowledgeable professionals who could then assume the role of trainers, thereby leading to sustainability of the project. This pattern of inviting subject experts into China to provide topic-specific training has continued at a steady pace. Symposia and workshops are routinely organized to disseminate training in IACUC function, veterinary care roles, environmental enrichment, and other related subjects throughout their research community. Examples include the annual China Pharmaceutical R&D Summits; the "Shanghai Laboratory Animal Welfare Sharing Conference" co-hosted by the Office of Shanghai Administrative Committee for Laboratory Animals, Global Research Education & Training (GR8) and AAALAC International; Peking University; and others.[28]

ANIMAL TRANSPORTATION

MOST guidelines[9] stipulate that the cages used to transport animals must provide adequate ventilation, they must be clean, and the animals must be secure. Mixing of animal species or lines in the same cage is prohibited. Food and water, as well as veterinary care, must be available if the duration of transport is long. In addition, Beijing Municipality regulations[10] stipulate that the transfer facilities and transportation equipment must support the pathogen status of the animal.

Animal rights activists have successfully pressured Chinese airlines to cease shipments of nonhuman primates overseas. In February 2012, Hainan Airlines stopped a specific shipment of monkeys from China to Canada, and announced it would cease "its engagement in such shipments." In April 2012, the U.S. Department of Agriculture (USDA) cited Air China for improper transport procedures after a monkey was injured in transit from Beijing to New York. In May 2012, the USDA again cited Air China for two violation of the U.S. Animal Welfare Act based on the escape of a monkey on a flight from Beijing to New York City and for handling the monkey in a way that posed harm to the animals. A subsequent e-mail and telephone campaign by animal rightists directed at Air China resulted in the company also ceasing shipment of nonhuman primates internationally.

HOUSING AND ENRICHMENT

Cage size standards for common species of laboratory animals used in China are contained in the standard GB 14925-2001.[29] These are summarized in Table 5.1 (small laboratory animals) and Table 5.2 (large laboratory animals and chickens). Standards for other species are being developed. For example, a standard for zebrafish established by Hunter Biotechnology, Inc. has been adopted as the local provincial standard in Zhejiang, and is being discussed for adoption as the national standard. For institutions accredited by AAALAC International, cage size recommendations in the *Guide* apply.[30]

Concern for animal welfare in Chinese society is increasing across multiple sectors of animal "use". As an example, Davey and colleagues have documented this both in student attitudes toward animal welfare[31] and public attitudes surveyed at zoos.[32] In the first instance, he polled students about their level of concern for animal welfare across a variety of issues. His results affirmed that Chinese society has generally positive attitudes toward animal welfare initiatives. In the latter study he and his colleagues evaluated the impact on the Chinese public of zoo exhibits that included environmental enrichment. The enriched exhibit promoted greater zoo visitor interest, measured by the duration of the visit to the exhibit and other parameters. Specific to the laboratory animal environment, the *Guidelines on the Humane Treatment of Laboratory Animals* state that the facility and environment must provide for the animals' behavioral and physiological needs, including the provision

TABLE 9.1 Cage Sizes Required in GB 14295-2001 for Small Laboratory Animals

	Mouse (g)		Rat (g)		Guinea Pig (g)		Hamster (g)		Rabbit (kg)	
	<20	>20	<150	>150	<350	>350	<100	>100	<2.5	>2.5
Individual (m²)	0.0065	0.01	0.015	0.025	0.03	0.065	0.01	0.012	0.20	0.46
Mother with litter (m²)	0.016		0.08		0.09		0.09		0.93	
Minimum height (m)	0.13	0.15	0.18	0.18	0.18	0.22	0.18	0.18	0.40	0.45

TABLE 9.2 Cage Sizes Required in GB 14295-2001 for Large Laboratory Animals and Chickens

	Cat (kg)		Dog (kg)			Nonhuman Primate (kg)			Mini Pig (kg)		Chicken (kg)	
	<2.5	>2.5	<10	10–20	>20	<4	4–6	>6	<20	>20	<2	>2
Individual (m²)	0.28	0.37	0.60	1.0	1.5	0.5	0.6	0.75	0.96	1.2	0.12	0.15
Mother with litter (m²)	—		—			—			—		—	
Minimum height (m)	0.76	0.8	0.9	0.9	1.5	0.6	0.7	0.8	0.6	0.8	0.4	0.6

of environmental enrichment.[9] Also, the topic of the animal's housing environment, including enrichment, is being discussed at conferences more often (such as http://www.ibclifesciences.com/china/overview.xml). Davey and Wu[32] assessed attitudes of Chinese university students about animal research. Of the university students surveyed, concern was strong regarding the use of animals for safety testing of cosmetics and household products and there was a tendency for students to disagree with the statement that "humans have the right to use animals as we see fit." As the authors noted, these students will enter the workforce and some may become directly involved in animal use. Trends regarding the treatment of laboratory animals in China will likely be shaped by their views, and thus topics such as the provision of environmental enrichment will be of increasing importance.

HUSBANDRY AND ENVIRONMENT

The National Standard of Laboratory Animals,[15] first published in 1991 and revised in 2001, 2006, 2008, and 2010 defines the quality of the animals and the conditions for their housing and care. The National Technical Committee 281 on Laboratory Animal Science of the Standardization Administration of China (SAC/TC281) was created in 2005 and expanded the process of implementing standards. The standards address the pathogen status and genetic monitoring of the animals, environment and housing facilities, as well as feeding and nutrition. The standards apply to both breeding and research institutions and require that the facility accommodate the animals' behavioral and physiological needs, to include the provision of environmental enrichment. The current number of national standards is 93; however, MOST is evaluating the implementation of additional standards, such as regarding the transportation of animals. The Beijing Municipality regulations[10] require that "Production environments and facilities of laboratory animals shall comply with standard requirements of laboratory animals of different classes. According to the corresponding standards, laboratory animals of different classes and breeds shall be managed in different environments and facilities. Up-to-standard supplies such as feedstuffs, cages and bedding materials etc. shall be used." The regulations (Section IV, Article 24) also prohibit conducting experiments in the same room on animals that may interfere with each other (behaviorally or due to their pathogen status).

Cages should not contain toxic materials; should resist corrosion and high temperatures; should be impact resistant; and should be easy to clean, disinfect, and sterilize. Bedding material should provide good absorption, should have little dust, have no "foreign smells," should be nontoxic, and contain no grease or impurities. Feed should only be purchased from institutions with "Licenses for the Breeding of Laboratory Animals." The feed should be accompanied by a certificate of conformity. Drinking water provided to the animals should conform to the standards for domestic potable water in the city, or in the case of barrier housed animals, water must be sterilized.

VETERINARY CARE

Veterinary medical training in China has traditionally been associated with agriculture and the production of healthy animals to feed the country's growing population, and has a more technical emphasis.[33] The veterinary medical education program is typically 3–5 years postsecondary education and addresses a wide range of relevant topics. In addition, the cultural roots of Traditional Chinese Veterinary Medicine remain a key component of contemporary veterinary medical education in China. However, it has been suggested that students do not learn to synthesize the individual courses into a cohesive approach to evaluate, diagnose, and treat the patient due to a gap in clinical training.[34] In March 2012 the Chinese Veterinary Medical Association (CVMA) and the International Veterinary Collaboration for China, a consortium of veterinary schools in the United States, United Kingdom, and Zoetis (formerly Pfizer Animal Health), formalized a memorandum of understanding to enhance veterinary education in China by supporting Chinese students in preveterinary courses for one year, followed by an opportunity to apply to veterinary school at a participating university. Funding for the students is made available by the China Scholarship Council and covers all of their expenses, including travel back to China after they complete their doctor of Veterinary Medicine education. This represents a significant step in advancing the quality of veterinary medical practice and teaching that will be made available in China, and is in accord with a recommendation made for input from international educators several years ago.[33]

In addition, the Chinese Ninth Five-Year Plan called for government actions to strengthen institutions of higher education, specifically with the goal of assisting universities "reach the advanced international standards for the overall quality of teaching, scientific research, and the training of professional manpower, so as to establish their international prestige and position among universities in the world," referred to as Project 211.[27] The National Science Foundation has reported that research and development (R&D) expenditures in China increased by a record 28% in 2008–2009—a time when R&D growth stopped and decreased in the U.S. and Europe—and the 2010 data show an additional 22% increase.[3] Therefore, China's investment in intellectual development is well placed and strategically sound.

The CVMA, established in 2009, is the first professional organization representing veterinarians in China, and it is taking a leadership role in improving the quality of education provided to veterinary students. As China's population grows, there is increasing pressure for well-trained veterinarians to support greater numbers of food and companion animals, and these initiatives represent a central recognition of the demand for a better educated professional to meet these socioeconomic demands.

Concomitantly, there is an amplified need for veterinarians who are skilled and knowledgeable in the specialty of laboratory animal medicine. The inadequacy of specialty training in the Chinese veterinary medical education program is recognized within and outside of China,[28] and numerous training sessions have been held for Chinese veterinarians to begin to close the gap. Bayne et al.[6] have proposed a laddered system of training options (ranging from a simple mentor system to board certification) to address inadequacies in veterinary training and qualifications around the world, and follow-up actions are taking hold. For example, CALAS provides educational opportunities, and certificates of education are offered by several universities that provide training in laboratory animal science topics, to include medicine.[8]

Despite the changes occurring in the education system for veterinarians, other obstacles must still be overcome. Prominent among these is the limited number of pharmaceutical-grade anesthetics and analgesics available in China.[34] Agents that may result in human addiction are not routinely accessible, thus the choice for optimal drugs is severely restricted. Permits for the importation of certain drugs may be applied for, but the process can be quite time consuming and challenging.

The perception of the veterinarian as a professional is increasing, due in part to demands from Western clients working with contract research organizations or pharmaceutical companies based in China. Western clients express clear expectations that the veterinarian will be a partner in the research enterprise, will be trained and competent in the management and care of the species used, and will have a credible role on the IACUC. In addition, institutions that seek accreditation of their animal care and use programs by AAALAC International must conform with the standards for veterinary care in the *Guide*[30] and with AAALAC's Position Statement on the roles and responsibilities of the veterinarian (http://www.aaalac.org/accreditation/positionstatements.cfm#vetcare).

CONDUCT OF EXPERIMENTAL PROCEDURES

Researchers using animals must obtain a personal license issued by MOST, through the PDSTs, that is based on attending a training course and examination, as well as supervision. The license is valid for 5 years. In addition, an IACUC must review the proposed work and must approve it before the work can begin. The institution where the research is conducted must also be licensed for animal use (and animal production facilities undergo a separate licensing process) and, similarly, this license is valid for 5 years, though annual reviews and site visits are performed. It is illegal for procedures to be conducted without a license. The "Regulations of Beijing Municipality for the Administration of Affairs Concerning Laboratory Animals," Section IV, Article 26[10] states that "Individuals dealing with animal experiments shall design experiments based on the principles of substitution, reduction and optimization." In this manner, it is stipulated that the Three Rs must be addressed in the proposed research. As a corollary to the consideration of the Three Rs, in accordance with MOST guidance,[9] humane endpoints should be used whenever possible for reducing or eliminating unnecessary pain or distress.

The draft "Animal Protection Law of the People's Republic of China"[12] recommends that "The State encourages the sharing of experimental data and material domestically and internationally, in order to reduce the numbers of laboratory animals used; the State also promotes alternative [replacement] experimental methods, in order to reduce the number of unnecessary animal experiments; and refinement of experimental methodology, technology, content and procedures in order to avoid causing animals unnecessary suffering and harm." The draft law further suggests (Article 97), "If intense pain is going to be caused to laboratory animals, anesthesia, painless methods or other methods which avoid pain, suffering, distress or lasting harm, should be used. In cases where owing to the particular aims of the experiment, anesthesia cannot be used or there is no practicable way of using anesthesia in the experiment, the requirements of the basic guidelines on animal experimentation should be followed."[12]

EUTHANASIA

In accordance with MOST guidance,[9] euthanasia should be carried out in a manner that avoids pain and other animals should not be present when euthanasia is performed. The draft "Animal Protection Law of the People's Republic of China" states in Article 98, "On conclusion of the experiment, the health of the laboratory animals should be examined immediately.[12] If as a result of the experiment an animal has lost part of a limb or organ or is injured, and in continuing to live it will have to endure lasting pain, suffering, distress or harm, in compliance with the prerequisites of the basic principles or guidelines on animal experimentation, the animal may be promptly destroyed using a humane method."

EQUIPMENT AND FACILITIES

Project 985 (named for the year (1998) and month (May) of its implementation) was launched to promote the development of China's higher education system, to include the construction of new research centers and the improvement of facilities. To that end, China has invested in the construction of animal facilities that provide physical environments for the conduct of sophisticated animal research, though older facilities also remain in use.[29] In 2010, Kong and Qin[8] stated that there were 320 production facilities and more than 1530 facilities licensed for laboratory animal use. Approximately 100 standards have been issued by the Chinese government to promote quality control, and animal housing and use facilities are addressed in these standards. In addition, the "Regulations of Beijing Municipality for the Administration of Affairs Concerning Laboratory Animals," Section IV, Article 25 (The Utilization of Laboratory Animals)[10] states, "The use of up-to-standard laboratory animals and environments and facilities at the corresponding levels shall be taken as basic requirements in the research projects submitted for approval and in assessing the results of such projects, performances of verifications and inspections as well as in producing products with laboratory animals as productive materials. The results of animal experiments obtained by using non-conforming laboratory animals or in non-conforming laboratory environments are invalid. Moreover, products produced thereof shall not be sold."

In Beijing, construction requirements stipulate that interior walls should be smooth and even so that they can be cleaned and disinfected easily. Materials that are water resistant, shock resistant and "hard to peel off" are required for interior walls of the facility. The floor must not be slippery and should resist abrasion and corrosion. Its surface should be smooth and should not have cracks. Additional standards are imposed for barrier facilities that address the environment outside the facility as well as inside construction requirements. For example, the regulations state: "The outdoor environment shall be in good order and hygienic. No puddles, weeds, garbage, diluvial soil or breeding ground for mosquitoes and flies shall exist."[14] Insect and rodent pest control must be implemented outside the facility. The interior elements of the barrier are also described, to include walls, floors, ceilings, corridors, doors, power supply, activity areas for dogs and monkeys, air supply and exhaust, water supply, temperature, humidity, etc.

OVERSIGHT OF NONHUMAN PRIMATES IN RESEARCH

China is one of richest nonhuman primate habitats in the world. Of the more than 200 living species of primates in the world, four families, seven genus, 23 species, and 39 subspecies are found in China. Six subspecies of *Macaca mulatta* are distributed in more than 20 provinces in China. For example, in the Yunnan province, 15 species of nonhuman primates are found, making it the richest primate resource in China, which in turn facilitated it becoming the earliest province to undertake primate research.

During the mid twentieth century, Chinese scientists began investigating the diversity of nonhuman primate species, their geographic distribution, habitat, ecology, morphology, etc. In the 1960s–1970s, scientists started to breed non-human primates for research in physiology as well as disease prevention and treatment. In the early 1980s, laboratory nonhuman primates, such as rhesus monkeys, were bred at a large scale at the Shanghai Laboratory Animal Center, Chinese Academy of Sciences. In 1982, MOST organized the first meeting on laboratory animals and as a result four National Laboratory Animal Centers were set up; one of the four centers was a primate breeding center. Since the late 1980s, more than 40 monkey farms or nonhuman primate resource centers that breed rhesus and cynomolgus monkeys have been established. Those breeding farms are located mostly in the Guangdong, Guangxi, and Yunnan provinces. In 2006, there were approximately 108,940 monkeys in China, of which 86,200 were cynomolgus and 22,740 were rhesus monkeys. However, the number of nonhuman primates has increased over time. To date, there are more than

TABLE 9.3 National Microbiological Standards for Nonhuman Primates

	Bacteria	Viruses	Parasites
Conventional	Salmonella spp. Pathogenic dermal fungi Shigella spp. M. tuberculosis	Herpes B	Ectoparasites Toxoplasma gondii
Specific pathogen free (conventional plus...)	Yersinia enterocolitica Campylobacter jejuni	Simian retrotype D (SRV) Simian immunodeficient virus (SIV) Simian T-lymphotropic virus type-1 (STLV-1) Simian pox virus (SPV)	All helminths Entamoeba spp. Plasmodium spp. Flagellates

200,000 cynomolgus and 30,000 rhesus monkeys in China. All animals are in a closed breeding environment; the animals used in biomedical research and drug discovery are F2 animals with a well-defined microbiological status (Table 5.3). As of 2011, it was estimated that 25,000 monkeys were used domestically and about 25,000 were exported to overseas such as the United States and European countries.[35]

Primate facilities must strictly follow the national standard (GB 50447-2008) for architectural and technical requirements.[29] Monkey breeding and transportation, export of animals as well as tissue and blood are conducted under a permit from the Ministry of Forest in China and are also subject to MOST regulation. Each monkey farm is allowed to provide a specific, limited number of animals for export and domestic use yearly. Violation of this limit is subject to severe punishment, including a significant fine and even imprisonment. Enforcement of these limits is taken seriously. For example, in 2005 an individual was sentenced to prison for 14 years plus a fine of 2 million RMB due to the lack of a permit for animal breeding and transportation.

OTHER

China has established a network of national laboratory animal production centers to supply national needs. The goal is to "scientifically protect and manage the resources of laboratory animals... so as to achieve an assurance of the quality of breeds."[36] To participate in the National Resource Centre for Laboratory Animals, the institution must go through a formal application process and meet certain standards for quality control.

In addition, attention is paid to genetically modified animals through the Statute on the Administration of Transgenic Organism Biosafety in the Agriculture System issued by the Ministry of Agriculture (MOA).[37] This statute requires the establishment of a cross-ministry collaborative network to address the biosafety of transgenic organisms. MOA has also issued related regulations covering safety evaluation, export and import supervision, labeling systems, product review, and advertising. Further oversight is achieved by the Regulation on Sanitation Oversight of Transgenic Food[38] and the Regulation on the Import and Export of Transgenic Products[39] issued by the Ministry of Health and the General Administration of Quality Supervision, Inspection, and Quarantine of China, respectively.

CONCLUSIONS

China has established a complex matrix of regulations at the state, provincial, and local level to address laboratory animal welfare. It is clear that China intends to establish an environment for the conduct of biomedical research that is on a level with the international community. Yet, like the oversight framework in any country, establishment of regulations is just a first step. Implementation of the regulations and guidelines must be enforced with tangible consequences (e.g. suspension of funding) should noncompliance be detected. In many countries this is achieved by inspections from government representatives and the submission of annual reports that summarize the institution's use of animals (e.g. types of research, species and numbers of animals used, etc.). While the regulatory environment is important, regardless of how the standards are developed and implemented by oversight bodies, daily compliance by personnel at the institution is key. The institutional "culture" has a profound impact on the degree of compliance with requirements as well as attention to quality science and animal welfare. Klein and Bayne[40] propose that the establishment of a culture of care, conscience, and responsibility is as or more important than the regulatory framework itself because it promotes an environment of workplace integrity, ethics-based decision making, good communication of institutional expectations, clear lines of authority, and a system for continuous development and improvement of the animal care and use program. When this institutional culture is established, regardless of the details of the regulatory framework, institutional personnel will ground their decisions and actions in an ethical context that will promote quality science and animal welfare.

China is a very diverse region overflowing with energy, talent, and potential. There is no doubt that China is increasingly becoming a global focus for the conduct of animal-based biomedical research. Kong and Qin[8] estimate that 19 million laboratory animals were produced and 16 million animals were used in China in 2010. It has been estimated that over 100,000 people work in the field of laboratory animal science. Likely, those numbers have increased in the intervening time and will continue to grow. It is important that collaborations with Chinese colleagues continue with the goal of promoting training and competency in laboratory animal medicine and science for IACUC members, veterinarians, researchers and other staff involved in the production and use of laboratory animals as one way to harmonize standards in laboratory animal welfare and quality science.

REFERENCES

1. World Health Organization. World Health Statistics 2012. Available at: http://www.who.int/gho/publications/world_health_statistics/2012/en/index.html.
2. *Health in the Americas*. 2007 ed. vols 1 and 2. ISBN 9275116229. Pan American Health Organization/World Health Organization. Available at: http://www.paho.org.
3. National Science Foundation. Science and Engineering Indicators 2012. Available at: http://www.nsf.gov/statistics/seind12/c4/c4h.htm.
4. Foroohar R, Liu M. It's China's world; we're just living in it. *Newsweek Magazine* March 22, 2010:36–9.
5. Bayne K, Bayvel ACD, Williams V. Laboratory animal welfare: international issues. pp. 55–76. In: Bayne K, Turner PV, editors. *Laboratory animal welfare*. New York: Elsevier; 2014.
6. Bayne K, Bayvel D, MacArthur Clark J, Demers G, Joubert C, Kurosawa T, et al. Harmonizing veterinary training and qualifications in laboratory animal medicine: a global perspective. *ILAR J* 2011;**52**(3):393–403.
7. National Research Council. *International animal research regulations: impact on neuroscience research: workshop summary*. Washington, DC: The National Academies Press; 2012.
8. Kong Q, Qin C. Laboratory animal science in China: current status and potential for the adoption of Three R alternatives. *ATLA* 2010;**38**:53–69.
9. Ministry of Science and Technology (MOST). Guideline of humane treatment of laboratory animals; 2006. Available at: http://www.lascn.net/policy/ShowClass.asp?ClassID=4.
10. MOST. *Guideline of the Beijing municipality on the review of welfare and ethics of laboratory animals*. Beijing, People's Republic of China: Beijing Municipality Administration Office of Laboratory Animals (BAOLA); 2005. Available at: http://bjxkz.lascn.com/system_manager/news_manager/UploadFile/250/250_3.doc.
11. Kong Q, Qin C. Analysis of current laboratory animal science policies and administration in China. *ILAR e-Journal* 2010;**51**:E1–0. Available at: http://dels-old.nas.edu/ilar_n/ilarjournal/51_1/PDFs/v51(e1)Kong.pdf.
12. Animal Protection Law of the People's Republic of China. Available at: http://www.actasia.org/uploads/file/Feedback%20on%20draft%20APlegislation.pdf.
13. MOST. A Regulation on the Management of Laboratory Animal License System; 2002. Available at: http://www.lascn.net/policy/law/nationlaw/200805/17.html.
14. Beijing Municipal Government, Committee of Science & Technology. Regulations of Beijing Municipality for the Administration of Affairs Concerning Laboratory Animals; 2005. Available at: http://www.baola.org/ContentDetail/zhengceFG.aspx?CategoryId=48&contentId=7.
15. National Bureau of Quality Monitoring, Inspection & Quarantine. National Standards for Laboratory Animals; 2001, 2006, 2008, 2010. Available at: http://www.baola.org/ShowContentList2.aspx?CategoryId=24.
16. Guideline and Detailed Criteria for the Review of Provincial Laboratory Animal Monitoring Centers. Available at: http://www.lascn.net/policy/ShowClass.asp?ClassID=4.
17. Hennig W. Bioethics in China. *EMBO Rep* 2006;**7**(9).
18. Law of the People's Republic of China on Prevention and Control of Infectious Diseases. Available at: http://www.chinacdc.cn/flfg/wsfl/200507/t20050725_41129.htm.
19. Laboratory Bio-safety Management Regulation for Microbiological Pathogens. Available at: http://www.moa.gov.cn/fwllm/zxbs/xzxk/bszl/201104/t20110416_1970325.htm.
20. Classification List of Animal Pathogens. Available at: http://www.moa.gov.cn/fwllm/zxbs/xzxk/bszl/201104/t20110416_1970333.htm.
21. Fire Control Law of the People's Republic of China, Sixth, Article 14, Chapter 2. Available at: http://www.mps.gov.cn/n16/n1282/n3493/n3763/n4198/434859.html.

22. First International Symposium on Cosmetics. *Alternatives to animal experimentation for cosmetics*. Beijing, China: China Cosmetics Research Center (CCRC) of the Beijing Technology and Business University with major support from the Food Licensing Department of the SFDA; 2011.

23. China sFDA. Comment Letter on Alternative Method of Phototoxicity Study through 3T3 Neutral Red. Letter no. 45; 2012.

24. China sFDA. Comment Letter on the Management of Ordinary Cosmetics. Notice no. 57; 2012.

25. China sFDA. Comment Letter on Amended Management of Ordinary Cosmetics. Notice no. 263; 2012.

26. China Ministry of Health. Hygienic Standard for Cosmetics; 2007.

27. China Education and Research Network. Project 211: a brief introduction. Available at: http://www.edu.cn/20010101/21852.shtml.

28. Bayne K. Animal welfare in Asia: the AAALAC international experience. *ALN World* 2011;4(1):11–9.

29. Architectural and technical code for laboratory animal facility (GB 14925-2010). Available at: http://www.lascn.net/policy/ShowClass.asp?ClassID=4.

30. National Research Council. *Guide for the care and use of laboratory animals*. Washington, DC: The National Academies Press; 2011.

31. Davey G. Chinese university students' attitudes toward the ethical treatment and welfare of animals. *J Appl Welf Sci* 2006;9:289–97.

32. Davey G, Wu Z. Attitudes in China toward the use of animals in laboratory research. *ATLA* 2007;35:313–6.

33. Yin JC, Li GX, Ren XF. An overview of veterinary medical education in China: current status, deficiencies, and strategy for improvement. *J Vet Med Educ* 2006;33(2):238–43.

34. James A, Kelly H. East meets west: the reformation of veterinary education in China. *ALN Mag* May 7, 2012. Available at: http://www.alnmag.com/article/east-meets-west-reformation-veterinary-education-china?page=0,4.

35. China Laboratory Primate Breeding and Development Association; 2012. Available at: http://clpa.org.cn/organization.asp.

36. MOST. Guideline for National Resource Centre for Laboratory Animals; 1998. Available at: http://www.la-res.cn/la-res/website/publicinfo/showArticleInfo.jsp?articleId=7021b060-c7de-4a31-8daf-3c7074532678.

37. Ministry of Agriculture (MOA). Statute on Administration of Transgenic Organism Biosafety in Agriculture System; 2001. http://www.agri.gov.cn/xzsp_web/bszn/t20031105_133969.htm.

38. Ministry of Health (MOH). Regulation on Sanitation Administration of Transgenic Food; 2002. Available at: www.fsi.gov.cn/news.view.jsp?id=17596.

39. General Administration of Quality Supervision, Inspection and Quarantine of China (AQSIQ). Regulation on the Entry and Exit Transgenic Product; 2004. Available at: http://www.aqsiq.gov.cn/xxgkml/jlgg/zjl/20032004/200610/t20061027_12236.htm.

40. Klein HJ, Bayne KA. Establishing a culture of care, conscience, and responsibility: addressing the improvement of scientific discovery and animal welfare through science-based performance standards. *ILAR J* 2007;48:3–11.

Chapter 10

Laws, Regulations, Guidelines, and Principles Pertaining to Laboratory Animals in Far East Asia

Tsutomu Miki-Kurosawa[1], Jae Hak Park[2], and Chou-Chu Hong[3]

[1]*The Institute of Experimental Animal Sciences, Osaka University Medical School, Suita-shi, Osaka, Japan,* [2]*Laboratory Animal Medicine, College of Veterinary Medicine, Seoul National University, Seoul, Korea,* [3]*AAALAC International, New Taipei City, Taiwan*

Chapter Outline

Laboratory Animals. http://dx.doi.org/10.1016/B978-0-12-397856-1.00010-6

JAPAN

Background and Current Situation

The importance of laboratory animal science was becoming more recognized in the early 1950s in Japan. One of the oldest national associations in Asia is the Japanese Association of Laboratory Animal Science founded in 1951. Biomedical scientists have contributed to establish the national legislation system in their own countries by the reflection of the public concern of animal welfare. The Japanese government does not formally publish its legislative documents in English so therefore, there is a paucity of Japanese laboratory animal welfare information in other countries. However, the advancement of translation technology through the Internet may help us understand the Japanese legislative framework in this chapter.

The Law of Humane Treatment and Management of Animals (Law No. 105, 1973) was published in 1973.[1] The Law was influenced by western countries, in particular the UK. The Law was revised several times and the most recent revision was made in 2012. The Law already included 3Rs tenets at the revision in 2005. Before the revision in 2005, Refinement only had been stated, but Reduction and Replacement were added to the Law in terms of laboratory animals. Under the Law, the Standards Relating to the Care and Management of Experimental Animals (Notice No. 6 of the Prime Minister's Office 1980) were published.[2] The Standard was revised several times and the new revision is under consideration in 2013. At the revision in 2006, the title of the Standard was changed and the current title is the Standards Relating to the Care, Management, and Relief of Pain of Experimental Animals.[3] The euthanasia of laboratory animals is regulated by the Standards Relating to the Care and Management of Experimental Animals (Notice No. 6 of the Prime Minister's Office 1980) and Guidelines on Methods of Sacrificing Animals (Notice No. 40 of the Prime Minister's Office, July 4, 1995).[4] The use of genetically modified laboratory animals is restricted by the Law Concerning the Conservation and Sustainable Use of Biological Diversity through Regulations on the Use of Living Modified Organisms, which is in accord with the Cartagena treaty.[5] This Law may be the first law to restrict the usage of animals for research in Japan. There are many reported violations of this Law due to the paucity of practical information for the compliance in research settings. The importation of nonhuman primates (NHP) is restricted by the Invasive Alien Species Law and the Law Concerning the Prevention of Infectious Diseases and Medical Care for Patients with Infectious Diseases.[6,7] The institutions using NHP should be registered and the importers of NHP need a license for importation and quarantine. The Japanese government restricts the countries from which NHP can be imported to Japan.

With these Laws and Standards, the Ministry of Education, Culture, Sports, Science and Technology and the Ministry of Health, Labor and Welfare compiled the Fundamental Guidelines for Proper Conduct of Animal Experiment and Related Activities in Academic Research Institutions under the jurisdiction of the Ministry of Education, Culture, Sports, Science and Technology and the Basic Policies for the Conduct of Animal Experimentation in the Ministry of Health, Labor and Welfare. The Ministry of Agriculture and Fisheries also issued the Guideline for the Proper Conduct of Animal Experimentation.[8]

The Science Council of Japan issued the Guidelines for Proper Conduct of Animal Experiments (the Guide)[9] to serve as reference material or a model when research institutions compile their own regulations for animal experimentation in accordance with the above fundamental guidelines and basic policies in 2006.

These guidelines indicate the responsibility of the President of every research institution to carry out animal experimentation. The President of each research institution is asked to establish institutional regulations to control animal experiments and to find the Institutional Animal Care and Use Committee (IACUC). However, the duty of the IACUC is limited to the review of protocol, but does not extend to the oversight of the entire animal care program. The involvement of the veterinarian as a member of the IACUC is not stated. There are no descriptions of veterinary care and the laboratory animal veterinarian's responsibilities to look after laboratory animal's well-being in these guidelines. Generally speaking, so-called self-regulation is emphasized, as the legal restriction of scientific activities is thought to demolish the research, the legal system for laboratory animal welfare is not stringently enforced in Japan.

Scope

The Law of Humane Treatment and Management of Animals covers all aspects of the use and management of animals in Japan.[1] Under the Law, the Standards Relating to the Care and Management of Experimental Animals regulates the use of animals for scientific purposes including research, education, testing, and manufacture of biological products.[3] In the Law and Standard, animals are defined as mammals, birds, and reptiles. Neither fish nor cephalopods are included as laboratory animals.

The Principles

The Law of Humane Treatment and Management of Animals focuses on the use, as well as management of animals, but does not emphasize laboratory animal welfare. The relief of pain of laboratory animals is mandatory as Refinement but Replacement and Reduction are both just suggestions. Laboratory animal welfare and the registration of research institutions using animals are not prescribed.

Ethical Review

The IACUC is a body for the ethical review of the protocol for animal experimentation indicated by the various guidelines. The President of each research institution is responsible for oversight of animal experimentation. The IACUC members are nominated by the President and include researchers conducting animal experiments, laboratory animal specialists, and other persons of knowledge and experience. Ethical review is limited to protocol review, and does not extend to oversight of the entire animal care and use program.

An IACUC should be established in each institution. The President of the institution requests the IACUC to review the animal experiment protocols submitted by principal investigators based on scientific rationale and in consideration of animal welfare. The President of the institution then approves or does not approve the protocol based on the report of the IACUC. After completion of the animal experiment, the President of the institution examines the results obtained and instructs the principal investigator and manager to make improvements based on the advice of the IACUC.

Institutional and Designated Personnel Responsibilities and Qualifications

Following the Guide by the Science Council of Japan, the President of the institution bears the final responsibility for all experiments conducted at the institution.[9] The President of the institution has the responsibility to prepare the facilities considered necessary for proper care and management of the laboratory animals and the proper and safe conduct of the animal experiments, appoint the manager, and appoint a person with knowledge and experience related to laboratory animals as the laboratory animal manager. The President of the institution also has to provide education for related persons including the researchers and animal technicians with the cooperation of the manager and laboratory animal manager to inform them of the related laws and policies. However, the competencies and qualifications of the personnel are not stated.

In each institution, in-house regulations, including the authority and responsibilities of the President of the institution, standard operating procedures (SOP) for the conduct of animal experiments, proper care and management of laboratory animals, and methods of maintenance and management of facilities should be established.

Personnel Training, Education, and Competency Requirement

The Guide of Science Council of Japan states that the President of the institution should endeavor to provide separate education and training of the laboratory animal manager, researcher(s), and animal technicians as required.[9] Education and training should be provided prior to engaging in animal experiments and should also be provided later as required.

Education and training should be conducted in accordance with in-house regulations, and the dates of instruction, educational content, and the names of the instructor and those receiving instruction should be recorded and retained.

Education and training content should be specified in the in-house regulations, taking into account activities undertaken in the institution. From the viewpoint of proper conduct of animal experiments, the following items should be included in education and training:

1. items related to pertinent laws and ordinances, by laws, guidelines, in-house regulations;
2. items related to animal experiments, and the handling of laboratory animals;
3. items related to the care and management of laboratory animals;
4. items related to safety assurance;
5. items related to the use of facilities.

Animal Transportation

Laboratory animal transportation is not regulated by the Law with an exception for genetically modified animals. The transportation of genetically modified animals is regulated by the Law Concerning the Conservation and Sustainable Use of Biological Diversity through Regulations on the Use of Living Modified Organisms.[5] The prevention of the escape of animals is a main concern. Animal welfare concern during transportation of laboratory animals is briefly stated in the standard.

Physical Plant, Housing, and Environmental Conditions

According to the Guide of the Science Council of Japan, there are some conditions to be met.[9] Respecting the opinions of the laboratory animal manager, the manager should provide the conditions necessary to meet the requirements for conduct of the research for animal physiology, ecology and behavior, and sanitary management, while establishing and administrating the facilities. Cleaning and disinfecting floors, inside walls, ceilings, and auxiliary equipment in the facilities should be easy, and the structure should facilitate maintenance of sanitary conditions to eliminate the possibility of laboratory animals being injured by projections, holes, depressions, inclined planes, etc. Inspections and maintenance of cages should be conducted to prevent laboratory animals escaping, to keep laboratory animals physically comfortable, and to make hygiene management and daily operations easy.

The following items should be considered when setting up facilities as per the Guide of the Science Council of Japan:

1. equipment for the care and management of laboratory animals, sanitation equipment for cleaning and disinfecting materials and devices, and experimentation equipment;
2. assurance of structures and reliability to prevent entry of wild animals;
3. assurance of structures and reliability to prevent laboratory animals escaping;
4. in facilities where animal experiments on infection with pathogens or using radioactive materials are conducted, effective equipment to prevent escape of infected animals, laboratory animals administered with chemicals and laboratory animals treated with radioactive materials;
5. facilities with the structure necessary to reduce odor and noise and storage facilities required for waste material;
6. within the limits that do not interfere with achieving the objective of the animal experiment, proper space, temperature, humidity, ventilation, lighting, etc., to assure laboratory animals are not subjected to excessive stress;
7. depending on the animal species and the objective of the experiment, air conditioning equipment necessary to maintain a constant environment in the facilities;
8. in facilities for rearing laboratory animals extremely susceptible to infectious diseases, sanitation and air conditioning equipment necessary to control microorganisms;
9. assurance of structures that enable researcher(s) and animal technicians to work without risk;
10. facilities and equipment such as safety cabinets, draft chambers, and localized air exhaust equipment to prevent work-related accidents as required. Also, education and training of researcher(s) and animal technicians to familiarize them with safe operating procedures;
11. when conducting gas sterilization of rearing equipment such as vinyl isolators, gas masks should be worn as required;

12. for autoclaves and ethylene oxide gas sterilizers, periodic inspections as indicated by laws and ordinances, as well as daily checks before starting work.

Behavioral and Environmental Management Enrichment

In most of the Laws, Standards, and Guidelines, behavioral and environmental management and enrichment were not clearly stated for laboratory animals in Japan.

Occupational Health and Safety

In Japan, we have stringent occupational health and safety laws and regulations for most industrial institutions, including biological hazard. However, the specific hazards in laboratory animal settings are not clearly defined.

Veterinary Care

There are no statements of veterinary care in laboratory animal regulations and guidelines in Japan.

Euthanasia

Euthanasia of laboratory animals is regulated by the Standards Relating to the Care and Management of Experimental Animals and Guidelines on Methods of Sacrificing Animals (Notice No. 40 of the Prime Minister's Office, July 4, 1995).[3,4] The overdose of anesthetics and carbon dioxide inhalation are recommended for most laboratory animals. For rodents, decapitation and cervical dislocation by experienced personnel is also recommended.

SPECIAL CONSIDERATIONS

Nonhuman Primates

The importation of NHP is restricted by the Invasive Alien Species Law and the Law Concerning the Prevention of Infectious Diseases and Medical Care for Patients with Infectious Diseases.[6,7] The institutions using NHP should be registered and the importers of NHP need a license for importation and quarantine. The Japanese government restricts the countries from which NHP can be imported to Japan.

Genetically Altered Animals

The use of genetically modified laboratory animals is restricted by the Law Concerning the Conservation and Sustainable Use of Biological Diversity through

Regulations on the Use of Living Modified Organisms which is in accord with the Cartagena treaty.[5]

Reuse of Animals

There are no restrictions for reuse of laboratory animals in Japan. This should be revised by the IACUC during protocol review.

Annual Reports

An annual report of animal experimentation is recommended by various guidelines including the number of protocols, number of laboratory animals used, and the results of animal experimentations.

Care and Use of Fish

Because fish are not defined as laboratory animals, there are no descriptions of care and use of fish for research in Japan.

SOUTH KOREA

General Framework

In Korea, there are two kinds of acts about animal experimentation. One is the Animal Protection Act (APA) which is managed by the Ministry of Agriculture, Food and Rural Affairs (MAFRA) and the other is the Laboratory Animal Act (LAA) which is managed by the Ministry of Food and Drug Safety (MFDS).[10,11] The Acts require that animal facilities must manage the IACUC. Registration in government is the duty of both laboratory animal breeder and user institutions. Personnel must be certified for training by the government. In addition, the research protocols of animal experiments must be approved by IACUC. Government started to designate excellent animal testing facilities with the introduction of certification schemes to entice the ethical and scientific value of animal experiments. The experimental animal facility creates its own guidelines, based on the APA and LAA, to perform the ethical operation and maintenance of the animal facility. In this section, ethical use of experimental animals, based on APA, LAA, and regulations for the Seoul National University IACUC are described.

The Korean Association for Laboratory Animal Science (KALAS) was established in 1985 in order to discuss the knowledge of the experiment, and promote the science and technology in laboratory animal sciences. From about the mid-1970s, experimental animal sciences were introduced in Korea. Accordingly, researchers and institutions felt the need for exchange and cooperation. This led to the establishment of KALAS on May 1, 1985. Presently, KALAS has about 4800 individual members, 55 affiliates, and 105 organization members (www.kalas.or.kr). To perform basic and applied research to develop a new test method that can replace the use of laboratory animals, and ultimately minimize the number of experimental animals and suffering, the inaugural meeting of the Korean Society for Alternative to Animal Experiments (KSAAE) was held February 23, 2007 (www.ksaae.org). The Korean College of Laboratory Animal Medicine (KCLAM) was established on October 25, 2006 for the purpose of (1) enhancement of humane technique of animal experiments; (2) development of knowledge in the fields of disease of experimental animals, operation, anesthesia, pain relief, animal welfare, and animal protection; (3) training for the development of laboratory animal veterinarians; and (4) establishment of technical standards. Laboratory animal veterinarian certification review shall consist of qualification review and written examination. A candidate must achieve specified standards to be certified (www.kclaam.org). By the LAA, the Korean Association for Laboratory Animals was established in 2009. The main roles are the education of administrators of the laboratory animal facilities, laboratory animal suppliers, laboratory animal handlers, the standardization of experimental animal products, and consultancy on management (www.kafla.kr).

The purpose of the APA is to provide adequate protection and management of animals including experimental animals by preventing their mistreatment and to guide citizens in the care, safety, and respect of animals.[10] To respond to the consultation of the animal welfare policy of the MAFRA, an animal welfare committee must be established in the MAFRA. This committee is different from the IACUC in the animal facilities. The animal welfare committee will carry out the following works: (1) establishment and enforcement of a comprehensive plan on animal welfare; (2) guidance and supervision of animal experiments such as the configuration of the IACUC; (3) establishment of a policy related to animal welfare husbandry and certification; and (4) matters relating to animal welfare and protection such as the rescue of animals and the prevention of abuse of animals.

The animal welfare committee shall consist of 10 people including one chairman. The Minister of MAFRA can impose the role on people who fall under each of the following categories: (1) a veterinarian who has long experience and knowledge of animal welfare and animal protection; (2) a person who has received the recommendation from nongovernmental organizations with relevant knowledge and experience related to animal welfare; (3) a person with expertise in animal welfare, and other people who meet the eligibility criteria specified by the Ordinance of MAFRA.

In the same Act, to prohibit animal abuse, no one shall commit the following actions toward animals: (1) killing by brutal methods, such as hanging; (2) killing animals in public or in the presence of other animals; (3) any other act of killing without valid reason set by the decree from MAFRA, such as veterinary concerns or threat to human life, health, and finance due to animals. Also, no one shall commit the following abusive acts toward animals: (1) harming an animal with instruments or substances (however, this is permitted in certain circumstances if approved by the IACUC and prescribed by decree of MAFRA, such as for pre-venting or curing disease); (2) harming an animal's body, extracting bodily fluids or installing devices to extract bodily fluids while it is alive (however, this is per-mitted in certain circumstances approved by the decree of MAFRA, such as for preventing or curing disease); (3) harming an animal for the purpose of gambling, advertisement, entertainment, or amusement (however, there are exceptions, such as for traditional games, by the decree of MAFRA); (4) Any other act of killing without valid reason set by the decree from MAFRA, such as veterinary care or threat to human life, health, and finance due to animals. No one shall capture and sell or kill any abandoned animal. If one attempts to perform experiments with animals without IACUC approval, he will be charged by a fine of more than 10 million KRW or imprisonment of up to one year. Also, the owner or caretaker of an animal shall not abandon his or her animal. If so, the owner of the aban-doned animals shall be imposed with a fine of 1 million KEW or less.

The MFDS manages the LAA whose purpose is to contribute to the develop-ment of life sciences and the improvement in national health by enhancing the ethics and reliability on animal testing through the appropriate administration of laboratory animals and animal testing.[11] The duties of MFDS for LAA are: (1) formulation and promotion of policies concerning the use and administra-tion of laboratory animals; (2) support for the establishment and operation of animal testing facilities; (3) support for the maintenance, conservation, and development of laboratory animals in animal testing facilities; (4) research sup-port for the improvement in quality of laboratory animals; (5) support for the collection and management of information, and education in connection with laboratory animals; and (6) formulation and promotion of policies concerning the development and approval of methods which can substitute animal testing.

The Korean Center for the Validation of Alternative Methods (KoCVAM; http://www.nifds.go.kr/en/inter/kocvam.jsp) was established based on this sentence in November 2009 within the National Institute of Food and Drug Safety Evaluation (NIFDS). KoCVAM seeks alternative methods through various activities to build cooperative relationships with both domestic and foreign organizations and to review and validate proposed alternatives. KoCVAM also intends to keenly respond to these global trends by globally promoting alternative test methods developed by Korean organizations.

The Principles

In the APA, the principles of animal protection are: (1) animals must be main-tained with the original prototype of habit and body and normal lifestyle; (2) free-dom from hunger and thirst by ready access to fresh water and a nutritional diet; (3) freedom to express normal behavior and not to experience inconvenience; (4) freedom from pain, injury, and disease; and (5) freedom from fear and distress.[10] These principles are absolutely against animal experiments. In the same Act, the principles of animal experiments different from the principles of animal protection were described as follows: (1) animal experimentation must be performed only after considering the promotion of human welfare and the dignity of animal lives; (2) when considering the use of animals in experiments, replacement must be considered as preferential; (3) experiments with animals must be performed by someone who possesses knowledge and experience in the ethical handling and scientific usage of the experimental animals, and only the minimum of the required number of animals should be used; (4) animals with low sensitivity should be used for experiments necessarily involving pain, and appropriate measures must be taken to lessen the pain, using veterinary methods that employ analgesics, sedatives, and anesthesia; (5) a person who has conducted experiments with animals shall inspect the subject animals immediately after the conclusion of experiments and if the animals will suffer lasting pain or injury, then such animals shall be euthanized as quickly as possible in the most painless way.

Scope/Applicability

The term "animal testing" used in APA and LAA means testing conducted on laboratory animals for scientific purposes, such as education, testing, research, and production of biological medicines.[10,11]

In APA, animal testing facilities are defined as facilities set in the national institutions and local governments; government-funded research institutes; institutions for safety and efficacy test of drugs, cosmetics, and medical instruments; universities; agricultural and fishery cooperatives; institutions testing potentially hazardous chemicals and pesticides; and feed, food, or food additives, health functional food manufacturing institutions.

While APA covers all the animal testing facilities, LAA limits the coverage to the animal testing facilities related with development, and safety/effective tests of foods, functional health foods, medical and pharmaceutical products, nonmedical and pharmaceutical products, biomedicines, medical appliances, cosmetics, and narcotics. So, the animal facilities defined by LAA are those that are related with manufacturing, importing or selling food, health supplements, drugs or quasidrugs, cosmetics, narcotics, medical devices, or that are delegated from MFDS to do the same jobs. The management systems of the animal facilities have been defined by LAA. The manager can deal either with animal testing facilities or laboratory animals. A manager of animal testing facilities shall be responsible for the formulation of guidelines on the scientific use and administration of laboratory animals, the education of those who carry out and engage in animal testing, the preferential consideration of methods that can substitute animal testing, the formulation of plans concerning the appropriate disposal of waste matter of animal testing, and the safety of workers. In this scheme, laboratory animal management is the responsibility of the IACUC which should be set within animal facilities. IACUC shall be appointed by the manager of the animal facilities, while IACUC members based on APA shall be appointed by the CEO, not by the manager or Institutional Official in an institution.

The term "laboratory animal" in LAA was defined as any vertebrate used or raised for the purpose of animal testing. APA defines animals as cattle, horse, swine, dog, cat, rabbit, chicken, duck, goat, sheep, deer, fox, mink, and other species as designated by APA that includes mammalian, birds, fish, and reptiles. Therefore, all kinds of vertebrates are regarded as laboratory animals. However, APA prohibits animal experiments on abandoned animals and animals that have served for humans, such as seeing eye dogs or guide dogs, rescue dogs, police dogs, military dogs, and drug and explosive detection dogs, except cases when it is necessary to conduct research on the species' health or disease. A person who performs animal experimentation with abandoned or service animals shall be punished by a fine of not more than 500,000 KRW. The exceptional cases are the diagnosis and treatment or research on zoonotic diseases or pest control from these animals. These experiments should be done after IACUC approval.

Authorization of User-Breeding Institutions/Inspection/Penalties by Competent Authorities and Institutional and Designated Personnel Responsibilities

Any person who intends to establish animal testing facilities by LAA shall register it with the MFDS.[11] The manager, who should be qualified as a person with a university degree and who has worked in animal facilities for over three years shall be assigned to animal testing facilities to manage the relevant facilities and laboratory animals. The MFDS needs documents to prove the qualifications of the manager and the status of the structure, the layout of the facility, and the area. The detailed requirements are that animal rooms must be separated by type of animals and from laboratory or waste storage room; floor and wall finishes must be constructed with materials convenient to clean and disinfect; the temperature and humidity must be able to be adjusted; and laboratory, surgery or autopsy rooms, waste storage rooms, and animal carcasses rooms should be fully equipped. Also a standard operations practice for disinfection and animal husbandry shall be prepared.

When a researcher uses laboratory animals in the animal testing facilities, LAA asks the researcher preferentially to use laboratory animals produced in animal testing facilities or excellent laboratory animal production facilities. When a researcher intends to use laboratory animals imported from abroad, they shall use laboratory animals of which qualifications have been certified by the government or institutions.

The MFDS may, for the appropriate use and administration of laboratory animals, designate animal testing facilities which have appropriate human resources and facilities, and the administration condition of which are excellent, as the excellent animal testing facilities. Any person who intends his/her animal testing facilities to be designated as excellent animal testing facilities shall apply for designation. Designation criteria of excellent animal testing facility include having one veterinarian manager and an animal expert with over three years' experience. Facilities shall meet the standards for the following items: offices, animal rooms, laboratories, quarantine rooms, operating rooms, necropsy rooms, cleaning rooms, warehouses, shower, and waste storage rooms. Each breeding room must be equipped with disinfection instruments and systems for adjusting the room environment such as temperature, humidity, ventilation, and mechanical devices or equipment to prevent the influx of pollutants from the outside, and a separating wall preventing cross-contamination. Animal rooms should be equipped with a protective device to prevent contamination from the lighting system, and easy-to-clean floors and walls. The MFDS may advise the relevant business manager who uses laboratory animals or a person who carries out research services to conduct such business in excellent animal testing facilities. Any person who has registered facilities as animal testing facilities or excellent animal testing facilities shall be guided and supervised by the MFDS.

Any person who intends to engage in the business of production, importation, or sale of laboratory animals shall register to the MFDS. However, this shall not apply to cases where he/she supplies laboratory animals produced in the process of maintenance or research in animal testing facilities. A supplier of laboratory animals shall manage laboratory animal production facilities and laboratory animals so that no harm may be caused to health and hygiene and the safety may be secured; when he/she transports laboratory animals, he/she shall do so by a method suitable to the ecology of such laboratory animals.

The MFDS may, for the improvement of quality of laboratory animals, designate laboratory animal production facilities which have sufficient human resources and facilities, and the administration conditions of which are excellent, as excellent laboratory animal production facilities. The requirements for being an excellent laboratory animal production facility are almost the same as those criteria for an excellent animal testing facility. Any person who has registered as a supplier of laboratory animals or has been designated as an excellent laboratory animal production facility shall be guided and supervised by the MFDS.

A person who has failed to register animal testing facilities or a person who has attached a mark of excellent laboratory animal production facilities or a mark similar or has publicized it shall be punished by a fine for negligence not exceeding 1 million KRW.

Nonhuman Primates

In South Korea, there are several nonhuman primate research facilities, such as the National Primate Research Center of KRIBB in Ochang (http://www.primate.re.kr/) and KIT of KRICT in Daejeon (http://www.kitox.re.kr/). There is no national guideline for the care and use of laboratory nonhuman primates in Korea. Seoul National University hospital has prepared the guideline for laboratory nonhuman primates. The guideline includes purposes and definition, laboratory primate husbandry, veterinary care, behavioral and psychological management of primates, animal care and health, planning and execution of animal experimentation. Also, the enrichment programs including social behavior and positive reinforcement training were considered in the facility.

In addition, Seoul National University enforce researchers who want to manage the primates to attend education on primate experiments with regard to animal protection and understanding of animal protection policies, animal rights theory and international trends, primate ethical production, management, use, and postprocessing, the ethical and scientific experiments using primates, evaluation of planning primate experiments, and primate laboratory safety guidelines.

Genetically Altered Animals: Special Considerations

A person who wants to develop or import genetically modified organisms must report or get permission from the head of the relevant central administrative, according to the act of genetically modified organisms.

Oversight and Ethical Review Process

By APP, an IACUC must be established in animal facilities in order to oversee the protection and ethical treatment of experimental animals.[10] Heads of institutions which have animal facilities must subject research proposals to the IACUC. When Heads of institutions that have animal facilities carry out animal experiments without going through the deliberations of the IACUC, they can be imposed with a fine of 1 million KRW or less.

IACUC is to perform the following functions: (1) deliberations on the proposals about animal experiments; (2) the Committee will direct animal experimentation so that it will be carried out in accordance with the principles of animal experiments; (3) the Committee can request that the Heads of institutions that have animal facilities take necessary measures for the protection and ethical treatment of experimental animals.

A committee member who is taking part in the experiments may not take part in that review. A member of the committee must not disclose or fraudulently use the information learned during his duty. Members of IACUC who steal or leak the secrets of proposals shall be punished by a fine of not more than 5 millions KRW.

The committee will be comprised of three to 15 members including one chairman of the committee. Members of the committee are to be appointed as specified by the President of the institution having the animal facility, and the chairman is to be elected from among the members. It is requisite that one member of the IACUC is a veterinarian who is specialized in laboratory animals (KCLAM) or who has worked for at least one year in a laboratory animal facilities, or who has completed the training in accordance with the provisions of the ordinance of MAFRA. Another requisite member of IACUC is a person who is very learned and experienced in animal protection, recommended by a nongovernment organization selected by presidential decree, and who meets the following criteria: (1) has been engaged in business for more than one year in relation with the animal protection and animal welfare; (2) has undertaken education on animal protection, animal welfare, and animal experiments carried out at school in accordance with the "Higher Education Act" or at corporations and organizations; (3) has more than one year experience on an institutional research board (IRB) committee; (4) has undertaken education on animal welfare and animal experiments, animal protection at the Animal and Plant Quarantine Agency (http://www.qia.go.kr/listindexWebAction).

At least one veterinarian and one person who is very learned and experienced in animal protection and recommended from a nongovernment organization must be included in the committee. At least one-third of the members must not be employed by the research institution or be affiliated with it. The term of membership of a committee is two years. Heads of the institution having experimental animal facilities, which do not establish the IACUC, shall be imposed with a fine of 1 million KRW or less.

LAA asks the animal facilities to set up a laboratory animal management committee, which shall be established and managed within the animal unit in order to review the proposals and secure the ethics, safety, and reliability of animal testing.[11] The member organization of the laboratory animal management committee is almost same as the IACUC defined in APA.

As an example of the implementation of applicable regulations, this is the organization, role, and protocol review procedure of Seoul National University IACUC:

1. Organization of the IACUC and its Authority

   The President of the University established the IACUC to enable the regulation of the Seoul National University, to give proper advice and supervision to researchers and provide administrative support for the IACUC to let it work appropriately. The IACUC is comprised of 15 members without a separate committee for each college that the central committee supervises.

   The IACUC is responsible for the supervision and evaluation of the program for the management and use of animals. Its functions include: inspection of facilities; evaluation of the program and animals' sphere of activity; submission of reports from each facility supervisor; examination on the use of suggested animals in research experiments and education (protocol); and establishment of a system to know the interest of each facility in the management and use of animals.

   The IACUC has meetings as frequently as possible to perform its duties and a meeting is held at least once in a month. Minutes of the IACUC meetings and results from meetings must be kept for five years. It must examine the animal management program at least once every six months and inspect animal facilities and their sphere of activity. A report upon such examinations and inspections with majority approval of the IACUC must include conditions of the program for the management and use of animals and other activities, and must be submitted to the president.

The IACUC is comprised of nonscientific and nonaffiliated members as well as scientific/affiliated members, to ensure a balance of professional expertise and community perspective. All IACUC members must participate in the review process. A quorum for each meeting must be equal to or greater than one half of regular voters where at least one nonaffiliated member must be present. The IACUC members must perform a follow-up inspection at least once a year in addition to the regular check-ups where they meet with researchers and staff of animal facilities to verify accordance between the use of animals and protocol and drugs and materials used in protocols and animal records.

2. The IACUC's Confidentiality

The IACUC members must not use information from their committee activities for their personal benefit or release such information to any other committee. Information requiring confidentiality can be related to an agreement of researchers, research process and progress, statistical information that requires the security of research facilities, researchers' source of income, profit and loss, and sales. The above information requires confidentiality. If anyone breaks this rule, (a) he/she will be removed from the IACUC; and (b) will be charged with a fine of 5 million KRW.

3. Submission of the Protocol

For an animal experiment using vertebrate animals, a protocol form must be completed and filed before its beginning phase and it must be approved by the IACUC regardless of types of animals and the experimental purpose.

4. Protocol Review

The review and approval process of a protocol begins with the completion of the form. Consultation with each facility's veterinarian in charge is a prerequisite prior to the submission of protocol. The review by the IACUC is progressed in three different forms, which are full review, expedited, and executive reviews.

A review is carried out with the following steps.

Step 1: Submission of the Protocol. Investigators submit completed protocols to administrative staff at the IACUC office online. Step 2: Initial Review. Upon the submission of the protocol, a type of review will be determined upon predicted pain and anxiety of animals and protocol methods. Modification of the protocol may be examined here. Step 3: Review. The executive administrator of IACUC considers if protocols require full IACUC review, expedited reviews, or executive reviews. Possible outcomes of the IACUC's review include approval, approval pending modification and/or clarification, and rejection.

a. Full Review—All IACUC members will review the protocols and vote.

b. Expedited Review—Two members from the IACUC are assigned. The primary reviewer is always a scientific representative of the IACUC and is responsible for presenting a summary of the protocol at the meeting along with any concerns or points requiring clarification or modification. The secondary reviewer adds any additional concerns. The protocol is then open for discussion by the IACUC. If no further comments are received and there is no request for a full IACUC review, a letter requesting clarifications and/or modifications is sent to the investigator or if a recommendation to approve outright has been made, a Verification of Approval is sent.

   c. Executive Review—If the purpose, hypothesis, and laboratory animals used in the experiment are the same as the previously approved protocol—in other words, if it was a review of a continuous project, the IACUC chair alone reviews and approves protocols falling in this category. This category includes continuing reviews, previously approved protocols that have been resubmitted or identical protocols submitted to different funding agencies, or protocols with no direct animal use, e.g. the use of shared animal products or slaughterhouse materials.

Step 4: Investigator notification. If a protocol was approved by IACUC voting, the investigator is provided with a Verification of Approval and letter certifying the approval. In cases where the IACUC requires clarification or modification, the investigator is notified by the IACUC chair in a letter. In such cases, the approval is issued following receipt of an acceptable response from the investigator. In cases of deferred or disapproved protocol, the investigator is notified by the chairman and advised as to the available options.

5. Criteria for Review

   Protocols may contain potential procedures that can potentially cause severe or uncontrollable pain or distress. Such procedures may include physical restraint, multiple surgeries, limits to the access to food and water, the use of adjuvants, termination upon the completion of protocols, use of hazardous stimuli, stimulation on skin and the cornea, induction of excessive sized tumor, blood-collecting from the heart or suborbital sinus, and housing in an abnormal environment. General information on research methods and purposes must be searched from texts, other scholars or references that inform of influences of such studies on animals. If an experiment is performed via an unknown method, a qualified preliminary examination is advised to be carried out under the supervision of an IACUC member to estimate possible impact and outcomes on animals. The following queries will be provided to the investigators who want to submit the protocols.

   Activities—All activities involving animals must be in accordance with this regulation and policy for the welfare of animals.

   Pain/Distress—Investigators must describe how to avoid/minimize discomfort/distress/pain. If pain/distress is caused, appropriate sedation, analgesia, or anesthesia will be used. Use of paralytics is prohibited. Animals with chronic/severe unrelievable pain will be painlessly killed.

Alternatives—The protocol investigator considers alternatives to procedures that may cause more than momentary or slight pain or distress to the animal and has provided a written narrative description of the methods and sources. These alternatives include methods that interfere less with animals, selecting another species, cultivating cell or organs and the use of computer simulation.

Rationale and Methods—All proposals must include: identification of the species and the approximate number of animals to be used; a rationale for involving animals and for the appropriateness of the species; and numbers of animals to be used.

A description of procedures designed to assure that discomfort and pain to animals will be limited to that which is unavoidable for the conduct of scientifically valuable research, including provision for the use of analgesic, anesthetic, and tranquilizing drugs appropriate to minimize discomfort and pain to animals; and a description of any euthanasia method to be used.

Duplication—Investigators must provide assurance that activities do not unnecessarily duplicate previous efforts.

Surgery—Investigators must meet requirements for sterile surgery and pre/postoperative care. They cannot use one animal for several major operative procedures from which it will recover, without meeting specific conditions.

Euthanasia—Euthanasia method must be consistent with the policy that is performed in a humane way.

Housing/Health—Animal living conditions must be consistent with standards of housing, feeding and care directed by veterinarian or scientist with appropriate expertise. Medical care must be provided by a qualified veterinarian.

Qualifications—Personnel must be appropriately trained and qualified.

Safety of the working environment for the personnel—In case of protocols dealing with hazardous materials, safe environment, and safety standards for the personnel must be provided.

Continuing Review Process—Upon the recommendation of the reviewer, the protocol is approved outright for a period of time up to three years, and the Committee reviews it at least once a year.

Three-Year Renewal—At the end of the third year of a protocol, the investigator must resubmit it to IACUC in order to continue research activities. A new animal use protocol form must be submitted; this form undergoes the same review process as any new protocol.

Modifications or amendments to approved protocols—Modifications of approved protocols must be documented appropriately, reviewed, and approved. The method for obtaining approval for a modification or amendment is similar to that for a complete protocol. A letter requesting the modification including an explanation of the rationale for the change, any amended animal use and protocol pages resulting from this change should be submitted to the IACUC. The chair, in consultation with the veterinarian of each facility if necessary, determines if the modification is "minor" or "significant." Minor modifications may be approved administratively by the chairman and the veterinarian of each facility without full review. A major modification may entail a large change in numbers of animals being used or requested, an increase in invasiveness, a change in species, an increase in pain or discomfort, or a change in the method of euthanasia. Major modifications require review by the full committee members. A written description of the significant changes in the protocol is provided to the IACUC.

Termination of protocols—It is the responsibility of the investigator to notify the IACUC when a project is completed. Projects that have been completed, withdrawn or terminated are closed immediately upon notification. The administrative office of the IACUC is notified by the chairman. All closed projects are kept in the administrative office of the IACUC for a five-year period from date of closure.

Reuse

The decision may be different depending on the IACUC. There is no legal provision for animal reuse. Generally it is assumed it is not a good idea to cause pain to animals more than twice. In addition reuse is rare because most of the experiments will be ended with the pathological examination of animals after euthanasia.

Setting Free/Rehoming

If pain is not involved after the experiment is finished, or infectious pathogens to humans and other animals are not found, rehoming of the companion animals that were used in experiments is encouraged. For example, if the animals were only blood sampled during nutritional studies, such animals can live a normal life through adoption.

Occupational Health and Safety

When a manager or administrator of animal testing facilities conducts animal testing using substances or pathogens which may cause harm, he/she shall take the necessary precautionary measures to ensure the prevention of any harm to humans and animals. When a disaster caused by animal testing facilities and laboratory animal production facilities is deemed detrimental to the public, the manager or administrator shall take the necessary measures immediately, such as closure, or disinfection, and then report the results to the government.

When the responsible manager or administrator of the animal testing facility intends to use any biological harmful substances prescribed by Ordinance of the Ministry of Health and Welfare for animal testing, he/she shall report it in advance to the government. A manager or administrator of animal testing facilities or a supplier of laboratory animals shall dispose of the carcass of the laboratory animals from the relevant facilities lest it should be taken outside and reused, or a disaster occurs. Waste matter, such as carcasses of laboratory animals from animal testing facilities and laboratory animal production facilities shall be disposed of pursuant to the Wastes Control Act.

Education, Training, and Competence of Personnel

The government has developed regular training for IACUC members. The IACUC members recommended from nonprofit organizations should participate in the training offered by MAFRA and obtain the certification. In addition, the following persons should participate in educational programs offered by the MFDS: (1) a person who has established animal testing facilities; (2) an administrator; (3) a supplier of laboratory animals; (4) a person who conducts animal testing. Persons who want to conduct research in each animal experimental facility must attend the training for the management and ethical handling of animals offered by each institution before they perform the experiments.

Transport

According to APA, a person using a vehicle to transport all kind of animals must comply with the following: to supply feed and water suitable for animals; avoid shock and injury caused by activities such as abrupt acceleration or braking; animal transport vehicles will be designed as a structure that can minimize the pain caused by sudden temperature changes and dyspnea; young or pregnant or sick animals or suckling animals should be kept out of contact with other animals; the animal shall be carefully treated so as not to impact and injure when animals are shipped or unloaded; do not use an electric tool for the transportation.[10]

On the other hand, LAA described that suppliers of laboratory animals must use the transportation vehicles which maintain adequate environmental conditions such as temperature, humidity, ventilation, and space dependent on the species, and secure the health and safety of laboratory animals from cross-infection between animals.[11]

Husbandry, Housing, and Environment

Morphological, physiological and biochemical characteristics of experimental animals, and microbial and drugs susceptibility, may be affected by genetic factors and environmental factors. Therefore, the regulation strictly sets the standards for genetic factors and environmental factors in experimental animals. The APA and LAA SOP related to environmental factors and animal breeding and management and veterinary care was made by referring to the Guide for the Care and Use of Laboratory Animals (NRC), the 1996 US National Institute of Health Office of Laboratory Animal Welfare (OLAW), and AAALAC International standards.[12-14] The SOP describes the appropriate environmental factors such as temperature, humidity, air flow, wind speed, etc.;

physico-chemical factors such as ventilation, dust, odor, noise, illumination, etc.; enclosures such as residential buildings, cages, floors, feeders, drinkers, etc.; and nutrition, food, water, etc.; biological factors such as homogenous biological factors of animals such as social rank, competition, stocking density; heterogenous biological factors such as microorganisms, people, and other species, animals, etc.

APA and LAA SOP asks for the self evaluation of management programs. The operation of the facility, if possible, should be evaluated once every six months. This is to periodically check the matters relating to the welfare of animals in facilities to ensure compliance with the provisions of the SOP. OLAW inspection forms are suggested as standard inspection forms. The results must be submitted to MAFRA and MFDS, whose delegates regularly inspect the animal facilities.

Veterinary Care

Veterinary medical care is an essential part of an animal care program to provide animals with preventive medicine and treatment, humane and ethical management, and to perform effective animal experiments. It is especially important to take the animals into consideration and not cause unnecessary pain to them in terms of animal care and well-being. It is highly desirable to have an appropriate veterinary manager under the supervision of an attending veterinarian with wide experience in laboratory animal science. In addition, one must keep in mind that medical examination and treatment on dogs, cats, and other animals must comply with the Veterinary License Act.

APA and LAA SOP states that directors of animal facilities should ask veterinarians to practice the proper management of all the animals.[10,11] The veterinarian has the following rights and responsibilities: professional education and training of other veterinarians and researchers on experimental conduct; the authority for the responsible veterinary care of the health and welfare of the animals in facilities; the purchase and import of animals including the establishment of isolation and quarantine procedures; the animal health management program; cooperating with scientific techniques and research progress; ensuring and verification of the safety for the experimental environment (occupational health and safety responsibilities); gathering the latest information related to laboratory animals' medicine and welfare. The SOP describe the important fields managed by the veterinarian such as decisions on the use of anesthesia and sedatives and euthanasia, choice of appropriate animal species for the study, and preoperative, operative, and postoperative management.

Euthanasia

Generally, most institutions follow the American Veterinary Medical Association (AVMA) guidelines.[15]

TAIWAN

General Framework

The APA was approved by the Legislative Yuan and promulgated by the President of Republic of China in 1998.[16] The Council of Agriculture (COA) of the Executive Yuan is the authority of the Government that supervises the implementation of the Act. The APA consists of seven chapters and 40 articles. The Act defines the "animal" as any vertebrate that is fed or kept by people, including economic animals, laboratory animals, pet animals, and other kinds of animals. "Laboratory animal" is defined as an animal that is fed or kept for the purpose of scientific application. According to the Article 16, Chapter III of the Act, institutions that perform animal experimentation shall organize an IACUC to oversee the animal care and use program for animal experiments. In 2012, there were 208 organized IACUC in Taiwan.

The COA conducts site visits for institutions with animals used in research, teaching, and testing since 2003. Forty institutions are selected each year for the assessment by a multidisciplinary team consisting of laboratory animal science professionals, animal protection inspectors and animal welfare interested group.[17] The COA also published the book *A Guidebook for the Care and Use of Laboratory Animals* as the reference for the institutions.[18] Reports with mandatory and/or suggestions for improvement will be forwarded to the institutions following the visits. Any institution in violation of the Act will be subjected to the fine according to the Act.

A Guidebook for the Care and Use of Laboratory Animals has 25 chapters including an introduction, animal care and use programs, animal husbandry, veterinary care, physical plans, animal breeding, occupational health and safety programs, disaster plan and emergency, and special chapters for pigs, fish and amphibians, dogs, cats, and nonhuman primates.

The annual report of the institution usage of laboratory animals in the animal use and care program should be submitted to the COA for review. The report must include the types and sizes of animal facilities, the numbers of animals and species used in the programs, the sources of animal procurement, the euthanasia methods, and animal numbers subjected to it and the methods and quantity of carcass disposal.

In 2012, there were 208 institutions which had organized the IACUC and were registered in the COA in Taiwan, of which 55% were research institutions and universities/colleges, and 45% were hospitals, contract research organizations (CRO), and various types of pharmaceutical companies and nonprofit medical research institutions.

Annually, the IACUC held a total of 423 meetings and reviewed 5178 animal study protocols for animal experimentations. The IACUC approved 3994 animal study protocols without revision (77.13%), 1137 with revision (21.96%), and rejected 47 of them (0.91%). Among the IACUC activities, the most typical were semiannual inspections of animal facilities, followed by consultation on animal breeding, experimentation, and care, while termination and/or suspension of animal experimentation was seldom executed. Further analyses of the annual reports show that the numbers of IACUC members were moderately correlated with the amount of work load of the IACUC and frequency of the IACUC meeting, but not with the numbers of amendments to animal use protocols.

The sizes of the animal facilities varied widely among institutions: 55 institutions (27.5%) had a facility size less than 20 square meters, 59 institutions (29.5%) were between 20 and 65 square meters, 25 institutions (12.5%) had 170–350 square meters, 17 institutions (8.5%) had 1700–3300 square meters, and 12 institutions (6%) were larger than 3500 square meters. The numbers of laboratory animals used among institutions also differed considerably from less than 50 to more than 20,000 annually. Forty-nine institutions (34.5%) used less than 50 animals, 14 institutions (7%) used 51–100 animals, 43 institutions (21.5%) used 101–500 animals, 38 institutions (19%) used 1001–5000 animals, 15 institutions (7.5%) used 5001–10,000 animals, 10 institutions (5%) used 10,001–20,000 animals, and 10 institutions (5%) used more than 20,000 animals. Regarding the number of animal species used, 32 institutions (16%) did not use any animals at all this year (2011), 84 institutions (42%) used one species only, 44 institutions (22%) used two species, 17 institutions (8.5%) used three species, and 14 institutions (7%) used five to 10 species. The numbers of laboratory animals used in different types of institutions in 2011 indicated that 9.1% of animals were by CRO and pharmaceutical institutions, 42% by research institutions, 38% by universities and colleges, and less than 2.2% by other categories.

Seventy percent of the animals used in 2011 were produced by animal breeding centers or commercial breeders domestically, 26% from institutional in-house breeding, 1.8% from local market (mainly fish and amphibians), and less than 2.2% from importation or wildlife. A total of 1,526,725 animals were used for scientific application. Rodents accounted for 79%, fishes 14%, and farm animals 7%. In term of animal species, laboratory mice accounted for 53%, laboratory rats 25%, chicken/embryo 2.4%, and rabbits 2.0%. The fish used in the research were *Rachycentron canadum, Tilapia nilotica, Epinephelus fuscogusttatus, Danio rerio, Anguilla japonica, Cyprinus carpio,* zebra fish, and others. The methods of euthanasia were included: 78% of rodents were euthanized by carbon dioxide exposure, while 22% by the combination of carbon dioxide exposure and physical methods. Rabbits, mini-pigs, guinea pigs, and dogs were euthanized by injection of pentobarbital solution.

The carcasses of rodents, mini-pig, dogs, amphibian, and others were mainly disposed by incineration through licensed commercial medical waste companies, while some were by in-house incineration.

A brief retrospective examination of the previous years reveals that there was constant decline in the number of laboratory animals used in research, teaching and testing, suggesting a substantial change of laboratory animal practice.

Thus, there is a better understanding of the 3R principles and trend of laboratory animal use in Taiwan.

The Principles

In its general provision, the APA (Act) states it was enacted to respect the lives of animals and protect them.[16] Animal protection shall be implemented in advance with regulations provided in this Act. Like most national systems of oversight of scientific use of animals, the Act incorporates the principles of the "Three Rs" in the policy.[19] The Act also states: the scientific application means the application for the purpose of teaching, experiment, the manufacturing of biological products, laboratory merchandise, pharmaceutics, or organ transplant.

The Act states that the competent authority at central government level shall select experts, scholars, competent agencies and registered civic animal protection organization representatives to periodically formulate animal protection policy and review the implementation of this Act. The number of experts, scholars, and registered civic animal protection organization representatives, who do not have the capacity of governmental representatives, shall not be less than two thirds of the total number. In Articles 6, 7, and 8 of the Act, it is mandated that no one shall be allowed to harass, maltreat, or hurt an animal; an owner shall prevent his animal from infringement of the life, body, freedom, or property of others; and the competent authority at the central government level shall post a public notice about the animals that are prohibited to be raised, exported, or imported.

The Act also states that owners must provide necessary medical treatment to the animals that are injured or sick. The medical treatment or surgery of animals, based on the need for the health or management of animals, shall be operated by veterinarians. The Act also sets requirements for meeting animals' physical and psychological requirements.

The number of live animals used in scientific application shall be avoided or reduced to a possible minimum when usage is absolutely necessary. The application shall be done in a way that inflicts the least amount of pain or hurt on the animals.

Scope/Applicability

The Act implements the policies to all institutions using animals for scientific purposes.[16] This includes academic institutions, government research institutions, and private pharmaceutical, biomedical companies. The institutions that perform the scientific application of animals shall organize an IACUC to supervise the scientific application of animal experiments. By the end of each calendar year, the IACUC must submit the reports of the animal care and use programs, the usage of animals and semiannual facility inspection reports to the COA for the compliance of the Law.

As the animal study protocol review is the responsibility of the IACUC, the COA provides guidelines on the review process, includes number of animals, pain and distress, anesthesia and analgesics, euthanasia, and disposal of refuse. The COA also provide two workshops annually related to the operation and management of IACUC to new animal users, IACUC members and/or Institutional Officials (IO) of institutions in Taiwan. The content of the course "Laboratory Animal Humane Management" includes the following topics: (1) Animal Protection Act and laboratory animal related regulations; (2) laboratory animal science application and auditing consultation; (3) IACUC roles and functions; (4) laboratory animal use - 3Rs; (5) laboratory animal facility planning and management; (6) small laboratory animal care and transportation; (7) laboratory rodent diseases; (8) laboratory animal occupational health and safety; (9) laboratory animal pain assessment and anesthesia; and (10) IACUC practical questions and discussion.

The Act declares that schools below the level of senior high school shall not teach any lesson that may cause the injury or death of any animal which is not included in the approved curriculum by the competent authority in charge of educational administration.

Authorization of Institutions/Inspection/Penalties by Competent Authorities

Any institutions which uses animals for scientific research must register in COA in Taiwan or will be fined according to the Act.[16] An active IACUC is required, whose composition, authority, responsibilities, and functioning are defined in a written term of reference based on a supplemental document issued by the COA. The COA inspection teams visit at least 40 institutions of registered organizations annually, the inspection team will review the IACUC function and the animal care and use programs, to check if: (1) the overall structure of the animal care and use program is sound, effective and well supported; (2) the IACUC remains active and functional, meeting at least twice a year, visiting the animal facilities at least once a year and fulfilling all of the responsibilities described in the COA supplemental document; (3) the veterinary and animal care services continue to meet institution needs and *Guidebook* standards; (4) the training and continued education program is active and provided for the personnel; and (5) the animal facilities meet the institutional needs and *Guidebook* standard.

All the registered institutions are required to submit animal use data to the COA on an annual basis by the end of March of each year. The report includes the types and sizes of animal facilities, the numbers of animals and species used in the programs, the sources of animal procurement, the euthanasia methods and animal numbers for euthanasia, and the methods and quantity of carcass disposal. This information permits the COA to publish reasonably comprehensive data on an annual basis. Annual statistics and some analysis of data are published by

COA and available for the institution and related agencies. The annual reports includes; (1) all vertebrates including fish used for research, teaching, and testing; (2) animals held, if assigned to a protocol; (3) the termination of the animals used in the research and experiments; (4) the usage of animals in the scientific purpose, i.e. teaching, research/development, biomedical, agricultural, pharmaceutical, and CRO services; and (5) the distribution of animal usage in the research, teaching, and testing services. The information of the statistical analysis can be found on the website (http://www.coa.gov.tw).

Oversight and Ethical Review Process

The supplemental document issued by the COA states that the structure and resources of the IACUC programs, the composition, functioning and effectiveness of the IACUC, and the appropriateness of animal care and use practice, procedure and facilities shall be implemented.[18] The principal responsibilities and functions of the IACUC under the "Act" include the coordination and review of the animal study protocols and animal care and use programs, such as (1) the activities and procedures related to the care of animals; (2) the standards of care and facilities for animals; (3) the training and qualification of personnel that are engaged in the care of animals; and (4) procedures for the prevention of unnecessary pain including the use of anesthetics, analgesics and euthanasia.

Education, Training, and Competence Personnel

In Taiwan, all personnel from institutions involved with the use of animals in scientific research should be competent and adequately trained in the principles of ethical care and use of animals. Animal users (including principal investigators, researchers, graduate students, postdoctoral fellows, research staff, and study directors, etc.), animal health professionals, IACUC members, and IO have different education and training needs. The institution should formulate programs which could provide the suitable training and education programs for them. The Chinese-Taipei Society for Laboratory Animal Sciences (CSLAS) provides three different levels of training programs for researchers, animal technologists, and animal technicians annually. The training programs are funded by the COA and executed by the CSLAS.

Responsibilities of Laboratory Animal Veterinarian

The main responsibilities of attending veterinarians in the laboratory animal facilities have been outlined by the COA.[18] Duties include: (1) supervision of disease detection and surveillance, prevention, diagnosis, and treatment; (2) supervision of handling and restraint, anesthetics, analgesics, and tranquilizer drugs; and methods of euthanasia; (3) supervision of surgical and post surgical care; (4) supervision of the promotion and monitoring of animals' physical and psychological well-being; (5) oversight of the adequacy of the husbandry program; (6) involvement in the review and approval of all animal care and use, e.g. via a role on the IACUC; (7) training of institutional staff in the care and use of laboratory animals; (8) assistance in establishment and/or monitoring of the occupational health and safety program; (9) monitoring for zoonotic diseases; (10) advise on and monitor biohazard control policies and procedures relevant to the animal care and use program; and (11) monitoring experimental animal care and usage through Postapproval Monitoring (PAM) and daily runs. However, the duties and responsibilities of the laboratory animal veterinarian should correspond with the size and program of the individual institution.

Animal Husbandry, Housing, and Enrichment

In Taiwan, the animal husbandry, housing and enrichment are described in the *Guidebook*.[18] For the general environmental parameters such as room temperature, ventilation, heating, ventilation, and air conditioning, etc., basic recommendations are included in the *Guidebook* as well. More specific information relevant to individual animal species is included in the specific chapter of the *Guidebook*. Meanwhile the *Guide for the Care and Use of Laboratory Animals* published by NRC in the USA serves as the best reference resource.[20]

Euthanasia

The chapter of euthanasia in the *Guidebook*[18] is based on the recommendation made by ICLAS Working Group on Harmonization, and the 2007 AVMA "Guidelines on Euthanasia."[21,22] An overview of acceptable method of euthanasia in the *Guidebook* for common species used for research, teaching, and testing is included. The methods of euthanasia were reported in the COA annual statistical analysis in 2011: 78% of rodents were euthanized by carbon dioxide exposure, while 22% by the combination of carbon dioxide exposure and physical methods. Rabbits, mini-pigs, guinea pigs, and dogs were euthanized by injection of pentobarbital solution. The carcasses of rodents, mini-pig, dogs, amphibian, and others were mainly disposed by incineration through licensed commercial medical waste companies, while part of them were by in-house incineration.

Animal Facility and Special Animal Biosafety

The Act requires that animal facilities for the care and use of all animals in research, teaching, and testing shall be conducive to the well-being and safety of the animals, provide an appropriately safe workplace for personnel, and establish a stable research environment.[16] The *Guidebook* provides the guidelines to promote optimal levels of animal care and facilitate good research.[18] The guidelines are intended to be viewed as a tool for achieving acceptable standards and not as mandatory instructions. The *Guidebook* also encourages the idea of adopting barrier systems for reducing or minimizing cross-contamination.

Taiwan Center for Disease Control also issues a specific requirement and guideline to govern the animal biosafety research laboratory in the institution including ABSL 2, 3, and 4 level facilities.

REFERENCES

1. Japan, 1973. The Law of Humane Treatment and Management of Animals (Law No. 105, 1973).
2. Japan, 1980. The Standards Relating to the Care, Management and Relief of Pain of Experimental Animals (Notice No. 6 of the Prime Minister's Office 1980).
3. Japan, 2006. The Standards Relating to the Care, Management and Relief of Pain of Experimental Animals (Revised in 2006).
4. Japan, 1995. Guidelines on Methods of Sacrificing Animals (Notice No. 40 of the Prime Minister's Office, July 4, 1995).
5. Japan, 2007. The Law Concerning the Conservation and Sustainable Use of Biological Diversity through Regulations on the Use of Living Modified Organisms.
6. Japan, 2004. Invasive Alien Species Law.
7. Japan, 1998. The Law Concerning the Prevention of Infectious Diseases and Medical Care for Patients with Infectious Diseases.
8. Japan, 2006. Ministries of Education, Culture, Sports, Science and Technology; Health, Labor and Welfare; Agriculture and Fisheries. The Guidelines for the Proper Conduct of Animal Experimentation.
9. Japan, 2006. Science Council. Guidelines for Proper Conduct of Animal Experiments (the Guide).
10. South Korea, 1991. Ministry of Agriculture, Food and Rural Affairs. Animal Protection Act 1991 (Revised 2008).
11. South Korea, 2009. Ministry of Food and Drug Safety. Laboratory Animal Act.
12. Institute of Laboratory Animal Resources. *Guide for the care and use of laboratory animals.* National Research Council; 1996.
13. http://olaw.nih.gov.
14. http://www.aaalac.org.
15. American Veterinary Medical Associations. *Guidelines for the euthanasia of animals: 2013 edition*; 2013. https://www.avma.org/KB/Policies/Documents/euthanasia.pdf.
16. Taiwan Legislative Yuan, 1998. Animal Protection Act. Promulgated on November 4, 1999.
17. Taiwan, 2000. Enforcement Rules of Animal Protection Act. Promulgated on January 19, 2000.
18. Taiwan Council of Agriculture, 2010. *Guidebook for the care and use of laboratory animals.* 3rd ed. (Chinese version) (ISBN 957-30437-2-6).
19. Russell WMS, Burch RL. *The principles of humane experimental technique.* London UK: Special edition published by Universities Federation for Animal Welfare—UFAW, 1992; 1959.
20. National Academy of Sciences. *Guide for the care and use of laboratory animals.* Washington, D.C.: National Academy Press; 2011.
21. Demers G, Griffin G, De Vroey G, Haywood JR, Zurlo J, Bedard M. Harmonization of animal care and use guidance. *Science* 2006;**312**(5774):700–1.
22. American Veterinary Medical Association. *Guidelines on euthanasia.* Schaumburg IL: AVMA; 2007.

Chapter 11

Laws, Regulations, Guidelines, and Principles Pertaining to Laboratory Animals in Southeast Asia

Montip Gettayacamin[1], Richard Grant[2], Imelda Liunanita Winoto[3], Dondin Sajuthi[4], Yasmina Arditi Paramastri[5], Joanna Debby Khoo[6], Pradon Chatikavanij[7], Jason Villano[8], and Abdul Rahim Mutalib[9]

[1]AAALAC International, Samutprakarn, Thailand, [2]SNBL (Cambodia) Ltd, Khan Daun Penh, Phnom Penh, Cambodia, [3]Primate Research Center, Bogor Agricultural University, Bogor, Indonesia, [4]Primate Research Center, Veterinary Teaching Hospital, Bogor Agricultural University (IPB), Bogor, Indonesia, [5]Department of Pathology, Microbiology and Immunology, Vanderbilt University Medical Center, Nashville, TN, USA, [6]Agri-Food and Veterinary Authority of Singapore, Singapore, [7]National Research Council of Thailand (NRCT), Bangkok, Thailand, [8]Unit for Laboratory Animal Medicine, University of Michigan Medical School, Ann Arbor, MI, USA, [9]Department of Veterinary Pathology and Microbiology, Faculty of Veterinary Medicine, University Putra Malaysia, Serdang, Malaysia

Laboratory Animals. http://dx.doi.org/10.1016/B978-0-12-397856-1.00011-8

INTRODUCTION

Southeast Asia is a subregion of Asia, consisting of the following countries: Brunei or the Nation of Brunei; Cambodia or the Kingdom of Cambodia; Christmas Island (a territory of Australia); Indonesia or the Republic of Indonesia; Laos or the Lao People's Democratic Republic; Malaysia; Myanmar; Philippines or the Republic of the Philippines; Singapore or the Republic of Singapore; East Timor or Timor-Leste or the Democratic Republic of Timor-Leste; Thailand or the Kingdom of Thailand; and Vietnam or the Socialist Republic of Vietnam. All these countries are members of the World Organisation for Animal Health (OIE). The OIE *Terrestrial Animal Health Code* (the *Terrestrial Code*) sets out standards for the improvement of animal health and welfare and veterinary public health worldwide, including standards for safe international trade in terrestrial animals (mammals, birds, and bees) and their products. This is achieved through the detailing of animal health measures to be used by the veterinary authorities of importing and exporting countries to avoid the transfer of agents pathogenic for animals or humans, while avoiding unjustified trade barriers. The *Terrestrial Code* is a reference document for use by veterinary authorities, import/export services, epidemiologists and all those involved in international trade. Section 7 of the *Terrestrial Code* covers animal welfare and Chapter 7.1 is called, "Introduction to the Recommendations for Animal Welfare." There is no specific overarching legislation applied for animal welfare relevant directly to laboratory animals for research, testing, and teaching in Southeast Asia. This chapter describes current laws, regulations, guidelines, and principles pertaining to laboratory animals used for research, teaching, and testing in some countries of Southeast Asia region including Cambodia, Indonesia, Philippines, Thailand and Singapore. Table 11.1 summarizes legal requirements, country and institutional oversight bodies, and special requirements pertaining to laboratory animals in these countries.

CAMBODIA

Background and Current Situation

Cambodia is primarily an agricultural country with very little infrastructure for animal experimentation. More than 95% of Cambodians are Buddhists and the humane care and treatment of animals is regarded as important in the context of those beliefs. The Royal University of Agriculture graduates students with Bachelor and Masters Degrees in Veterinary Science and Animal Husbandry. Research being conducted at the Royal University of Agriculture, the Royal University of Phnom Penh and the National Veterinary Research Institute consists of surveys of wild animals and agricultural species for disease surveillance. Foreign research institutions exist in Cambodia, namely the Pasteur Institute and the United States Naval Medical Research Unit (NAMRU), but those institutions report that no laboratory animal experimentation takes place. If animal experiments were to be performed, these foreign institutes report that they would follow their home-country requirements for animal care and use.

The Association for Assessment and Accreditation of Laboratory Animal Care (AAALAC) International lists one accredited cynomolgus monkey (*Macaca fascicularis*) breeding program in Cambodia. Additional breeding centers reported that they will seek AAALAC International accreditation in 2013 or 2014.

Summary of Regulations or Legal Requirements

In Cambodia, guidelines for animal care are primarily focused on agricultural production and enforced by the Ministry of Agriculture, Forestry, and Fisheries (MAFF). Governmental and academic research institutions generally follow the OIE *Terrestrial Animal Health Code*.[1] Chapter 7.1 of the *Terrestrial Code* called *Introduction to the Recommendations for Animal Welfare,* generally recognizes the "Five Freedoms" and the "Three Rs" principles.[2] There is no specific regulation or guideline covering the care and use of laboratory animals in research, teaching, and testing. No formal protocol review or prestudy approvals are required for the animal studies.

Exportation of Nonhuman Primates from Cambodia

In Cambodia, the primary policy-making and permit-issuing for exporting nonhuman primates (NHPs) is the authority of MAFF involving the CITES office, the Department of Animal Health and Production (DAHP), the National Veterinary Research Institute (NaVRI), and the Forestry Administration (FA). There is currently a moratorium on exportation of wild-caught NHPs, enforced by officials at FA and DAHP. The veterinary authorities of these two departments regularly inspect breeding facilities to ensure that housing and transport conditions are acceptable, and only the first generation offspring animals (F1)

TABLE 11.1 General Legal Requirements Pertaining to Laboratory Animals, Country Institutional Oversight Bodies, and Special Requirements in Some Southeast Asia Countries

Countries	Laws and Regulations	Guidelines and Principles	Legal Oversight Body	Institutional Oversight Body	Remarks
Cambodia	Not applicable	Not applicable	For exporting NHPs: Ministry of Agriculture, Forestry and Fisheries (MAFF)	Not applicable	Laboratory animals are mammals (species covered in the OIE *terrestrial animal health code*[1])
Indonesia	Law of the Republic of Indonesia No. 18 (2009)	The National Guidelines on Health Research Ethics (2011)[8] Teaching guidebook for ethics on health research (2011)[6]	National Bioethics Committee	Not required But some institutions maintain Institutional Animal Care and Use Committee (IACUC). Some institutions merge the IACUC with the Health Research Ethical Committee	Conservation Law No. 5 (1990)[13] covers capturing and handling methods of animals from the natural habitat
Malaysia	The Animal Welfare Act—pending approval in 2013	*The Malaysian Code of Practice for the Care and Use of Animals for Scientific Purposes*— pending approval in 2013[9]	Department of Veterinary Services The National Bioethics Council of Malaysia	Institutional Animal Care and Use Committee (IACUC)	
Philippines	The *Republic Act 8485* (the Animal Welfare Act of 1998)[16]	*Administrative Order No. 40 (AO 40), Series of 1999* (Department of Agriculture)[21] PALAS Code for the Care and Use of Laboratory Animals in the Philippines (COP)[22]	Bureau of Animal Industry (BAI), Department of Agriculture[17]	Institutional Animal Care and Use Committee (IACUC)	Authorization to conduct animal research (renewed annually) is granted to institutions by the BAI based on the acceptability of the animal care and use program, including accreditation by the Philippine Association for Laboratory Animal Science (PALAS), the presence of IACUC and animal care personnel training program

Singapore	The Animals and Birds Act (Chapter 7) • The Animals and Birds (care and use of animals for scientific purposes) Rules (2004)[24]	NACLAR guidelines (National Advisory Committee for Laboratory Animal Research guidelines[27] on the care and use of animals for scientific purposes) • The guiding principles for the care and use of animals for scientific purposes[25] • The Institutional Animal Care and Use Committee (IACUC) guidelines • The training guidelines	AVA (Agri-Food and Veterinary Authority of Singapore)	Institutional Animal Care and Use Committee (IACUC)	Any research facility which intends to use animals for scientific purposes must obtain a license from the AVA
Thailand	Not applicable	Ethical principles and guidelines for the use of animals for scientific purposes (2006), National Research Council of Thailand[32]	National Research Council of Thailand	Institutional Animal Care and Use Committee (IACUC)	

or greater are exported. The following steps are the documents in chronological order required for exporting NHPs:

1. CITES Permit obtained from CITES Management Authority, MAFF.
2. NaVRI Animal Health Certificate obtained from NaVRI, DAHP, MAFF.
3. Veterinary Certificate obtained from DAHP, MAFF.
4. Transit Permit(s) as needed obtained from the transit country if animals will move from one plane to another.
5. Cambodian Transportation Permit obtained from DAHP, MAFF.
6. Cambodian Trucking Permit obtained from FA, MAFF after FA officials perform facility inspection.
7. Inspection Certificate (optional for exporting to Japan only) obtained from DAHP, MAFF.
8. Veterinary Health Certificates of Cynomolgus Monkeys: Monkeys' Data and the Records of Exporting Nonhuman Primate issued by the exporter and signed by the facility's Attending Veterinarian.

INDONESIA

Background and Current Situation

Governmental interference in animal related issues in Indonesia has been established since its very early days, as indicated by the presence of Law No. 432 2012,[3] which regulated transmissible diseases control, animal husbandry improvement, and veterinary health and hygiene. However, it focused mostly on farm animals.

In 1967, Law of Republic Indonesia No. 6, Article 22[4] gave specific attention to animal welfare by regulating the following aspects: housing, care, transportation, use, euthanasia methods, and humane treatment of animals. Article 22 also indicated that the use of animals should be done with respect to animal welfare. Holding site and housing requirements were regulated by local government authorities at district level. Attempts should be made to limit disturbance to public facilities and compounds. The housing requirements were regulated within achievable criteria to support production growth. Efforts should be made to ensure humane treatment of animals as God's living creatures. During transportation, animals should be kept in normal position, in adequate space with good ventilation. Animals should not be abused or overly used above their maximum tolerances or capabilities. Euthanasia should be done with respect to religious procedure requirements with minimal pain and distress.

In 2009, this law was replaced by Law of the Republic of Indonesia No. 18 that covers Veterinary Public Health and Animal Welfare.[5] With regard to laboratory animals, veterinary authority in the use of animals as laboratory animals and in comparative medicine is discussed in Chapter VII, Veterinary Authority, Article 74.[6]

A laboratory animal program was initially established by the United States Naval Medical Research Unit No. 2 (NAMRU-2) in Indonesia in January 1970 by a request of the Indonesian health authorities (the Ministry of Health of Republic Indonesia) following their success assisting the Indonesian government in controlling a Bubonic plague outbreak in Central Java. NAMRU-2 animal facility was the first AAALAC accredited facility in Indonesia in 1996, followed by three other institutions (a university research center, a primate breeding facility, and a pharmaceutical company) in 2006, 2007 and 2012. Currently, laboratory animals are extensively used in teaching, testing, and research in many universities, pharmaceutical companies, vaccine companies, and research institutions in Indonesia.

Summary of Regulations or Legal Requirements

Scope

Law No. 18[5] covers animal welfare in general. Chapter VI describes Veterinary Public Health and Animal Welfare and Part 2 governs Animal Welfare. Articles 66 and 67 of the law define that "animal" means any vertebrate animal and some invertebrate animals which can feel pain such as a crab. The law covers all aspects of the efforts to support animal welfare. This includes capturing animals from the wild, handling and restraint, housing and care, transportation, euthanasia methods, and humane treatment of animals. In addition, Veterinary Authority in the use of laboratory animals in comparative medicine is also covered under Chapter VII, Veterinary Authority; Article 74 of this Law.[6]

The principles of the ethical treatment on the use of animals in comparative medicine are described in the Government of the Republic of Indonesia's Regulation No. 95.[7] Detailed guidelines on the use of laboratory animals in biomedical research are developed by the Health Research Ethics Committee, the Ministry of Health, Republic of Indonesia in the National Guidelines on Health Research Ethics[8] and the Teaching Guide book for Ethics on Health Research.[9] However, the Guidelines[8] focus on the use of animals in biomedical research for human health, but do not include the use of animals in research for animal health, industrial or military purposes.

For the use of fetuses in health research, fetuses at postimplantation stage are treated as adult animals in regard to use of appropriate anesthetics and analgesics. If a surgical procedure is conducted on any fetus and as a result the fetus cannot grow to become a self-supported individual, euthanasia must be done after birth. And for the use of endangered species, a permit must be obtained from the Forestry Department of Indonesia prior to the use of animals in health research.[8]

Applicability

There is neither an animal facility registration, authorization nor personnel licensing system applied in Indonesia.

The Principles

Law No. 18[5] does not address the Three Rs (3Rs) explicitly but emphasizes the importance of the Five Freedoms (5Fs). However, both the Guidelines[8] and the Guide Book[9] address the 3Rs and the 5Fs: (1) Freedom from hunger and thirst; (2) Freedom from discomfort; (3) Freedom from pain, injuries, and diseases; (4) Freedom from fear and distress; and (5) Freedom to express natural behaviors.

Ethical Review

Article 74[6] covers the requirements of competent veterinarians to oversee the animal care and use program, according to the ethical treatment of animals and animal welfare.

The National Guidelines on Health Research Ethics[8] cover ethical aspects of animal use for health research, protocol review procedures by ethics committees, authority and responsibility of ethics committees, and special considerations on study protocols reviews to include the following topics:

- Risk-benefit analysis
- Animal use justification
- Pain categories and alleviation
- Study design and statistical analysis
- Veterinary care
- Animal procurement, transportation, husbandry, sanitation and housing[4]
- Restraint
- Food and water restriction
- Behavioral management including physical activity and social housing of the animals
- Humane endpoints and humane treatments
- Multiple major survival surgeries
- Use of anesthetics, analgesics, and neuromuscular blocking agents
- Euthanasia methods
- Personnel qualifications and trainings
- Annual reports and final reports
- Occupational health and safety (OHS) program to include annual medical check ups and vaccinations, personal protective equipment (PPE), physical plant and engineering controls, and waste management
- Use of fetuses in health research
- Use of endangered species in health research
- Animal importation and exportation procedures.

The National Committee of Ethics on Health Research also developed the Teaching Guide Book for Ethics on Health Research.[9]

In 2004, the National Bioethics Committee was established by inter-departmental decrees of three ministries: the Ministry of Research and Technology, the Ministry of Health and the Ministry of Agriculture of the Republic of Indonesia.[10] The National Bioethics Committee membership consists of experts from various fields, including religious representatives and humanists. The main functions of the committee are:

- To improve review on matters regarding bioethical principles
- To provide recommendations to the government on the bioethical aspects of research development and application of the biological sciences and technology
- To disseminate public understanding on bioethics.

Both the Guidelines[8] and the Guide Book[9] state that in order to be ethically justified, animal use for research, teaching, and testing must be covered in a protocol that is approved by the ethical committee of the institution where the activities will be done.

A few institutions have established Institutional Animal Care and Use Committees (IACUC) in Indonesia since this practice was initiated for the first time by the United States Naval Medical Research Unit No. 2 (NAMRU-2) in 1991 and introduced to other institutions in 1993. This oversight body has been established in academic institutions and universities (at either departmental level or campus-wide program), governmental institutions, research institutions, pharmaceutical companies, and contract research organizations using animals for research, teaching and testing. Some institutions merge the oversight responsibilities of the IACUC with the Health Research Ethical Committee. For example, the Indonesian National Institute of Health Research and Development (Badan Litbangkes Republik Indonesia) Health Research Ethical Committee also serves as the IACUC.

Each institution has their own animal use protocol format to fulfill their specific requirements, but basically they cover the following items to be reviewed by the ethical committee:

- Purpose of the project (research, teaching, breeding, antibody production)
- Animal specifications (species, stock or strain, sex, age or bodyweight, number)
- Housing method
- Length of the project
- Lay summary of the project
- Hypothesis
- Literature search to confirm no duplication of efforts and the least painful procedures are chosen
- Project description and procedures
- Ethical consideration (species justification and animal number justification)
- Pain category, the use of analgesia, justification for not using pain alleviation
- Animal use procedures:
  - If the project involves anesthesia: drugs, doses, routes of application, frequency, duration, methods of monitoring
  - Administration of drugs, reagents, vaccines, cells, agents, substances, etc.
  - Surgery (survival/non survival, number of survival surgery)
  - Antibody production (the use of adjuvant)
  - Behavioral testing
  - The use of hazardous agents
  - Blood sampling
  - Food and water deprivation
  - The use of paralytic agents
  - Euthanasia method
  - Humane endpoint

- Restraint method and duration
- Tumor induction
- Toxicity testing
- Tissue collection
- Personnel qualifications (educational degree, role in the project, training or experience)
- Genetically engineered animal use/production
- Adverse effect (unexpected outcomes).

Institutional and Designated Personnel Responsibilities and Qualifications

The National Guidelines on Health Research Ethics[8] indicate that the institution must appoint a veterinarian with expertise and experience in laboratory animal medicine and science to be in charge of and oversee the care and use of laboratory animals.

Personnel Training, Education and Competency Requirements

The Guidelines[8] state that researchers and research supporting personnel must be competent and adequately trained in the care and handling of laboratory animals. Regular and periodic trainings are required for each research staff prior to the proposal submission or beginning of the study.

Animal Care and Use Areas Addressed by the Regulations

Animal Transportation

Law No. 18[4] states that transportation of animals should be done in a way that the animals are free from pain and distress. Detailed explanations of the regulation are provided by the Government Regulation No. 95 on Veterinary Public Health and Animal Welfare,[7] as follows:

Application of animal welfare principles on transportation must be done at minimum in the conditions that:

- Do not harm, prevent injury and do not cause stress to the animals.
- Use clean and proper transportation means or vehicles according to the capacity.
- Provide adequate food and water to fulfill physiological need of the animals.

If animals are transported in cages, the space provided must allow the animals to move freely in the cages, as well as prevent the presence of predators and incompatible animals. Animals must be protected from direct heat and rain.

Transportation must be conducted under supervision and/or recommendations of an attending/authorized veterinarian.

Physical Plant and Environmental Conditions

There is no specific guideline or regulation for these aspects.

Housing or Enclosures

Animal holding and housing areas must be built in a way to ensure the animals' freedom to express their natural behaviors. The regulation does not specify minimum space requirements for each animal species, but the Guidelines[8] indicate that minimum space requirements stated in the ILAR *Guide for the Care and Use of Laboratory Animals*[11] should be adopted. The following guidelines are included in the Government Regulation No. 95[7]:

● Holding and housing enclosures should provide no harm, and do not cause injury and/or stress to the animals.
● Facilities and equipment do not cause harm, injury and/or stress to the animals.
● Providing separate housing of different species or allow physical separation by species.
● Use of clean cages that provide enough space to allow free movement, provide protection from predators and incompatible animals, as well as provide protection from direct heat from sunlight and rain.

Daily Care Activities

There is no specific guideline or regulation on food and water provision but animal care should be done in a way to ensure the animals are free from hunger and thirst and fulfill their physiologic requirements. There is no specific guideline on sanitation procedures, but animal husbandry should be done in a way to ensure the animals are free from diseases, pain, and distress.

Behavioral and Environmental Management Enrichment

The Guidelines[8] indicate that animals used in research must have opportunities to express their natural behaviors and display normal activities according to the biology of the species. Social species are required to have social experience. Therefore, group composition, housing structure and placement of the animals must be considered and allow visual, auditory and olfactory contact among their conspecifics. Territorial species could be distressed by the presence of other animals within their territorial boundary. Housing of multiple species in the same room may cause fear and distress.

Occupational Health and Safety

An OHS Program as indicated in the Guidelines[8] consists of:

● Annual medical evaluation and vaccination relevant to medical risk assessment and analysis
● PPE as needed, i.e. masks, gloves, boots or shoe covers, caps, goggles, face shields, and lab coats
● Physical facilities or rooms in conformity with safety and ergonomic requirements to minimize work related injuries and accidents
● Proper waste management to prevent environmental pollution.

Veterinary Care

The institution must appoint a veterinarian with adequate knowledge and experience in laboratory animal medicine to be involved, responsible for, and oversee the care and use of laboratory animals.[8]

Euthanasia

As indicated in the Guidelines,[8] euthanasia policy is referred to the American Veterinary Medical Association (*AVMA*) *Guidelines on Euthanasia*.[12] Euthanasia methods should be appropriate and without pain and distress and done by professional, qualified or adequately trained personnel; and death following euthanasia must be confirmed. Euthanasia should be done without the presence of other animals in the same room or location.

Special Considerations

Nonhuman Primates: Capturing and handling methods of animals from the original habitat must be in compliance with Conservation Law.[13]
Genetically Altered Animals: There is no specific guideline or regulation.
Reuse of Animals: There is no specific guideline or regulation.

MALAYSIA

Background and Current Situation

The use of laboratory animals in Malaysia (or Malaya as it was called then) began in the early 1900s when the Institute for Medical Research (IMR) was established in Kuala Lumpur by the British colonial government. The Institute's main function was the study of tropical diseases, and laboratory animals were used as animal models. In 1948, the Veterinary Research Institute was established in Ipoh, the capital of the state of Perak, and laboratory animals were used mainly as diagnostic tools and for vaccine testing for animal diseases. To date many universities, especially those with medical, veterinary, and life science faculties, and research institutions have laboratory animal facilities. Although many of the facilities are of the conventional type, at least four of the newer ones were designed to accommodate specific pathogen free rodents. At the moment none of the animal facilities are AAALAC accredited. However, these four facilities (one government research institute and three public universities) are working toward AAALAC International accreditation.

The Laboratory Animal Science Association of Malaysia (LASAM) established in 1994 has as one of its main activities, among others, to promote the humane use of animals in research through recognition of ethical principles and scientific responsibilities. To that effect, LASAM has organized and carried out yearly seminars, workshops and conferences to educate the scientific community in Malaysia.

Summary of Regulations or Legal Requirements

Two acts are in existence in Malaysia where animals are concerned. The Laws of Malaysia Act 647 also known as the Animal Act 1953, revised in 2006[14] is mainly concerned with acts of cruelty to livestock and pets with penalties or fines imposed by the courts if found guilty. This act is in the process of being revised and updated at the moment of writing this document (March 2013) by the Department of Veterinary Services. The other act is the Wildlife Conservation Act 2010, Laws of Malaysia Act 716.[15] This Act stipulates, among others, that a permit issued by the Department of Wildlife and National Parks must be obtained for research activities using wildlife.

The Malaysian government, through the Department of Veterinary Services, is drafting a bill called the Animal Welfare Act[16] which is scheduled to be tabled in parliament during 2013. A section of the bill is solely dedicated to the use of animals for scientific purposes with a set of regulations attached to it. This regulation, proposed to be called *The Malaysian Code of Practice for the Care and Use of Animals for Scientific Purposes* (the *Code of Practice*[17])," was drafted by the Laboratory Animal Science Association of Malaysia (LASAM) taking into consideration established guidelines and codes of practices from several countries. Due to the absence of governmental guidelines on the care and use of animals for research, testing, and teaching purposes at the moment, many research institutions and universities have adopted the Code of Practice drafted by LASAM as their official guide and endorsed by their IACUCs. For the interim period, while waiting for the proposed Animal Welfare Act to be passed in the parliament and become law, the National Bioethics Council of Malaysia will adopt this Code of Practice document as an official guide for those who want to use animals for scientific purposes which include research, teaching and display.

The *Code of Practice* is an exhaustive document which promotes animal welfare and the ethical use of animals for research, teaching, and display. It emphasizes the 3Rs principles when planning a research project using animals. It describes in detail the composition, the terms of reference, and the IACUC function. It also describes in detail the specifications and conditions for animal housing, cage dimensions, room temperature, and ventilation rates among others. It also emphasizes the need to train personnel involved in the care, use, and management of animals and in techniques of handling, restraint, and carrying out procedures on the animals used for scientific purposes.

The proposed Animal Welfare Act will require that all animal facilities be licensed and subjected to yearly inspection. Enforcement of the Act will be carried out by the Department of Veterinary Services under the Ministry of Agriculture and Agro-based Industry. The Act also stipulates that institutions wanting to use animals for scientific purpose must establish an IACUC or Animal Ethics Committee (AEC) and all research proposals using animals must be vetted and approved by the IACUC/AEC before the study can be carried out. Currently there are 14 IACUCs/AECs in various institutions of higher learning and research institutions in Malaysia.

PHILIPPINES

Background and Current Situation

Laboratory animal research in the Philippines started as early as the eighteenth century when Spanish conquistadors sent a mission to the country to study smallpox. The mission resulted in the foundation of the Board of Vaccination later renamed Bureau of Science under the American rule.[18] Animals have since been used primarily for the production of antivenins. To date, laboratory animals are used mainly by various governmental and academic institutions such as the Research Institute for Tropical Medicine and the University of the Philippines. In 1988, the Philippine Association for Laboratory Animal Science (PALAS) was founded to promote laboratory animal science in the country and has since conducted annual scientific convention and animal technician training.[18] Private institutions that conduct research include two AAALAC International-accredited programs that export Philippine cynomolgus macaques (*M. fascicularis*).

It is important to note that although the Philippines has its own sets of rules and regulations on animal research, the framework is mainly based on internationally recognized references such as the *International Guiding Principles for Biomedical Research Involving Animals*, Council for International Organizations of Medical Sciences (CIOMS)[19]; *Guide for the Care and Use of Laboratory Animals (Guide)*[11]; and *Biosafety in Microbiological and Biomedical Laboratories*.[20]

Summary of Regulations and Legal Requirements

The Republic Act 8485,[21] also known as "The Animal Welfare Act of 1998" protects and promotes animal welfare in the country by supervising and regulating the establishment and operations of facilities used for breeding, maintaining, keeping, treating or training of all animals either as objects of trade or as household pets.

Pursuant to the Act, Administrative Order No. 40 (AO 40),[22] Series of 1999 (Department of Agriculture) was issued to specifically institute the rules and regulations that govern animal research in the country. Applicable guidelines and principles considered as support documents include those from the *Guide*, the PALAS Code of Practice for the Care and Use of Laboratory Animals in the Philippines (COP), and the CIOMS International Guiding Principles for Biomedical Research Involving Animals. The government can also issue additional guidance on other specific issues.

The Philippine Association for Laboratory Animal Science (PALAS) Code of Practice for the Care and Use of Laboratory Animals in the Philippines

(COP), a relatively comprehensive reference document endorsed by local regulatory agencies, is used as the basis for accreditation.[23] The COP uses performance standards as the basis of an animal care and use program's effectiveness. However, it also provides strict guidelines on certain aspects, especially relating to engineering and husbandry standards (e.g. 10–20 air changes/h, weekly cleaning of mouse and rat cages). It highlights the importance of occupational health and safety programs, training of animal care personnel, and the role of the IACUC and its policies. It discusses transportation of animals, especially in conjunction with quarantine and acclimation of those that are newly received, and strongly advocates the use of standards established by the International Air Transport Association (IATA). The document also addresses other topics including animal health monitoring, disposal of animal carcasses, provision of environmental enrichment (especially to NHP), and animal housing with the relevant environmental controls. It encourages the rational reuse of animals based on the severity of the procedures and the ability of the animal to fully recover from the procedures.

Scope

The AO 40 only applies to research using live vertebrate animals used in biomedical research; teaching and instruction; product (food, drugs, agrochemicals and cosmetics) testing; and production of antisera or other biologicals. Manipulation is defined as any technique that interferes with normal physiological, behavioral or anatomical integrity of the animal. It includes exposing the animal to any parasite, microorganism, drug, chemical, biological product, radiation, electrical stimulation, or environmental condition; subjecting it to enforced activity, unusual restraint, abnormal nutrition or surgical procedures; and depriving it of usual care. Exempted from the AO 40 requirements are clinical tests on animals for evaluating veterinary products in accordance with regulatory requirements or standard procedures; recognized veterinary procedures such as therapy, prophylaxis, diagnostic or disease surveillance procedures necessary or desirable for animal welfare; and recognized agricultural practices (e.g. castration, genetic engineering or embryo manipulation, unless they form part of an experiment).

Authorization of User-Breeding Institutions

The Act promulgates that the Director of the Bureau of Animal Industry (BAI), Department of Agriculture shall supervise and regulate these facilities, and that animal transport should provide maximum comfort to the animals while in transit and prevent animal sickness, death and cruelty.

Authorization to conduct animal research (valid for two years and renewed annually thereafter) is granted to institutions by the BAI based on the acceptability of the animal care and use program, including adherence and congruency to the principles of the PALAS COP, and the presence of an IACUC and animal

TABLE 11.2 Pain Categorization of Experimental Procedures as Outlined in Administrative Order No. 40 in Philippines[22]

Category	Pain Level	Reviewer/s	Examples of Procedures
Category 1	Low or mild pain	IACUC chairperson	Venipuncture, superficial tissue biopsy, laparoscopy
Category 2	Medium or moderate with reliable postprocedural analgesia and care	IACUC chairperson and 1–2 members	Most surgical procedures, toxicity studies avoiding lethal endpoints
Category 3	High or substantial severity	Quorum of the IACUC	Acute toxicity procedures with significant morbidity or death as endpoint; some models of disease and major surgery where significant postoperative suffering may result

care personnel training program. The AO 40 relies on the animal care program being accredited by PALAS and/or AAALAC International. It requires the institution to describe its animal care and use program based on personnel training and qualifications (including continuing education opportunities), husbandry care and provisions, the animals' physical environment, and veterinary care. Authorization is revoked when there is a failure to subscribe to or conduct the animal technician training program, or when there is misrepresentation of any information in the program description and reports. Such revocation may be appealed to BAI's Animal Welfare Committee. Re-application for authorization can only be done after one year from the date of revocation.

Ethical Principles and Ethical Review

The AO 40 clearly outlines various IACUC responsibilities. It states that the IACUC should be guided by Russell and Burch's 3Rs and records and documents should be archived for five years. Members are appointed by the Institutional Official and should include a veterinarian, a scientist who works with animals, and a nonaffiliated public member. Any additional member may be appointed provided the person possesses the aforementioned qualifications. If necessary, the chairperson may request an expert to attend the committee meeting for protocol review.

Protocol reviews, including the number of reviewers involved in the process, are based on pain categories (Table 11.2). An IACUC member who has a conflicting interest (e.g. direct involvement in the project) may only provide information requested by the IACUC and may not participate in the review and approval of the protocol. The AO 40 gives an example of a protocol form, and required information includes: qualifications of personnel, significance of the study, experimental design including animal details, and consideration of alternatives. Multiple major survival surgical procedures are also particularly reviewed, as the IACUC is to describe its policy on the conduct of such studies in its annual report to BAI. For protocol amendments, the principal investigator shall submit revised parts of the protocol to the IACUC Chairperson for evaluation. Approval of a protocol may be revoked by the majority of the IACUC quorum on the basis of noncompliance to the approved program and protocol. The Institutional Official in consultation with the IACUC shall review the reasons for revocation, take appropriate corrective action, and report that action and the full circumstances of the revocation to BAI's Animal Welfare Division.

The AO 40 dictates that the IACUC's responsibilities should monitor and review the animal care and use program, including conducting facility inspections. The PALAS COP and the National Research Council (NRC) Guide are used as references for review and inspections, and any deficiencies are classified as either minor or significant (those that pose a threat to animal health or safety) after careful deliberation with the facility manager and/or Institutional Official. The AO 40 frames a checklist for the review and inspection, and this checklist includes veterinary care; personnel qualifications and training; macro- and microenvironment including type, space, noise, illumination, temperature, and humidity; and husbandry including sanitation, food and water provision, and environmental enrichment program. If significant deficiencies are not resolved by the set correction dates, the IACUC must notify BAI's Animal Welfare Division within 30 working days. Actions taken for the identified deficiencies are included in the required annual report submitted to BAI, which then issues a Certificate of Approval.

Animal Care and Use Areas Addressed by the Regulations

Veterinary Care

The AO 40 briefly outlines the requirements for veterinary care in an animal care and use program. The veterinarian may be employed part-time or as a consultant, but the frequency and duration of visits should be described in the animal care and use program. Weekend, holiday, and emergency veterinary care should be available. The animal care and use program should describe quarantine and acclimation, disease surveillance, preventive medicine, analgesia, anesthesia, and euthanasia. The IACUC has a responsibility to provide necessary technical recommendations and guidelines/references (e.g. recommended doses for anesthetics and analgesics) to animal users. Finally, as discussed previously, the PALAS COP provides a more comprehensive reference for various aspects of veterinary care. For example, it lists various pathogens found on laboratory animals based on animal species, zoonotic and/or opportunistic potential, and the potential to confound research.

Euthanasia

A significant portion of the AO 40 lists euthanasia methods for various laboratory animal species. These include the use of carbon dioxide (for rodents placed in an uncrowded chamber), inhalant anesthetics, barbiturates (most recommended for dogs and cats), stunning, cervical dislocation (for mice, rats weighing less than 250 g, and rabbits weighing less than 1 kg), decapitation, and exsanguination (performed under anesthesia). Agents such as ether and chloroform are also listed as unsatisfactory and are discouraged. Whatever method is employed, euthanasia should always be done away from public view.

SINGAPORE

Background and Current Situation

Research involving the use of animals is regulated to safeguard laboratory animal welfare. Under the Animals and Birds (Care and Use of Animals for Scientific Purposes) Rules 2004,[24] any research facility which intends to keep or use animals for scientific purposes must obtain a license from the Agri-Food and Veterinary Authority of Singapore (AVA).

The National Advisory Committee for Laboratory Animal Research (NACLAR) was established in 2003 with the aim of promoting responsible and humane use of animals for research. Members included representatives from academia, research organizations, the AVA, veterinarians, as well as legal and ethical specialists. In 2004, NACLAR developed national guidelines known as the NACLAR Guidelines.[25] Research facilities are required to comply with the guidelines as part of licensing requirements.

The NACLAR Guidelines are modeled closely after best practices in countries such as Australia, Canada, New Zealand, the United States, and organizations such as the Council for International Organizations of Medical Sciences[19] and the European Convention for the Protection of Vertebrate Animals Used for Experimental and Other Scientific Purposes.[26] Consultation with local research communities, associations, government agencies, religious groups, animal welfare groups, and the general public were conducted in its formulation. Today, the NACLAR Guidelines set out responsibilities for all parties involved in the care and use of animals for scientific purposes, in accordance with widely accepted scientific, ethical and legal principles.

In Singapore, self-regulation is enforced. Research facilities are licensed, but not the individual researchers or the projects. However, requirements imposed through licensing make each research facility accountable for researchers and research projects under it. Each research facility has to put in place a program and a system of checks to ensure that animals are used responsibly. In this regard, the appointed IACUC of each facility acts as an internal approval and auditing body to evaluate all proposed use of animals for scientific purposes in compliance with the NACLAR Guidelines. The AVA maintains regulatory oversight and audits IACUCs to check for effective self-regulation.

Summary of Regulations or Legal Requirements

Under the Animals and Birds Act Chapter 7,[27] the Animals and Birds (Care and Use of Animals for Scientific Purposes) Rules 2004[24] governs the care and use of animals for scientific purposes. The Rules commenced in effect from November 15, 2004. The NACLAR Guidelines establish best practices in the care and use of animals for research.

The NACLAR Guidelines are presented in three parts:

a. The Guiding Principles for the Care and Use of Animals for Scientific Purposes (Guiding Principles).
b. The Institutional Animal Care and Use Committee (IACUC) Guidelines (IACUC Guidelines) detail how the IACUC should operate in carrying out its responsibilities.
c. The Training Guidelines provide the training scope and requirements for all personnel involved in the care and use of animals for scientific purposes.

Scope

The Guiding Principles cover all aspects of the care and use of animals for scientific purposes performed to acquire, develop or demonstrate knowledge or techniques in any scientific discipline. This includes the purpose of teaching, field trials, environmental studies, research, diagnosis, product testing, and the production of biological products.

In relation, animals are defined under the Animals and Birds (Care and Use of Animals for Scientific Purposes) Rules 2004 and the NACLAR Guidelines as any live vertebrate, including any fish, amphibian, reptile, bird, and nonhuman mammal.

Applicability

Research facility licensing requirements and the NACLAR Guidelines apply to all research which is conducted in the facility. The IACUC of the facility maintains overall responsibilities in ensuring that all animal research complies with relevant legislation as well as the NACLAR Guidelines.

The Principles

The 3Rs principles of Replacement, Reduction and Refinement are integral to the NACLAR Guidelines and form the basis of the Guiding Principles. Chapter 2: General Principles for the Care and Use of Animals for Scientific Purposes recommends an approach to animal use based on the 3Rs.

Proposals submitted to the IACUC must contain sufficient information to justify the proposed use of animals. Researchers must demonstrate in their proposals that procedures will avoid or minimize discomfort, distress and pain to the animals, and that alternatives have been considered. They must also ensure that experiments are not unnecessarily duplicated. The rationale for involving animals and the appropriateness of the species and approximate numbers of animals used must also be explained. In all experiments, animal welfare must be observed and maintained at all times.

Animal research facilities are further required to submit an annual report by the Chief Executive Officer (CEO) per calendar year. Assurance must be provided that all researchers and personnel comply with the Guiding Principles, and that any exceptions to this are explained, approved by IACUC, and recorded officially. Assurance must also be given that professionally acceptable standards governing the care and use of animals were maintained, with appropriate use of anesthetics, analgesics and tranquilizers where necessary, and that each researcher has considered alternatives to any procedure which can cause pain or distress to animals.

Ethical Review

The IACUC is the institutional oversight body appointed by the CEO of each animal research facility. The IACUC reports directly to the CEO, who is then responsible for acting on the IACUC's recommendations. The IACUC advises the CEO of the institution's compliance, establishes plans and schedules for correction of deficiencies, and makes recommendations regarding any aspects of the animal program, facilities or personnel training. This system ensures that management is fully apprised of the facility's care and use of animals for scientific purposes.

As such, the IACUC must be provided with adequate and appropriate facilities, powers, and resources to carry out its duty. It reviews and approves or rejects proposals involving animal use, taking into consideration ethical and welfare aspects as well as scientific or educational value. It is mandatory to review the animal and care use program semiannually, and inspect the housing and procedural rooms of the research facility annually.

IACUC considers these areas when reviewing proposals involving animal use:

- Species and numbers of animals to be used
- Rationale for using animals and justification of the appropriateness of the species and numbers to be used
- Complete description of the project
- Whether proposals avoid or minimize stress, distress or pain
- Whether the 3Rs have been complied with, i.e. replacement using nonliving systems, reduction of numbers or refinement of techniques
- That there is no unnecessary duplication of previous experiments
- That animals are not used repeatedly unless justified
- Adequate description and justification of humane endpoints
- Appropriate use of drugs (sedatives, analgesics or anesthetics) and techniques in consultation with the Attending Veterinarian
- That paralytics will not be used unless when combined with anesthesia
- Qualifications and training of personnel conducting procedures on animals
- Adequate and appropriate care and husbandry of animals

- Adequate and appropriate presurgical, surgical and postsurgical procedures, techniques and care
- Euthanasia methods.

The IACUC also reviews and investigates complaints about animal care and use, monitors compliance of approved projects involving animals, ensures animal users and caretakers are adequately trained, and that occupational health and safety standards are complied with. Additionally, the IACUC serves a key role in advising of potential security risks and vulnerabilities, and in conducting appropriate risk assessments to develop an emergency preparedness and disaster plan to deal with any potential or real threats.

The IACUC comprises at least five members, including one separate member from each of the following four categories:

a. A veterinarian with training or experience in laboratory animal science and medicine, and who has experience in the routine care of the species of animals used in the research facility
b. A person with appropriate experience in the use of animals for scientific purposes
c. A person who is not affiliated in any way with the research facility and is not a user of animals for scientific purposes
d. A person whose primary concerns or interests is in nonscientific areas.

The Attending Veterinarian (AV) must be appointed to the IACUC, but the CEO or equivalent cannot be appointed to the IACUC. Additionally, a maximum of three members is allowed from the same unit or department within the facility, to avoid influencing IACUC decisions.

Certain IACUC decisions, such as approvals or suspensions/withdrawals of projects require a quorum. In Singapore, a quorum is defined as more than 50 percent of the IACUC members with at least one representative from the "nonscientific" or "nonaffiliate" category.

Institutional and Designated Personnel Responsibilities and Qualifications

CEO: Defined in the NACLAR Guidelines as the Chief Executive Officer (or person of similar standing) of an Institution who is in the position to grant resources to the IACUC and to enforce the IACUC's recommendations. The CEO cannot be appointed to the IACUC.

Attending veterinarian: Defined in the NACLAR Guidelines as a veterinarian engaged formally on a full-time or part-time basis, to maintain oversight of the animal care and use program and to provide adequate veterinary care. All veterinarians including the Attending Veterinarian must have qualifications in veterinary science and are licensed by the AVA.

IACUC: Defined in the NACLAR Guidelines as the Institutional Animal Care and Use Committee constituted by institutions.

Investigator: Defined in the NACLAR Guidelines as a person who proposes or has approval to conduct a project involving the use of animals.

Staff: Defined in the NACLAR Guidelines as all persons involved in the housing, feeding, and general care of the animals, or who otherwise assist investigators. Staff should have appropriate veterinary or animal care qualifications or experience, as recommended in the Training Guidelines.

Personnel Training, Education, and Competency Requirement

Personnel training and education are described in the IACUC Guidelines, Chapter 2: Oversight of the Animal Care and Use Program. The research facility has a responsibility to ensure that all groups of personnel involved in the care and use of animals for scientific purposes (researchers, research technicians, animal technicians, and other personnel) receive appropriate training, and are adequately trained to recognize and alleviate signs of pain and distress, as well as conduct animal procedures without causing unnecessary harm. Training should also be extended to temporary staff such as students or visiting researchers, where necessary.

The NACLAR Guidelines recommend that training programs should include the knowledge of humane methods of animal handling and experimentation; basic needs of different animal species; proper handling and care; proper preprocedural, procedural and postprocedural techniques and care; aseptic surgical methods and procedures; and proper use of anesthetics, analgesics and tranquilizers. Occupational health and safety is also mentioned as a component. The NACLAR Guidelines encourage institutions to support training programs and to encourage active participation by personnel, especially where training is not mandatory in a facility. Training records are advised to be maintained, whether by the individuals, IACUC, departments or training coordinators.

Specific training requirements and recommendations are highlighted in the Training Guidelines for animal facility staff (caretakers, technicians, managers, and veterinarians), researchers, IACUC members, service personnel, and teachers (tertiary and nontertiary levels). There is also a recommendation that training be provided for any biohazard-related work involving genetically modified organisms, radiation, chemicals and general hazards.

IACUC members are required to undergo the Responsible Care and Use of Laboratory Animals Course, which provides an understanding of basic animal experimentation requirements, and teaches basic handling skills and manipulation techniques. This course is compulsory for all IACUC members within the first 12 months of their appointment to the IACUC.

At least 50% of IACUC members are required to receive formal IACUC training; nevertheless, all IACUC members are encouraged to undertake this training to better understand their individual roles and responsibilities within the various categories of IACUC membership. IACUC training introduces members specifically to the mandates, compositions, and functions of the IACUC as well as relevant regulations, policies, and operations. IACUC training must be undertaken within the first 12 months of appointment to the IACUC.

Animal Care and Use Areas Addressed by the Regulations

Animal Transportation

Animal transportation is addressed in the Guiding Principles, Chapter 4: Procurement and Transport of Animals. The chapter lists factors which can cause distress during transportation, such as confinement, movement, noise, environmental conditions, personnel, duration and mode of transportation, food and water restrictions, and stocking density.

Animals must be transported under species-appropriate conditions which meet welfare standards or international transport standards, with potential sources of distress identified and minimized. Sufficient shelter, food, water, nesting or bedding material, and space must be provided to the animal to express normal behaviors. At the same time, animals should be protected from sudden movements, extremes of temperature and humidity, and be prevented from escaping. Delivery and receipt of animals must be carried out appropriately and responsibly, with all animals accounted for.

Physical Plant and Environmental Conditions

Physical plant and environmental conditions are detailed in the Guiding Principles, Chapter 3: Animal Housing and Management. Information is also provided in the IACUC Guidelines, Chapter 2: Oversight of the Animal Care and Use Program. Animals should be provided with environmental conditions which suit their behavioral and biological needs where possible, with appropriate air exchange, temperature, humidity, noise, light intensity, light cycles, and ventilation. Recommended temperature and humidity ranges, ventilation, and lighting for common laboratory animal species are further listed in the NACLAR Guidelines, Appendix IID.

Environmental conditions should be monitored daily and recorded, with prompt investigation of any deviations or abnormalities. It is recognized that outdoors-housed animals face a different set of environmental conditions, especially in a tropical climate like Singapore's. Nevertheless, the conditions should be suitable for the species of animals.

Housing or Enclosures

Chapter 2: Animal Housing and Management of the Guiding Principles and Chapter 2: Oversight of the Animal Care and Use Program of the IACUC Guidelines recommend that animal housing facilities should be appropriately staffed, designed, constructed, equipped, and maintained to provide a high standard of animal well-being. Both outdoor and indoor housing facilities should suit the needs of the particular species, with pens, cages, and containers designed, constructed, and maintained to meet species-specific behavioral and environmental requirements.

Generally, pens, cages, and containers should be of durable materials, be maintained in a clean and good condition, provide safety and shelter to animals, and allow for expression of species-specific behaviors. At the same time, they should not allow or cause injury to the animals, and also prevent animals from escaping. Bedding or nesting materials should be provided where appropriate. Housing or caging systems should allow for easy observation of animals with minimal disturbance.

The Guiding Principles, Appendix II, Standards for Housing and Environmental Conditions provides cage size recommendations for the most common laboratory animal species. Recommendations for floor area and height of cages are based on individual animal body weights and are similar to the *Guide for the Care and Use of Laboratory Animals*, NRC 1996.[11]

The Guiding Principles specifically advise that rodents are not housed on wire-floor cages, unless absolutely necessary, and if then, only for short periods. A solid resting area should be provided for wire-floor cages. Social housing is encouraged as far as possible. Animals that must be housed individually should receive environmental enrichment or have the impact of social isolation minimized.

Housing facilities should be disinfected and cleaned appropriately and there should be a pest control program in place. Adequate contingency plans must be in place to cover emergencies such as flooding or fire, or the breakdown of lighting, heating, cooling or ventilation.

Daily Care Activities

Chapter 2: Animal Housing and Management of the Guiding Principles details guidelines for daily care activities, as does Chapter 2: Oversight of the Animal Care and Use Program of the IACUC Guidelines. Animals should receive sufficient species-appropriate, uncontaminated, palatable, and nutritionally adequate food to meet their needs for growth or maintenance. Food should be stored in appropriate storage areas off the floor to minimize deterioration and contamination with regular cleaning of feeding equipment performed.

Animals should have access to a constant and reliable source of drinking water, which should be clean and uncontaminated. Any variation to food or drinking requirements in nutrition-based trials must be evaluated and approved by IACUC and should be gradual.

Adequate and effective hygiene and sanitation practices must be maintained and monitored on a regular basis. This should include cleaning and disinfection of all individual pens, cages or enclosures, animal housing areas, procedural areas, feed and equipment storage areas etc. A pest control program should be in place.

Behavioral and Environmental Management Enrichment

Enrichment and environmental complexities are also detailed in the Guiding Principles, Chapter 2: Animal Housing and Management and the IACUC

Guidelines Chapter 2: Oversight of the Animal Care and Use Program. Wherever possible, animals should be provided with enrichment or environmental complexities that promote expression of normal behaviors.

Group housing is recommended for social animals as far as possible, in socially compatible settings. Where animals must be individually housed, the effects of physical isolation should be minimized by the use of noncontact communication, reflection, or by increasing environmental complexity as far as possible. A variety of enrichment types can be provided apart from social enrichment, such as occupational enrichment, physical enrichment, sensory enrichment, or nutritional enrichment.

Animal Use and Cultural Aspects in the Use of Animals

Research in Singapore is generally carried out by researchers of diverse nationalities, and cultural aspects have not tended to influence choices of animal species. Animals are generally chosen according to their suitability for the study. Commonly used animal species include rodents (rats, mice), guinea pigs, rabbits, pigs, goats, nonhuman primates, poultry, and fish.

Animal use should be undertaken only after due consideration of the project's value to human or animal health, or the advancement of knowledge, weighed against the potential effects on animal welfare. Animals used must be of an appropriate species and quality for the scientific purpose concerned. While a minimum number may be required to obtain scientifically valid results, this should not result in greater suffering of individual animals in a project or study. Animal use projects must be as brief as possible, with death as an endpoint avoided if at all possible.

Occupational Health and Safety

In general, the facility's occupational health and safety program must conform to relevant biosafety standards set out by the Ministry of Health, Singapore and the occupational health and safety legislation set out by the Ministry of Manpower, Singapore. Occupational health and safety is an important component of the facility's animal care and use program (Chapter 2: Oversight of the Animal Care and Use Program of the IACUC Guidelines). The IACUC participates in the facility's occupational health and safety program by incorporating reviews of research activities that may be hazardous to personnel into the IACUC protocol review process. The IACUC additionally should ensure that personnel have received appropriate training.

Veterinary Care

A dedicated chapter of the Guiding Principles, Chapter 6: Veterinary Care describes standards of veterinary care required. The specific role of the veterinarian is detailed in Chapter 2: Oversight of the Animal Care and Use Program of the IACUC Guidelines.

Each facility must formally engage an Attending Veterinarian (AV) who is licensed by the AVA, with the appropriate authority to oversee the animal care and use program, and to provide veterinary care to the animals. The AV can be employed on a full-time or part-time basis with a formal schedule of regular visits to the facility. Interim arrangements must be made to ensure constant and ready access to veterinary care when the AV cannot be present.

Components of veterinary care include:

- Availability of appropriate facilities, personnel, equipment, and services which comply with the Guiding Principles
- Use of appropriate methods to prevent and control diseases, diagnose and treat diseases and injuries
- 24 hour emergency, weekend and holiday care
- Monitoring of animals' health and well-being
- Provision of training and guidance to researchers or other personnel involved in the care and use of animals in handling, immobilization, anesthesia, analgesia, tranquilization, surgical manipulations, and euthanasia
- Adequate preprocedural, surgical, and postprocedural care
- Preventative medicine programs
- Attention to both physical health and psychological well-being
- Provision of advice to researchers on the design and implementation of study proposals.

There must be effective communication channels between the staff and the veterinarian, such that the veterinarian is notified of any case requiring his/her attention. The veterinarian must be able to provide prompt care and treatment, and to keep the staff informed of any further monitoring to be made, or daily treatments to be carried out.

Euthanasia

Euthanasia is described under Chapter 8, Responsibilities of Investigators in the Guiding Principles and under Chapter 3, Review of Proposals of the IACUC Guidelines. Examples of acceptable, conditionally acceptable (requires IACUC approval of scientific justification) and unacceptable methods are listed in Chapter 3: Review of Proposals of the IACUC Guidelines. There are three references on euthanasia (Appendix I): *Euthanasia of Animals Used for Scientific Purposes* by the Australian and New Zealand Council for the Care of Animals in Research and Teaching,[28] the Canadian Council *Guide to the Care and Use of Experimental Animals*, Chapter XII,[29] and The AVMA *Report of the AVMA Panel on Euthanasia*.[30]

Euthanasia of animals is recognized as necessary to relieve pain, distress or suffering to animals when humane endpoints have been reached. The choice of euthanasia method is dependent on factors such as species, age, availability of restraint, personnel skill, and others. Any method used must avoid distress, be reliable, and produce rapid loss of consciousness without pain till death occurs.

Euthanasia should be performed in a quiet, clean environment, preferably away from other animals, with death determined prior to carcass disposal. Any dependent neonates must have alternative care provided, or also be euthanized, where applicable. Only personnel who have been designated by a veterinarian, or demonstrated to a veterinarian their competency, should perform euthanasia on animals.

Generally, IACUC should review and approve the choices of euthanasia methods based on minimum pain, distress, anxiety or apprehension caused; minimum delay until unconsciousness; reliability and irreversibility; personnel safety; emotional effect on personnel; compatibility with research goals, requirements and purposes; compatibility with species, age, and health status; and drug availability and human abuse potential.

Special Considerations

Nonhuman Primates

Special considerations for the care and use of NHPs for scientific purposes are indicated in Appendix III, Additional Information on the Care and Use of Nonhuman Primate for Scientific Purposes of the Guiding Principles. NHPs are regarded as being similar to humans, with highly developed mental and emotional capacities. As such, environmental enrichment is especially important. Group housing is preferred due to their complex social hierarchy and interactions, but the potential for problems such as dominance hierarchies, wounding, and disease transmission should be considered. Housing enclosures should be designed in ways such that NHPs can express specific behaviors such as gathering food, perching, swinging as well as vertical flight. Appendix III also advises that while interaction is encouraged, it should not be forced; direct physical contact should be minimized to prevent bilateral disease transmission; as well as to avoid emotional attachment.

Genetically Altered Animals

The Genetic Modification Advisory Committee (GMAC) oversees and advises on issues relating to genetic modification and genetically modified organisms. The GMAC Guidelines[31] details general considerations in transport of transgenic animals, with two critical principles:

- The animals are prevented from escaping.
- The need for arrival to the intended delivery, with proper identification processes and accounting for all animals by a competent biologist.

The aim of both principles is to ensure that transgenic animals do not escape into the environment and potentially interbreed with feral populations. It may also be necessary for the Institutional Biosafety Committee of the facility to make additional rules or conditions, and inspect transport arrangements to ensure compliance with the above two principles. In addition to transportation of animals adhering to transportation guidelines detailed in the Guiding Principles of the NACLAR Guidelines, the GMAC Guidelines also stipulate that the transport carriers should comply with International Air Transport Association (IATA) criteria. They must be escape-proof and allow for easy inspection without the carriers being opened.

Special considerations for the use of transgenic animals are described in Chapter 3: Review of Proposals of the IACUC Guidelines. The IACUC has a responsibility in determining if any mutant gene can result in a severely debilitating phenotype, and mitigating measures which can be put into place to address this. For example, modified husbandry measures or housing conditions can be of use.

General criteria for humane endpoints should be included in the project proposal. Specific clinical abnormalities which are known or suspected to occur in the development of a new mutant model should either be included at the start, or made known to the IACUC when the information becomes available. It is recognized that experimental endpoints may differ from mutant animals as compared to "normal" animals where the phenotype involves clinical abnormalities.

Personnel performing genetic manipulations should be qualified and trained in these procedures as well as related procedures such as aseptic surgery.

Reuse of Animals

Reuse of animals is addressed in the Guiding Principles under Chapter 8: Responsibilities of Investigators. Generally, animals should not be used in more than one experiment (whether in the same project or a different one) without evaluation and approval by the IACUC. While reuse of animals may reduce the total number of animals used in a project, the general principles are that animals should not be used in unnecessarily repeated or duplicated scientific activities, and that a reduction of animals used should not result in greater suffering of individual animals thereafter.

Animal reuse should not be advocated as a reduction strategy, and reduction should not be a rationale for reusing animals. Thus, refinement and reduction goals should be balanced on a case-by-case basis. In all circumstances, humane endpoints should be observed to determine when animals can be removed from the study, treated or euthanized.

The IACUC is advised to consider if experiments involving reuse of animals contain procedures which cause animals' pain or distress; whether there are potential long-term or cumulative effects of pain or distress; the total time an animal will be used; the amount of pain, distress or biological stress likely to be caused by next and subsequent procedures; and whether animals are allowed to recover fully from the first study prior to being used in subsequent studies. If the IACUC approves animal reuse, it may then decide to require that the previously approved protocol be amended to include the approved changes, or a separate protocol be developed to cover the reuse of animals in another study.

Animals are specifically not permitted for use in multiple major survival surgeries (MMSS) unless scientifically justified by the researcher in writing, such as if the MMSS is a related component of the same study; or if it is required by the AV as a procedure for the animal's health and well-being; or unless there are any other special circumstances which have been duly evaluated and approved by the IACUC.

Other Areas

Annual Reports

Each research facility is required to prepare and submit an annual report per calendar year, i.e. 1 January to 31 December. The report must be signed off and certified by the CEO of the facility. The categories of information required in the report fall under:

Assurance

Assurance should be provided that the facility has adhered to the NACLAR Guidelines. Any exceptions must be explained by the researcher and approved by the IACUC. Each researcher must have considered alternatives to painful procedures, and professionally acceptable standards must have been applied to the care and use of animals.

Background Information and Statistics

The composition of the IACUC and the name of the AV are stated. Locations of all facilities where animals were housed or used should be indicated. Species types and numbers of animals used in different pain categories should also be listed. The different pain categories are:

a. Scientific purposes involving no pain or distress, or with the use of pain-relieving drugs
b. Scientific purposes involving accompanying pain or distress for which appropriate anesthetics, analgesics or tranquilizers were used
c. Scientific purposes involving accompanying pain or distress for which appropriate anesthetics, analgesics or tranquilizers were not used due to potential adverse effects on the procedures or results, and which an explanation is provided by the researcher and accepted by IACUC
d. Animals being bred, conditioned or held for scientific purposes but not yet used for such purposes.

Self-Regulation

Dates of the semiannual animal care and use program review, and the annual facility inspection review are to be indicated. Deficiencies should be identified as significant or minor with relation to compromise of animal welfare, with corrective actions planned, and a time frame for rectification scheduled.

Care and Use of Fish

Appendix IV, Additional Information on the Care and Use of Fish for Scientific Purposes of the Guiding Principles describes special considerations for fish in terms of choice of species used; procurement; transportation and handling; acclimatization; quarantine; design and construction of facilities; water quality and temperature; illumination; stocking density and water flow; diet and feeding; health program and disease control; analgesia, anesthesia, and invasive procedures; euthanasia; and dangerous species and zoonoses.

THAILAND

Background Information and Current Situation

Laboratory animal science in Thailand was virtually unknown until 1961 when the Southeast Asia Treaty Organization (SEATO) Medical Research Laboratory set up a laboratory animal unit for conducting biomedical research on tropical diseases. Upon dissolution of SEATO in 1977, the Laboratory was renamed Armed Forces Research Institute of Medical Science (AFRIMS). AFRIMS operates as a joint American-Thai military medical research venture composed of the United States Army Medical Component (USAMC) and the Royal Thai Army Component. The laboratory animal program at the USAMC-AFRIMS has followed the US Animal Welfare Law and all other relevant regulations. It has earned and maintained AAALAC International accreditation since 1999.

In contrast, needs for laboratory animals for biomedical research in Thailand did not come to light until 1971 after series of ad hoc committee meetings which led to a memorandum of understanding to set up a national centre under Mahidol University's responsibility for breeding and supplying small laboratory animals to users in the country. At that time, Chulalongkorn University responded by including laboratory animal science subject in its curriculum at the Faculty of Veterinary Science. The development of laboratory animal care and use in Thailand has since slowly progressed and has the following landmarks:

- In 1973, the World Health Organization/United Nations development Program agreed to provide technical support for the national laboratory animal project and assigned Dr Stian Erichsen (the International Council for Laboratory Animal Science's (ICLAS) Secretary General), as the consultant to Mahidol University.
- In 1974, the National Economic and Social Development Board approved the project proposed by Mahidol University.
- In 1975, the National Laboratory Animal Centre of Thailand (NLACT) was established at Salaya Campus, Mahidol University.
- In 1978, the NLACT commenced its supply of rats and mice nationwide.
- In 1988, Thailand hosted the 9th ICLAS International Symposium in Bangkok.

- In 1999, the National Research Council of Thailand (NRCT) issued the *Ethical Principles and Guidelines for the Use of Animals for Scientific Purposes* (the *Ethical Guidelines*). The *Ethical Guidelines* were based on the CIOMS International Guiding Principles for Biomedical Research involving animals.[19]
- In 2001, the Thai Government Cabinet approved the NRCT's proposal for the National Committee for Research Animal Development (NCRAD).
- In 2002, the Thai Association of Laboratory Animal Science (TALAS) was founded.
- In 2006, the NRCT's *Ethical Guidelines* were revised and the English version was established.[32]
- In 2007, the Cabinet approved the National Strategic Plans for the Care and Use of Animals for Scientific Purposes and the setting up of the Secretariat Office for NCRAD (SONCRAD) to oversee the Strategic Plans development.
- In 2012, the 5th Asian Federation of Laboratory Animal Science Associations (AFLAS) was organized in Bangkok by TALAS.

There are at least 25 Thai governmental institutions with over 70 laboratory animal units throughout the country using laboratory animals for research, teaching, and testing. AAALAC International has accredited two of these institutional programs in 2010 and 2012 respectively. At least 10 more institutions should be able to seek and achieve accreditation within the next 2 years. The other institutions will gradually be managed to obtain funds for improving their animal facilities. The NLACT supplies approximately 300,000 rodents each year. About 90% are mice and the others are rats. Outbred rodents are kept under strict hygienic conventional system and the inbred animals are under specified pathogen free condition. Most laboratory animal facilities in Thailand are generally set up for biomedical research and the animals on demand are mice, rats, guinea pigs, rabbits, and hamsters. Rats, mice, and rabbits have been occasionally used for drug testing and biological products development. Dogs, cats, pigs, and monkeys are seldom used due to lack of sources and facilities. All animal users are required by the *Ethical Guidelines* to prepare protocols for using animals, and the protocols must be approved by the Institutional IACUC before using the animals.

Summary of Regulations, Guidelines, or Legal Requirements

Thailand has no laws concerning animal welfare and the care and use of laboratory animals for scientific purposes but has issued the NRCT's *Ethical Guidelines*.[31] Animal users in Thailand are expected to use animals only for acquiring new knowledge in biomedical sciences and improving the well-being of both humans and animals without offending the public by carefully selecting only quality animals, providing high standards of animal care and management, and using refined techniques to reduce animals' pain and distress.

In 2009, the NRCT has issued the *Fundamental Principles on Designing Animal Care and Use Facilities for Scientific Work*.[33] The Principles provide recommendations on the following areas including location, accessibility, building characteristics, functional and support areas, facility design criteria, air ventilation and management, air temperature, relative humidity and air pressure control and management, and safety standard.

Scope

The *Ethical Guidelines*[32] define that "an animal" means any vertebrate animal and cover educational institutions at all levels, state enterprises, governmental and private sectors.

The Principles

The 3Rs principles (Replacement, Reduction, and Refinement) are integral to the *Ethical Guidelines*. The *Ethical Guidelines* emphasize the five following Ethical Principles and recommend practical guidelines for the use of animals for scientific purposes:

1. Animal users are to be aware of the value of life of animals.
   Animals are to be used only for specific purposes, which have been carefully considered as beneficial and most necessary for the development of the quality of life of both humans and animals and/or the progress of science, and when it has been established that there is no other available equivalent option.
2. Animal users are to be aware of the accuracy of the research outcome using the minimal number of animals.
   Animal users are to be aware that the objective and goal of the project can be achieved accurately with the least number of animals depending on the genetics and health quality of the animals and the appropriate planning and techniques used in the experimental design.
3. The use of wild animals must not violate laws or policies for wildlife conservation.
   The use of wildlife is to be restricted to scientific research that cannot be replaced by any other kind of animals, and it is to abide by the laws and policies for wildlife conservation.
4. Animal users need to be aware that animals are living beings just as humans are living beings.
   Animal users have to be aware that animals experience a sense of pain and respond to their surroundings in the same way that humans do. Animals are to be treated with caution to avoid stress, pain and suffering by providing optimum conditions for transportation, animal husbandry, environmental enrichment, prevention of diseases and appropriate experimental techniques.

5. Animal users must keep detailed data and records of animal experiments. Animal users must strictly follow the protocol described in their proposal, and all the details of the experiments are to be recorded in full and made available for public release or investigation at all times.

Ethical Review

The NRCT appoints the National Committee for Research Animal Development Committee (NCRAD) to be responsible for monitoring and promoting the ethical use of animals in research, testing, production of biological materials, and teaching, through the *Ethical Guidelines*; and for amending the *Ethical Guidelines* as needed. The NCRAD, supported by the SONCRAD, promotes and reinforces that the use animals in Thailand follows the *Ethical Guidelines*,[32] supervising governmental and private institutions. The care and use of laboratory animals in all institutions is overseen by the SONCRAD. The Committee has the authority to investigate the internal affairs of an institute which is subject to public complaints or accusation by the people, public media, published documents, academic publications and research funding organizations for its violation of the Ethical Guidelines. The Committee cultivates awareness in the Budget Bureau and institutional budgeting units to support the necessity of the *Ethical Guidelines*, and to endorse requests from institutions for sufficient funding; and co-ordinates with research funding organizations to render their support to the projects endorsed by the Institutional Committees.

Every institution that uses animals in research, testing, production of biological materials, and teaching is advised to have at least one committee to manage and be accountable for the use of animals, so that the *Ethical Guidelines* are followed. The Institutional Committee should be pluralistic; and its membership should be diverse including members of the management of the institution, researchers, and lay people. The responsibilities of the Institutional Committee are as follows:

- To set up standard operational procedures according to the *Ethical Guidelines*.
- To review every project that involves the use of animals in research, testing, production of biological materials, and teaching at or outside the institution, and to forward it to the executive board of the institute for approval. Only projects that follow the *Ethical Guidelines* should be authorized to proceed or continue.
- To monitor the use of animals so that experimentation follows the *Ethical Guidelines*.
- To manage the animal unit of the institute so that it complies with the standards as set forth in the *Ethical Guidelines*.
- To support and ensure that the animal unit of the institute is sufficiently funded for facilitating the standard set up in the *Ethical Guidelines*.
- To provide personnel with continuing education and training to increase the knowledge in laboratory animal science.

In addition, the *Ethical Guidelines* recommend that editorial boards of academic journals should request the author(s) of the submitted research paper to provide detailed information concerning the genetic background and the number of animals used, animal care provided, and experimental protocols including the certificate of approval for the research project issued by the author's Institutional Committee. The manuscript should be rejected unless all the above requirements are fulfilled.

The *Ethical Guidelines* recommend that before using animals, the users should submit their protocol in detail, including the following information and considerations:

1. Research steps and plans, including the objectives and expected benefit to uplift the quality of life of humans or animals and/or the progress of science, and the accumulation of academic information.
2. The user should support the need to use animals with evidence and reasons that there is no other alternative. Animal users should reserve the use of animals for situations when there is unavoidable necessity or when there is no other available option.
3. The users should carefully study all information and related documents related to their research, and they should utilize all the information to make the most out of the use of the animals in their research.
4. Animal users should select the species and the breed of animal that fits in with the objectives and goals of the research.
5. A statistical method should be employed for estimating the least number of animals required while ensuring the most accurate and acceptable results.
6. Animal users should include in the protocol the set up of the experiment, the techniques and the materials to be used, and the care and management of the animals before, during and after experimentation. Animal users should employ only the most appropriate techniques, technology and statistical methods in planning and evaluating the project.
7. Animal users and animal caretakers should treat animals with kindness and should avoid any procedure or process that causes pain and stress to the animals. In case the situation is unavoidable, they should explicitly state the reasons that should be entirely based on academic grounds.
8. The experiment should be concluded before the animals die of extreme pain. At the end of each experiment, the users are responsible to euthanize all animals. In case the animals are to survive, the users must provide the reasons for such necessity in their proposals, and must be responsible for rearing the animals under conditions appropriate for the species. The animals should neither be released to nature, nor should they be abandoned at the animal unit without appropriate care.

The NRCT issued the *Standard for Institutional Animal Care and Use Committee*[34] in 2012. Any institution maintaining and using animals in scientific purposes should appoint one committee (IACUC) and show the IACUC reporting line and relationship to other sections clearly in its organization chart. The NRCT's Standard determines responsibilities of the IACUC to follow the *Ethical Guidelines*, and The NRCT's Standard describes IACUC compositions and functions.

The IACUC should include at least the following seven members and one member may have more than one role:

- An Institutional Administrator as the Chair of the Committee
- The Chief of the laboratory animal unit as a member and secretary
- The Attending Veterinarian
- At least two scientists with experience in research using laboratory animals
- A representative from a nonscientific discipline organization
- An expert or professional in others areas, i.e. a statistician, a lawyer or a librarian
- Must have at least one nonaffiliated member.

The IACUC has the following functions:

- Establish an annual budget plan to improve the animal care and use program
- Submit the annual budget plan to the institutional official (IO)
- Review animal use protocols
- Perform an oversight of the animal use protocol procedures
- Improve personnel
- Communicate continuously to the public and distribute knowledge
- Establish standard operating procedures (SOPs) and practices in the care and use of animals at the institution
- Make a routine annual institutional report and submit it to the IO.

Personnel Training, Education, and Competency Requirement

The *Ethical Guidelines* Item 4.4.2 recommends that animal units be staffed with a veterinarian or a well-trained academic who has sufficient knowledge and experience in laboratory animal science, and a well-trained team of animal caretakers.

The *Ethical Guidelines* Item 4.5.3 recommends that animal users should acquire the following basic skills regarding the animals before performing experiments: animal handling and restraining; identification; sexing; drug and substance administration; samplings of blood, stool, urine and tissues from living animals; anesthesia and anesthetizing procedures; euthanasia; and necropsy procedures.

The NRCT's Standard for IACUC Item 2.1[34] recommends that the IACUC members must receive a training in subjects relevant to the *Ethical Guidelines* and there must be at least one annual continuing education. In addition, Item 1.6 recommends that all personnel involved in the care and use of animals should receive education and training to enhance their knowledge and skill.

Animal Care and Use Areas Addressed by the *Ethical Guidelines*

Animal Transportation

The *Ethical Guidelines* Item 4.1 recommends that the means of transportation should provide safety for the animals and should have the least impact on the well-being of the animals. The animals should not be exposed to extreme environments. Adequate spaces and appropriate temperature and ventilation should be provided to avoid stress. Delivery boxes should be strong and well secured to avoid escape.

Animal Care

Physical Plant and Environmental Conditions

The *Ethical Guidelines* Item 4.2 recommends that the animal facility must be equipped with systems that can control infection, temperature, humidity, ventilation, lighting, and sound, to suit the needs of each species.

Housing or Enclosures

The *Ethical Guidelines* Item 4.3.1 recommends that animal cages must be strong enough to prevent the animals from escaping. The types and sizes depend upon the standard required for species, weight, and number of animals in the cage, but there is no specific recommendation on cage sizes and enrichment. The cages must be free from sharp edges or projections that could cause injury to the animals. They should be made of durable material that can withstand the chemical substances and the heat employed for disinfection or sterilization.

Daily Care Activities

The *Ethical Guidelines* Item 4.3.3 recommends that the animals must be fed daily with food and water, which are free from pathogens, toxic and carcinogenic substances. Their food has to be nutritionally adequate comprised of protein, fat, starch, vitamins, minerals, and fibers, in the proportions that are suitable for the requirements of the species. The bedding must be absorptive, and not disintegrate when wet. It must be free from sharp edges and contain no toxic substances and germs.

REFERENCES

1. The World Organisation for Animal Health (OIE). In: *Terrestrial animal health code*. 21st ed. vols 1–2. Available at: http://www.oie.int/en/international-standard-setting/terrestrial-code/access-online/. Accessed December 13, 2012.
2. The World Organisation for Animal Health (OIE). Chapter 7.1. Introduction to the Recommendations for Animal Welfare. In: *Terrestrial animal health code*. vol. 1. Available at: http://www.oie.int/index.php?id=169&L=0&htmfile=chapitre_1.7.1.htm. Accessed December 13, 2012.

3. Statsblad Law No. 432. 1912. Government Act (Ordonantie) Regarding Legal Review on Regulations Related to Government's Supervision on Animal and Veterinary Inspector (Ordonansi tentang Peninjauan Kembali Ketentuan-Ketentuantentang Pengawasan Pemerintahdalam Bidang Kehewanandan Polisi Kehewanan 1912 No. 432) (Herziening van de Bepalingen Omtrent het Veeartsenijkundige Staatstoezicht en de Veeartsenijkundige Politie, Staatsblad 1912 No. 432).

4. Law of the Republic of Indonesia No. 6. 1967. Chapter 22; Article 22 (Undang-Undang Republik Indonesia No. 6 Tahun 1967 Pasal 22).

5. Law of the Republic of Indonesia No. 18. 2009. Chapter VI, Veterinary Public Health and Animal Welfare; Part 2, Animal Welfare; Articles 66–67 (Undang-Undang Republik Indonesia Nomor 18 Tahun 2009. Bab VI. Kesehatanmasyarakat Veteriner. Bagiankedua: Kesejahteraan Hewan, Pasal 66–67).

6. Law of the Republic of Indonesia No. 18. 2009. Chapter VII, Veterinary Authority; Article 74 (Undang-Undang Republik Indonesia Nomor 18 Tahun 2009 Bab VII—Otoritas Veteriner, Pasal 74).

7. Government of Republic of Indonesia's Regulation No. 95. 2012. Veterinary Public Health and Animal Welfare(Peraturan Pemerintah Republik Indonesia. Kesehatan Masyarakat Veterinerdan Kesejahteraan Hewan Nomer 95 Tahun 2012).

8. The National Guidelines on Health Research Ethics. *Health research ethics committee*. Republic of Indonesia: The Ministry of Health; 2011.

9. Teaching Guide book for Ethics on Health Research. *Health research ethics committee*. Republic of Indonesia: The Ministry of Health; 2011.

10. National Bioethics Committee Report. 2008. The Ministry of Health, Republic of Indonesia.

11. National Research Council (NRC). *Guide for the care and use of laboratory animals*. Washington: National Academies Press; 1996.

12. American Veterinary Medical Association (AVMA). *AVMA guidelines on euthanasia*. Schaumburg, IL: AVMA; 2007.

13. Conservation Law No. 5. 1990. The Ministry of Forestry, Republic of Indonesia. (Undang-Undang Konservasi Hayati No. 5 Tahun 1990. Kementrian Kehutanan Republik Indonesia).

14. The Laws of Malaysia Act 647 Also Known as the Animal Act 1953 (Revised in 2006). Department of Veterinary Services. Available at: http://www.maqis.gov.my/c/document_library/get_file?uuid=51bf870c-537b-4681-bb03-d1ab4c7066db&groupId=29274. Accessed February 22, 2013.

15. The Wildlife Conservation Act 2010, Laws of Malaysia Act 716. Department of Wildlife and National Parks. Available at: http://www.wildlife.gov.my/images/stories/akta/Wildlife_Conservation_Act2010_Act716.pdf. Accessed February 22, 2013.

16. The Animal Welfare Act draft. Department of Veterinary Services. Available at: http://research.dvs.gov.my/survey/upload/survey. Accessed on February 22, 2013.

17. The Malaysian Code of Practice for the Care and Use of Animals for Scientific Purposes (Draft). The Laboratory Animal Science Association of Malaysia (LASAM). Available at: http://www.vet.upm.edu.my/IACUC%20-%20UPM%20Code%20of%20Practice.pdf. Accessed February 22, 2013.

18. Masangkay JS. 2008. The Status of Laboratory Animal Research in the Philippines. Proceedings of the 3rd AFLAS Congress and the 8th CALAS Annual Meeting in Beijing, China, September 27–29, 2008: 13–16.

19. CIOMS (Council for International Organizations of Medical Sciences). International Guiding Principles for Biomedical Research Involving Animals, 1985. Available at: http://www.cioms.ch/images/stories/CIOMS/guidelines/1985_texts_of_guidelines.htm. Accessed November 20, 2012.

20. Department of Health and Human Services. In: Richmond JY, McKinney RW, editors. *Biosafety in microbiological and biomedical laboratories*. 4th ed. Washington: Government Printing Office; 1999.

21. The Republic Act 8485. An Act to Promote Animal Welfare in the Philippines, Otherwise Known as The Animal Welfare Act of 1998. Available at: www.angelfire.com/ok2/animalwelfare/welfareact.html. Accessed October 6, 2012.

22. Administrative Order No. 40. Series of 1999. Rules and Regulations on the Conduct of Scientific Procedures Using Animals. Philippines Department of Agriculture. Available at: http://paws.org.ph/Portals/_Rainbow/images/default/AWA%20AO40-Rules%20and%20Regulations%20on%20the%20Conduct%20of%20Scientific%20Procedures%20Using%20Animals.pdf. Accessed October 6, 2012.

23. Philippine Association for Laboratory Animal Science (PALAS). *Code of practice for the care and use of laboratory animals in the Philippines*. 2nd ed. 2002.

24. Animals and Birds (Care and Use of Animals for Scientific Purposes) Rules, 2004. Agri-Food and Veterinary Authority of Singapore. Available at: http://www.ava.gov.sg/NR/rdonlyres/0CA18578-7610-4917-BB67-C7DF4B96504B/17786/19web_AB_CareandUseofAnimals forScientificPurposesR.pdf. Accessed November 20, 2012.

25. NACLAR Guidelines (National Advisory Committee for Laboratory Animal Research Guidelines on the Care and Use of Animals for Scientific Purposes), 2004. Agri-Food and Veterinary Authority of Singapore. Available at: http://www.ava.gov.sg/NR/rdonlyres/C64255C0-3933-4EBC-B869-84621A9BF682/13557/Attach3_AnimalsforScientificPurposes.PDF. Accessed November 20, 2012.

26. European Convention for the Protection of Vertebrate Animals Used for Experimental and Other Scientific Purposes, 1986. Available at: http://conventions.coe.int/treaty/en/treaties/html/123.htm. Accessed November 20, 2012.

27. Animals and Birds Act. Chapter 7, Agri-Food and Veterinary Authority of Singapore, 2002. Available at: http://www.ava.gov.sg/NR/rdonlyres/0CA18578-7610-4917-BB67-C7DF4B96504B/17773/7web_ABAct.pdf. Accessed November 20, 2012.

28. ANZCCART (the Australia and New Zealand Council for the Care of Animals in Research and Teaching). In: Reilly JS, editor. *Euthanasia of animals used for scientific purposes: a monograph*. SA: Glen Osmond; 1993.

29. CCAC (Canadian Council on Animal Care). Chapter XII 2nd ed. *Guide to the care and use of experimental animals*, vol. 1. Albert St., Ottawa, Ontario, Canada: CCAC, 1993; 1510–130, K1P5G4. Available at http://www.ccac.ca/Documents/Standards/Guidelines/Experimental_Animals_Vol1.pdf. Accessed February 23, 2013.

30. AVMA (American Veterinary Medical Association). Report of the AVMA panel on euthanasia. *J Am Vet Med Assoc* 2001;**218**(5):669–96.

31. GMAC Guidelines (The Genetic Modification Advisory Committee Guidelines), 2006. The Singapore Biosafety Guidelines for Research on GMOs. Available at: http://www.gmac.gov.sg/Index_Guidelines_Overview_on_GMAC_Guidelines.html. Accessed November 20, 2012.

32. Ethical Principles and Guidelines for the Use of Animals for Scientific Purposes, 2006. National Research Council of Thailand (NRCT). Available at: http://www.labanimals.net/images/stories/Ethics/Ethics%20on%20Animals%20.pdf. Accessed January 17, 2013.

33. Fundamental Principles on Designing Animal Care and Use Facilities for Scientific Work, 2009. National Research Council of Thailand (NRCT). Available at: http://labanimals.net/images/stories/Download/Design.pdf. Accessed January 25, 2013.

34. Standard for Institutional Animal Care and Use Committee, 2012. National Research Council of Thailand (NRCT). Available at: http://labanimals.net/images/stories/IACUC/IACUC%20Standard.pdf. Accessed January 25, 2013.

Chapter 12

Laboratory Animals Regulations and Recommendations for Global Collaborative Research: Australia and New Zealand

John Schofield[1], Denise Noonan[2], Yvette Chen[3], and Peter Penson[4]

[1]*J & L Consulting, New Zealand,* [2]*The University of Adelaide, SA, Australia,* [3]*The University of Melbourne, Australia,* [4]*Rosanna, Victoria, Australia*

Chapter Outline

Laboratory Animals. http://dx.doi.org/10.1016/B978-0-12-397856-1.00012-X

GENERAL FRAMEWORK—AUSTRALIA

Regulatory Framework

The Australian framework for the humane use of animals for research is underpinned by the principles of prevention of cruelty, duty-of-care, and Russell and Burch's 3Rs principles of humane animal use (Replacement, Reduction, Refinement). Similarities between the key principles of the Australian and New Zealand systems reflect the historic influence of United Kingdom animal welfare legislation upon Australia and New Zealand.

The Commonwealth of Australia consists of a federation of six states and two territories with a third tier of local government authorities. The regulatory framework for animal welfare and animal-based research reflects this federation. The Australian Government controls matters including those relating to gene technology, quarantine, animal export and trade in animal products, while all matters relating to scientific use of animals and animal welfare are subject to the animal welfare regulatory authority of each state or territory. Scientific uses or "Scientific Purposes" are defined by the *Australian Code for the care and use of animals for scientific purposes (2013) as: all activities conducted with the aim of acquiring or demonstrating knowledge or techniques in all areas of science including teaching, field trials, environmental studies, research (including the creation and breeding of a new animal line where the impact on animal wellbeing is unknown or uncertain), diagnosis, product testing, and the production of biological products.*[1] The State of New South Wales has a specific Act to regulate scientific use of animals, the *Animal Research Act 1985*.[2] Other states have sections within their animal welfare legislation that refer specifically to scientific use of animals: *Prevention of Cruelty to Animals Act 1986* (Victoria),[3] *Animal Care and Protection Act* (Queensland),[4] or *Animal Welfare Act* (Australian Capital Territory,[5] Northern Territory,[6] Western Australia,[7] South Australia,[8] and Tasmania[9]).

Regulatory responsibility for the welfare and scientific use of animals lies with different government departments. In New South Wales, Western Australia, Queensland, Victoria, and Tasmania, responsibility lies with Primary Industries Departments. In South Australia, it lies with the Department of Environment, Water and Natural Resources, and Local Government Departments are responsible in the Australian Capital Territory and the Northern Territory. Most states have appointed an Animal Welfare Advisory Committee, an expert panel drawn from members of the community that provides the relevant government minister with advice on animal welfare issues and on associated legislation, guidelines and Codes of Practice.

Although the specific legislation differs between the states and territories, all require the following: (1) prior licensing or registration of the scientific establishment that uses animals and/or premises where animals are bred, held or used for scientific purposes; (2) prior ethical review and approval, and ongoing oversight, of the animal care and use by an appropriately constituted Animal Ethics Committee; and (3) adherence to the principles and guidelines set out in the Guideline document entitled the *Australian Code for the care and use of animals for scientific purposes 2013* (8th edition; "Australian Code").[1]

Scope in Brief

The Australian legislative framework and relevant Codes for the use of animals in research apply to the use of animals for all types of research, as well as use of animals for the purposes of teaching. This means that in Australia, rather uniquely compared to parts of the rest of the world, similar requirements are in place for animals used for: (1) research and other "scientific purposes"—in medicine, biology, agriculture, veterinary, and other animal sciences, and in industries that use animals for product testing and the production of biological and environmental studies; (2) teaching (also considered to be a "scientific purpose" by definition in the Code) within universities and schools at primary, secondary, and tertiary levels; (3) in wildlife studies (including observational or some minimally invasive work); and (4) for veterinary clinical trials and field trials.

The species of animals protected by animal welfare legislation differ slightly between the states and territories; however, the Australian Code defines a (protected) animal as: *"any live nonhuman vertebrate (that is fish, amphibians, reptiles, birds and mammals, encompassing domestic animals, purpose-bred animals, livestock, wildlife) and cephalopods."*[1] In the State of Victoria, some decapod species are also included, following on from recent literature suggesting sentience in these species.[3]

State and territory animal welfare legislation makes provision for prohibition and restriction of particular animal procedures and uses. In accord with National Consultative Committee on Animal Welfare (NCCAW) guidelines, scientific use of the Draize test and the Lethal Dose 50 test is restricted, and in general requires approval from both the Animal Ethics Committee (AEC) and the government minister with responsibility for the animal welfare legislation.[10]

National Harmonization

There are mechanisms used to harmonize much of the regulatory and policy-setting framework in Australia.

- At the federal level, the Standing Committee on Primary Industries (SCoPI) and its Animal Welfare Committee bring together the New Zealand, Australian Commonwealth and state and territory ministers with agricultural responsibilities. SCoPI is the forum for the development of national agricultural policies, including welfare of animals.[11]
- The Australian Government Department of Agriculture, Fisheries and Forestry Animal Welfare Unit is responsible for animal welfare matters involving the Commonwealth of Australia. It performs a coordinating and support role with regard to the *Australian Animal Welfare Strategy* and *National Implementation Plan,* which provide national and international communities with a summary

of animal welfare arrangements in Australia and an outline of the plans for sustainable improvement of the welfare of all animals in Australia.[12]

At a day to day level, the Australian system is effectively a self-regulating system, since it is the locally-positioned and locally-run Animal Ethics Committees (AECs) which hold the authority to approve and oversee (and if need be, suspend) the care and use of these animals, rather than the government regulatory agency. At a higher level, the regulatory agencies have the authority to exercise their powers to license scientific institutions, inspect and audit and, as necessary, make recommendations for changes, issue penalties, suspend work or licenses, and prosecute. In addition, the Australian Code requires that all scientific establishments that use animals must participate in external reviews to ensure that regular, independent, and objective scrutiny of AEC procedures and practices takes place and that the AEC and institution are operating in accordance with the Australian Code.[1]

The Australian Code and Other Guidance Documents

The Australian Code has been prepared by a joint working party of the National Health and Medical Research Council (NHMRC), Commonwealth Scientific and Industrial Research Organisation (CSIRO), Australian Research Council, Universities Australia, representatives of Australian government departments, state and territory governments of Australia, and animal welfare organizations (Royal Society for the Prevention of Cruelty to Animals and Animals Australia).[1] The Australian Code has been recently revised and the 8th edition was endorsed and released during 2013. Authorship and review of the Australian Code is coordinated and overseen by the main funding body, the NHMRC. Representatives of the regulatory agencies in each state or territory participate in the revision process via the Jurisdictional Reference Group (previously known as the Code Reference Group).

NHMRC has produced supplementary policies and guidelines intended to be read in conjunction with the Australian Code. Compliance with these additional documents is particularly relevant to institutions receiving NHMRC grant funding.[13]

- Policy on the use of Non-Human Primates for Scientific Purposes (2003, currently under review by the NHMRC)
- Guidelines for monoclonal antibody production (2008)
- NHMRC Guidelines on the Care of Cats Used for Scientific Purposes (2009)
- NHMRC Guidelines on the Care of Dogs Used for Scientific Purposes (2009)
- Guidelines on the use of animals for training interventional medical practitioners and demonstrating new medical equipment and techniques (2009)
- Guidelines to promote the wellbeing of animals used for scientific purposes: The assessment and alleviation of pain and distress in research animals (2008)
- Guidelines for the generation, breeding, care and use of genetically modified and cloned animals for scientific purposes (2007, due for review by the NHMRC).[14]

In addition to the Australian Code, a large number of national Model Codes for animal welfare have been developed for livestock species.[15,16] State and territory animal welfare codes of practice have been developed from animal welfare legislation and Australian Model Codes. These are currently under review as part of the *Australian Animal Welfare Strategy*, and will be incorporated as standards and guidelines in Regulations under the animal welfare legislation in each state and territory.[12,16] The number of these is extensive, however a select list of Codes of Practice and guidance documents relating specifically to animals used for scientific purposes follows:

1. National Consultative Committee of Animal Welfare Guidelines[10] (N.B. NCCAW has been replaced by the Australian Animal Welfare Committee, AAWAC):
    a. Genetic manipulation of animals
    b. LD50 and Draize tests
    c. Pound sourced animals
2. New South Wales:
    a. Animal Research Review Panel Guidelines for the Care and Housing of Dogs in Scientific Institutions
    b. Animal Research Review Panel Guidelines for radio tracking in wildlife research
    c. Animal Research Review Panel Animal care guidelines for wildlife surveys
    d. Animal Research Review Panel Guidelines on opportunistic research on free living wildlife
    e. Animal Research Review Panel Guidelines on the use of feral animals in research
    f. Animal Research Review Panel Guidelines on collection of voucher specimens
    g. Animal Research Review Panel Guidelines on the use of pitfall traps
    h. Animal Research Review Panel Guidelines for the production of monoclonal antibodies
    i. Animal Research Review Panel guidelines on supply of dogs and cats for use in research
    j. Several other Animal Research Review Panel Guidelines for care and housing of specific laboratory animals[17]
    k. Animal Welfare Branch Fact Sheet 21: Supply of dogs and cats for use in research[18]
3. Victoria:
    a. Code of Practice for the Housing and Care of Laboratory Mice, Rats, Guinea Pigs, and Rabbits
    b. Code of practice for the use of animals from municipal pounds in scientific procedures[19]

4. South Australia—Wildlife Ethics Committee policies and guidelines:
   a. Policy on the euthanasia of research animals in the field
   b. Guidelines for the collection of hair and feather samples
   c. Guidelines for the use of live traps to capture terrestrial vertebrates
   d. Policy on the use of microchips for marking wildlife
   e. Guidelines for the use of tracking tunnels
   f. Guidelines for the transportation of live animals
   g. Policy on the collection of voucher specimens
   h. Policy on the collection of blood from wildlife.[20]

Where there is use of a species not specifically covered by one of the above research context-specific Guidelines, Codes or Standards pertinent to the care and use of that species in another context are used to provide general guidance on the minimum standards expected for the care of those species being held and used in a research environment. The Australian and New Zealand Council for the Care of Animals in Research and Teaching (ANZCCART) provides useful resource and reference material which is available at http://www.adelaide.edu. au/ANZCCART/resources/.[21]

GENERAL FRAMEWORK—NEW ZEALAND

The welfare of animals in New Zealand (NZ) is primarily provided for by the Animal Welfare Act 1999, (hereafter referred to as the AWA).[22] The purpose of the AWA is to reform the law relating to the welfare of animals and the prevention of their ill-treatment. In particular, the legislation requires owners of animals to attend to their proper welfare, it specifies conduct that is not permissible, and provides for approving the use of animals in research, testing and teaching (RTT). In addition, two national advisory committees were established: the National Animal Ethics Advisory Committee (NAEAC) and the National Animal Welfare Advisory Committee (NAWAC). The primary function of these committees is to advise the Minister on issues relating to codes of ethical conduct and animal welfare legislation. In support of these strategies, NZ has established a process to codify animal welfare. Currently there are codes of welfare and codes of recommendations and minimum standards (Table 12.1). These recommendations are voluntary guidelines that were produced before the Animal Welfare Act 1999 came into force. When individual codes of welfare are developed, they replace the older, corresponding codes of recommendations and minimum standards. Eventually, these older publications will be phased out completely.

The AWA is divided into eight parts that cover: care of animals, conduct towards animals, animal exports, advisory committees, codes of welfare, use of animals in RTT, provisions relating to administration and offenses. Part 6 of the AWA is devoted to the use of animals in research, testing and teaching (http://www.biosecurity.govt.nz/regs/animal-welfare).

TABLE 12.1 New Zealand Codes of Welfare and Codes of Recommendations and Minimum Standards

Codes of Welfare	Codes of Recommendations and Minimum Standards
Circuses	Animals in boarding establishments
Commercial slaughter	Animals at saleyards
Companion cats	Bobby calves (superseded by the code of welfare
Dairy cattle	for dairy cattle)
Deer	Deer during the removal of antlers
Dogs	Guidelines for the welfare of red & wapiti yearling
Goats	stags during the use of rubber rings to induce
Layer hens	analgesia for the removal of spiker velvet
Meat chickens	Guidelines for the welfare of yearling fallow deer
Painful husbandry procedures	during the use of rubber rings to prevent antler/
Pigs	pedicle growth
Rodeos	Emergency slaughter of farm livestock
Sheep and beef cattle	Horses
Transport within New Zealand	Ostrich and emu
Zoos	Sale of companion animals (also refer to code of welfare for companion cats and code of welfare for dogs)
	Sea transport of sheep from New Zealand

THE PRINCIPLES—AUSTRALIA

Justification, Responsibilities, and the 3Rs

Although the specific pieces of legislation relevant to animal use for scientific purposes differ between Australian states and territories, the key principles and framework are similar because they have all adopted the *Australian Code for the care and use of animals for scientific purposes 2013 (8th edition; "Australian Code"*; recently revised) into their respective legislation.[1] The way in which the guidance in this document is incorporated differs in subtle ways between states and territories; in some, mandatory compliance with the entire document is specified in the legislation; while in others, the key principles and terminology are embedded in the legislation itself. A detailed description and comparison of these differences are beyond the scope of this chapter, therefore, for simplicity the focus here is mainly on the requirements detailed in the Australian Code.

In its introduction, the Australian Code states that its purpose is to promote the ethical, humane and responsible care and use of animals for scientific purposes. The Code details the responsibilities of all those involved in the care and use of animals. This embraces a duty-of-care, an obligation to and support for the wellbeing of animals.

The Australian Code was the first, and remains the principal Australian guideline to mandate consideration of the principles of Replacement, Reduction and Refinement (3Rs) in the context of animal use for scientific purposes. The 3Rs must be applied at every stage of animal use, starting from the time of breeding or acquisition of the animals, for the duration of animal holding and use, and including the fate of the animals at the end of scientific use.

According to the Australian Code, a scientific or teaching activity can only be approved by an Animal Ethics Committee if the proposed activity has sufficient justification and that the work is essential for progress towards one of the following purposes: (1) to obtain and establish significant information relevant to the understanding of humans and/or animals; (2) for the maintenance and improvement of human and/or animal health and welfare; (3) for the improvement of animal management or production; (4) to obtain and establish significant information relevant to the understanding, maintenance or improvement of the natural environment; or (5) for the achievement of educational outcomes in science, as specified in the relevant curriculum or competency requirements. This prior approval pertains to all aspects of the proposed animal use, including the animals being used (includes the specification and justification of the requested species, strain, gender, age, number etc.), the procedures to be performed (including maximum permissible severity and intervention/endpoints), the investigators/personnel involved (including review of past experience and competence), and the institution and premises where the work will occur.

Ultimately, before approving the use of animals for scientific purposes the AEC must be convinced that the use is (1) essential; (2) has been appropriately justified by weighing the predicted scientific or educational value of the project against the potential effects on the welfare of the animals; and (3) that all ethical and animal welfare issues have been considered and the 3Rs have been adequately applied.

THE PRINCIPLES—NEW ZEALAND

The principles of the 3Rs have been recognized in New Zealand by adoption into the animal welfare legislation. In Part 6 of the AWA, Section 80 (2b) (1)–(4), these principles are included to promote efforts:[22]

1. "to reduce the number of animals used in research, testing, and teaching to the minimum necessary:
2. to refine techniques used in any research, testing, and teaching so that the harm caused to the animals is minimised and the benefits are maximised:
3. to replace animals as subjects for research, and testing by substituting, where appropriate, nonsentient or nonliving alternatives:
4. to replace the use of animals in teaching by substituting for animals, where appropriate, non-sentient or non-living alternatives or by imparting the information in another way."

SCOPE/APPLICABILITY—AUSTRALIA

State or territory animal welfare legislation applies to all animals in that jurisdiction. The general definition of an animal is a living nonhuman vertebrate; however, there are inclusions and exceptions of animal species that vary between the states and territories. Additional requirements and protection, as detailed in the legislation, apply to all animals used for research, testing and teaching purposes; however, common laboratory animal species such as rats, mice, guinea pigs, rabbits and non-human primates may be specifically addressed. As the *Australian Code for the care and use of animals for scientific purposes 8th edition* 2013 ("Australian Code"; recently revised) is adopted under state and territory legislation, the definition of an animal used in the Australian Code applies to all scientific use of animals nationally.[1] Terminology used in state legislation also varies, however the Australian Code defines the Scope and terminology thus:

Animal: *"any live non-human vertebrate (that is, fish, amphibians, reptiles, birds and mammals, encompassing domestic animals, purpose-bred animals, livestock, wildlife) and cephalopods." And "embryos, fetuses and larval forms [that] have progressed beyond half the gestation or incubation period of the relevant species, or they [have] become capable of independent feeding."*

Scientific purposes: *"all activities conducted with the aim of acquiring, developing or demonstrating knowledge or techniques in all areas of science, including teaching, field trials, environmental studies, research (including the creation and breeding of a new animal line where the impact on animal wellbeing is unknown or uncertain), diagnosis, product testing, and the production of biological products."*

Teaching: *"developing, imparting or demonstrating knowledge or techniques to achieve an educational outcome in any area of science as specified in the relevant curriculum or competency requirements."*

Scope as stated in the Australian Code:

"The Code encompasses all aspects of the care and use of animals where the aim is to acquire, develop or demonstrate knowledge or techniques in any area of science- for example, medicine, biology, agriculture, veterinary and other animal sciences, industry and teaching. It includes the use of animals in research, teaching with an educational outcome in science, field trials, product testing, diagnosis, the production of biological products and environmental studies."

"The Code applies throughout the animal's involvement in activities and projects, including acquisition, transport, breeding, housing, husbandry, the use of the animal in a project, and the provisions for the animal at the completion of their use."

"The Code covers all live nonhuman vertebrates and cephalopods. Institutions are responsible for determining when the use of an animal species not covered by the Code requires approval from an AEC, taking into account emerging evidence of sentience and the ability to experience pain and distress. Animals at early stages in their development—that is, in their embryonic, fetal and larval

forms—can experience pain and distress, but this occurs at different stages of development in different species. Thus decisions as to their welfare should, where possible, be based on evidence of their neurobiological development. As a guide, when embryos, fetuses and larval forms have progressed beyond half the gestation or incubation period of the relevant species, or they become capable of independent feeding, the potential for the experience of pain or distress should be taken into account."

All scientific research, teaching and testing uses of animals must have obtained prior ethical approval from an AEC before animal use can commence. There is no expedited review process for certain categories of animal use. All new animal use proposals or project applications must be considered at a quorate face-to-face committee meeting; however, minor modifications to approved projects may be considered by an executive subcommittee of the AEC. Decisions made by the executive subcommittee must be presented for ratification by the AEC at the next meeting.

In the ethical review process, a cost-benefit analysis approach is used as the tool for an AEC to consider and determine, by consensus, whether the likely or anticipated scientific or educational value outweighs the potential impact on the animals; whether the principles of the 3Rs have been applied in all respects of the care and use of the animals.

SCOPE/APPLICABILITY—NEW ZEALAND

The scope of the Good Practice Guide for the use of animals in research, testing and teaching[23] encompasses all aspects of the care and use of animals for scientific purposes in medicine, biology, agriculture, veterinary and other animal sciences, industry and teaching. It includes their use in research, teaching, field trials and product testing.

The key definitions from the AWA, which underpin the legislation, are as follows:

Animal

1. "means any live member of the animal kingdom that is—
   a. a mammal; or
   b. a bird; or
   c. a reptile; or
   d. an amphibian; or
   e. a fish (bony or cartilaginous); or
   f. any octopus, squid, crab, lobster, or crayfish (including freshwater crayfish); or
   g. any other member of the animal kingdom which is declared from time to time by the Governor."

Research, Testing, and Teaching

1. "any work (being investigative work or experimental work or diagnostic work or toxicity testing work or potency testing work) that involves the manipulation of any animal; or
2. any work that—
   a. is carried out for the purpose of producing antisera or other biological products; and
   b. involves the manipulation of any animal; or
3. any teaching that involves the manipulation of any animal

(2) The term defined by subsection (1) does not include any manipulation that is carried out on any animal that is in the immediate care of a veterinarian, if—

1. the veterinarian believes on reasonable grounds that the manipulation will not cause the animal unreasonable or unnecessary pain or distress, or lasting harm; and
2. the manipulation is—
   a. for clinical purposes in order to diagnose any disease in the animal or any associated animal; or
   b. for clinical purposes in order to assess the effectiveness of a proposed treatment regime for the animal or any associated animal; or
   c. for the purposes of assessing the characteristics of the animal with a view to maximising the productivity of the animal or any associated animal
3. The term defined by subsection (1) does not include any manipulation of an animal—
   a. which is carried out with the principal objective of—
      - assisting the breeding, marking, capturing, translocation, or trapping of animals of that type;
      - weighing or taking measurements from the animal; or
      - assessing the characteristics of animals of that type; and
   b. which is a manipulation of an animal that—
      - is carried out routinely; or
      - is a minor modification of a manipulation that is carried out routinely; within the term defined by subsection (1); and
   c. the term **teaching** means any teaching that comes within the term defined by subsection (1)."

Manipulation

1. "In this Act, unless the context otherwise requires, the term manipulation, in relation to an animal, means, subject to subsections (2) and (3), interfering with the normal physiological,behavioural, or anatomical integrity of the animal by deliberately—

a. subjecting it to a procedure which is unusual or abnormal when compared with that to which animals of that type would be subjected under normal management or practice and which involves—
  – exposing the animal to any parasite, micro-organism, drug, chemical, biological product, radiation, electrical stimulation, or environmental condition; or
  – enforced activity, restraint, nutrition, or surgical intervention; or
b. depriving the animal of usual care—and manipulating has a corresponding meaning.
2. The term defined by subsection (1) does not include—
a. any therapy or prophylaxis necessary or desirable for the welfare of an animal; or
b. the killing of an animal by the owner or person in charge as the end point of research, testing, or teaching if the animal is killed in such a manner that the animal does not suffer unreasonable or unnecessary pain or distress; or
c. the killing of an animal in order to undertake research, testing, or teaching on the dead animal or on prenatal or developmental tissue of the animal if the animal is killed in such a manner that the animal does not suffer unreasonable or unnecessary pain or distress."

Physical Health and Behavioral Needs

"In this Act, unless the context otherwise requires, the term **physical, health, and behavioural needs**, in relation to an animal, includes—

a. proper and sufficient food and water
b. adequate shelter
c. opportunity to display normal patterns of behaviour
d. physical handling in a manner which minimises the likelihood of unreasonable or unnecessary pain or distress
e. protection from, and rapid diagnosis of, any significant injury or disease—"

The framework in respect of animals used for research, testing and teaching is clear. AEC approval is required before any species (which meets the definition of an "animal") can be manipulated for the purpose of research, testing or teaching. This simple logic differentiates the use of animals for RTT from other animal use activities such as commercial farming, sport and recreation or assisting humans (e.g. guide dogs).

The NZ regulations defined by the AWA applies to all owners of animals (including parents or guardians of persons under 16 years of age who own animals), and to all persons in charge of animals. Hence, the regulations cover all institutions, colleges, schools, zoos, and commercial production units that use animals. It is important to note that the regulations which cover RTT, define "use" in terms of a "manipulation". Therefore, a teacher in school classroom, which maintains pet guinea pigs, is not subject to the RTT regulations. Nevertheless, the teacher, as person in charge, is required by the AWA to attend to the general welfare provisions of the AWA.

AUTHORIZATION OF USER-BREEDING INSTITUTIONS/INSPECTION/PENALTIES BY COMPETENT AUTHORITIES—AUSTRALIA

Scientific institutions that use animals for scientific purposes (research, teaching, or testing) are authorized to do so by licenses issued under the state or territory animal welfare or animal research legislation.[2–9] The breeding and supply of particular animal species for scientific use may also require authorization or licensing. The minister or government department with responsibilities for the animal welfare legislation issues these licenses, and administers any conditions imposed. Additional requirements and conditions usually apply to dogs and cats sourced from private owners and municipal pounds for scientific purposes.[10] Examples of additional state Codes and Guideline documents have been listed earlier in this chapter, under General Framework.[13,14,17–19]

Inspections of animal facilities, and the breeding and use of the animals are responsibilities of both the AECs and the relevant government department in most jurisdictions. Each licensed institution or breeding establishment is required to comply with the license conditions, as well as the responsibilities, procedures and practices stated in the *Australian Code for the care and use of animals for scientific purposes 8th edition* 2013 ("Australian Code"; recently revised). The AEC listed on the license must oversee the sourcing, supply, breeding, transport and husbandry of animals intended for scientific use. The external review of institutions, a requirement under the Australian Code, usually incorporates inspection of animal teaching and research areas, and animal holding and breeding facilities.

The government regulator receives annual statistics reports on animal use for scientific purposes from each licensed breeding establishment and research and teaching institution. The details of the statistical reporting are included in the relevant legislation, or in guidelines promulgated by the regulator.[2–9]

AUTHORIZATION OF USER-BREEDING INSTITUTIONS/INSPECTION/PENALTIES BY COMPETENT AUTHORITIES—NEW ZEALAND

There are no commercial rodent production facilities in New Zealand. Many universities and research institutions in New Zealand breed their own rodents/rabbits. These institutions are authorized through the Code of Ethical Conduct (http://www.biosecurity.govt.nz/animal-welfare/naeac/papers/naeaccec.htm), which is a government license to operate breeding facilities and use animals in research, teaching and testing. User or breeding institutions that perform "manipulations" are required to comply with Part 6 of the AWA.[22]

NON-HUMAN PRIMATES: SPECIAL CONSIDERATIONS AND RESTRICTIONS—AUSTRALIA

Non-human primates (NHP) are not indigenous to Australia. There are international treaties and Commonwealth, state and territory legislative restrictions upon their importation. Biocontainment holding, and exotic animal licensing and permit conditions are enforced by government regulatory authorities.[24] Additional ethical and animal welfare concerns apply to use of NHP in general and great apes in particular. These are addressed in animal welfare legislation, the *Australian Code for the care and use of animals for scientific purposes 8th edition* 2013 ("Australian Code"; recently revised), and additional guidelines such as the National Health and Medical Research Council *Policy on the use of Non-Human Primates for Scientific Purposes* (2003, currently under review by the NHMRC).[1,25] Currently great apes are not used for biomedical research or testing in Australia.

NHMRC Statement on the Use of Great Apes

"The species of great ape, gorilla, orangutan, chimpanzee and bonobo, are closely related to humans in evolution. Proposals to AECs requesting the use of great apes for scientific purposes may pose particular concerns. The Animal Welfare Committee of the NHMRC must be notified of proposals for the use of great apes for scientific purposes approved by the institutional AEC before the project can commence. Great apes may only be used for scientific purposes if the following conditions are met: Resources, including staff and housing, are available to ensure high standards of care for the animals; The use would potentially benefit the individual animal and the species to which the animal belongs; The potential benefits of the scientific knowledge gained will outweigh harm to the animal."

"Whenever non-human primates are used for scientific purposes, the investigator must justify the use of this species and weigh the scientific or educational value of the study against the potential effects on the welfare of the animal. Investigators are responsible for using animals humanely and treating them with respect as defined in the Australian Code. Before any research using non-human primates commences, approval must be obtained from the AEC where the work is to be undertaken. Research conducted overseas under the auspices of an Australian institution must conform to standards at least equivalent to those of the Australian Code of Practice and the NHMRC Policy on the use of Non-Human Primates for Scientific Purposes (2003)."

The NHMRC supports National Breeding Colonies for macaques (*Macaca nemestrina* and *Macaca fascicularis*), marmosets (*Callithrix jacchus*) and baboons (*Papio hamadryas*). These colonies have been established to centralize breeding, provide a consistently high standard of animal care and management, and to allow access to non-human primates for research.[25]

NHMRC Policy on the Care and Use of Nonhuman Primates for Scientific Purposes

"The NHMRC will only fund research using non-human primates which meets all of the following requirements:

1. No alternative to animal use or other species of animal is suitable for the particular research project and the predicted outcome of the project justifies the use of non-human primates.
2. The most appropriate species of non-human primate is chosen and detailed justification for such use is provided to the AEC.
3. The application to the AEC must state the fate of the non-human primates at the end of the project.
4. Whenever possible investigators obtain non-human primates from the NBCs.
5. Investigators must ensure that documentation of the source of each non-human primate and assessment of its behaviour, clinical history, and health status must accompany the animal and must be kept current.
6. To assist the AEC consider the case for importation from overseas in the context of the Code of Practice, investigators must provide the AEC with all available information regarding animal welfare at the source facility.
7. Non-human primates imported from overseas must not be taken from wild populations and must be accompanied by documentation to certify their status.
8. Confirmation of orders and dispatch of non-human primates do not occur before the institutional AEC has approved the project.
9. Investigators performing experiments overseas under the auspices of an Australian institution obtain approval from an Australian AEC that, of necessity, may include the delegation of authority to inspect sites and monitor projects at remote sites.
10. Animals are not bred in Australia outside the NBCs unless as a (sic) integral component of the research protocol with AEC approval.
11. Imported species of non-human primates are not used to establish a breeding colony outside the NBCs.
12. Non-human primates originally sourced from overseas are not exported from Australia.
13. Australian-bred non-human primates are not exported unless for a specific purpose, in which case exemption must be obtained from the AWC of the NHMRC.
14. The principal investigator and all animal care staff have training and experience specific for the species of non-human primate to be used.

15. Accommodation for non-human primates as described under Appendix 1 Point 4 is available before the animals are obtained. When new caging is required, personnel responsible for planning and financing the project first seek advice from NBC management. Before using non-human primates, investigators funded by NHMRC should be prepared to allow access to the AWC of the NHMRC to inspect the facilities where the animals will be housed and used.

16. Social interaction between experimental animals is accommodated. Animals should be held in appropriate social groupings unless the AEC has specifically approved individual caging as necessary for a particular project. When individual caging of animals is unavoidable, it is for a minimum time and the singly-housed animal is given auditory and visual contact with other animals in the colony.

17. When social isolation is unavoidable, attempts are made to increase the variety of environmental enrichment beyond that available for socially-grouped animals.

18. Daytime access to an outside enclosure is freely available to all non-human primates held for six weeks or longer. It must be ensured that the non-human primates are presented with as many choices of environment as possible. Outside enclosures are important to increase the variety of stimuli perceived. The external environment provides elements of enrichment. Access to an outside enclosure is recommended for all animals.

19. All personnel working with non-human primates are informed about the diseases transmissible between non-human primates and humans and are instructed on the measures needed to prevent disease transmission.

20. When non-human primates are killed for experimental or veterinary reasons, all possible attempts are made to inform other interested investigators of the availability of tissues and organs and their disease status."[25]

NON-HUMAN PRIMATES: SPECIAL CONSIDERATIONS AND RESTRICTIONS—NEW ZEALAND

The AWA prohibits the use of nonhuman hominids unless approved by the Director-General. These are defined to include the gorilla, chimpanzee, bonobo or orangutan. At the time of writing this review, there are few non-human primates used in research, testing and teaching in New Zealand.

GENETICALLY ALTERED ANIMALS—AUSTRALIA

The Commonwealth of Australia *Gene Technology Act* 2000[26] and *Gene Technology Regulations* 2001,[27] regulate genetically modified (altered) animals and other organisms (GMOs) in Australia. The aim of the legislation is to protect the health and safety of humans and the environment by identifying and managing risks posed by, or resulting from, gene technology. Biocontainment of GMOs is required, with the details of the physical containment determined by legislation and guidelines. This involves government accreditation of organizations generating or holding GMOs, certification of facilities, licensing of use of GMOs (dealings), monitoring and compliance activities, and maintenance of a "Record of GMO and GM Product Dealings" as required by the *Gene Technology Act 2000*.[26] The *Guidelines for the transport, storage and disposal of GMOs* (2011) applies to all genetically modified animals and other organisms regardless of their commercial, domestic or scientific use.[28]

The *Australian Code for the care and use of animals for scientific purposes 8th edition* 2013 ("Australian Code"; recently revised)[1] and the National Health and Medical Research Council Animal Welfare Committee *Guidelines for the generation, breeding, care and use of genetically modified and cloned animals for scientific purposes* 2007[29] address the animal welfare and ethical issues and requirements relating to genetically modified and cloned animals used for scientific purposes. These guidelines provide a brief overview of issues such as the potential for adverse animal welfare outcomes, the uncertainties associated with the genetic manipulation on the resultant genetic constitution (genotype) and biological characteristics or phenotype, and the large number of animals generated in order to obtain a small number with the desired genotype. The roles and responsibilities of the Animal Ethics Committee, the research scientist and the animal caregiver are detailed, and cross-referenced with the requirements of the Australian Code.

The Australian National Consultative Committee on Animal Welfare *Position statement on genetic manipulation of animals* (1991) states that "All stages of producing genetically modified animals, including developmental stages such as breeding-up prior to release, should be: Regarded as experimental under the *Australian Code for the Care and Use of Animals for Scientific Purposes*, and Covered by Animal Experimentation Ethics Committees (AEECs)."[10]

GENETICALLY ALTERED ANIMALS—NEW ZEALAND

The Hazardous Substances and New Organisms Act 1996[30] and the Biosecurity Act 1993[31] provides for controls on the use of genetically altered animals. The controls on the housing and use of genetically altered animals are detailed in the *Biosecurity Authority Standard 154.03.03 Containment Facilities for Vertebrate Laboratory Animals*.[32] These controls require the maintenance of registers, which record the numbers of genetically modified animals held in containment. The registers must be updated each month. Movement from secure containment in the animal facility to a research laboratory must also be documented. Government officials audit research, testing, and teaching institutions every six months to ensure compliance with the Standard.

INSTITUTIONAL AND DESIGNATED PERSONNEL RESPONSIBILITIES—AUSTRALIA

The state and territory animal welfare legislation states the licensing requirements and conditions that apply to scientific establishments that use animals for scientific purposes.[2–9] These conditions include, generally, the requirement to appoint one or more AEC, to report annually to the government regulator, and to fulfill the responsibilities detailed in the *Australian Code for the care and use of animals for scientific purposes 8th edition* 2013 ("Australian Code"; recently revised).[1] The state or territory legislation also usually stipulates that a particular individual, nominated by the institution, is named on the license for the scientific institution or establishment. This person is usually a senior manager with responsibilities that include oversight of the research, testing or teaching conducted by the institution, and with the authority to represent the institution on these matters to the government regulator.[2–9]

The Australian Code (Section 2.1) details institutional responsibilities and recognizes that matters of compliance with relevant legislation, and the Australian Code, lie with the governing body of the institution. The governing body of the institution ("the institution"), or its delegate, is held responsible and accountable for the activities of the personnel under their authority who use animals for scientific purposes, as well as other personnel and resources controlled by them on behalf of the institution. The institution is responsible for the effective function of the AEC/s listed on the license, is required to seek comments and reports from the AEC, and to resource it adequately so that it can fulfill its responsibilities and thereby assist the institution. In effect, the AEC performs a significant compliance management role on behalf of the institution.

The Chairperson of the AEC plays a pivotal role in the success of this arrangement, as communication between the AEC and the institution must be both frequent and effective. The Australian Code (Section 2.2.2) recommends thus:

"Institutions should consider appointing a chairperson who holds a senior position in the institution. If the chairperson is an external appointee, institutions must provide the chairperson with the necessary support and authority to carry out the role."

The responsibilities incumbent upon scientific researchers and teachers are similar to those detailed in the following New Zealand section. The responsibilities of institutions and other personnel follow.

Institutional responsibilities are detailed in Australian Code Section 2.1, and include:

- Establishing or accessing one or more AEC, to ensure that all scientific use of animals complies with the relevant legislation, Codes and guidelines;
- Provision of education, training, workshops and other information so that all animal users, animal care staff, and AEC members are aware of their responsibilities under the Code and legislation;

- Seeking comment from the AEC on all matters that may affect animal welfare, including building or modification of animals' facilities;
- Seeking AEC approval for all institutional animal care and use guidelines and emergency planning;
- Responding promptly to recommendations of the AEC;
- Establishing mechanisms to respond to enquiries or complaints concerning the use of animals within the institution;
- Ensuring adequate numbers of trained animal care staff and veterinary services are available;
- Establishing procedures for resolution of disputes between members of the AEC, animal users, and the institution;
- Ensuring that the AEC operation and report are reviewed annually by the institution, and at least every four years by an external review committee.

INSTITUTIONAL AND DESIGNATED PERSONNEL RESPONSIBILITIES—NEW ZEALAND

Institutions wishing to use animals are required to obtain a Code of Ethical Conduct (CEC), which is a license to use animals in research, testing and teaching. Details of the code requirements are specified above. Animal users are expected to comply with the policy document: *"Good Practice Guide for the use of animals in research, testing and teaching"*.[23] This defines responsibilities of users in terms of:

- AEC approval
- Planning projects
- Conduct of experiments
- Limiting pain and distress
- Animal welfare monitoring of pain or distress
- Study endpoints
- Repeated use of animals in experiments
- Duration of experiments
- Handling and restraining animals
- Completion of projects
- Euthanasia
- Autopsy
- Pre-operative planning
- Surgery
- Post-operative care
- Implanted devices
- Organ and tissue transplantation
- Neuromuscular paralysis
- Electro immobilization
- Animal models of disease
- Modifying animal behavior

- Toxicological experiments
- Experiments involving hazards to humans or other animals
- Experimental manipulation of animals' genetic material
- Cloning of animals
- Experimental induction of neoplasia
- Production of monoclonal antibodies
- Lesions of the central nervous system
- Withholding food or water
- Fetal experimentation
- Research on pain mechanisms and the relief of pain
- Animal welfare and animal health research.

Investigators are ultimately responsible for all matters relating to the welfare of animals under their control. However, the Good Practice Guide recognizes that in many institutions the duty of managing routine animal husbandry is delegated to professional animal care personnel on a daily basis. Therefore, strategies must be in place for effective communication between facility managers and investigators regarding animal welfare and research concerns.

OVERSIGHT AND ETHICAL REVIEW PROCESS—AUSTRALIA

The mainstay of the Australian system is the AEC, whose terms of reference, role, membership and mode of operation are comprehensively outlined in Section 2.2 of the *Australian Code for the care and use of animals for scientific purposes 8th edition* 2013 ("Australian Code"; recently revised).[1]

The AEC comprises a diverse membership consisting of at least one of each of the following categories of persons: a qualified veterinarian with experience relevant to the activities of the institution (Category "A"); a suitably qualified person with substantial recent experience in the use of animals in scientific or teaching activities (Category "B"); a person with demonstrable commitment to, and established experience in, furthering the welfare of animals, who is not employed by, or otherwise associated with the institution, and who is not involved in the care and use of animals for scientific purposes (Category "C"); and a person who is both independent of the institution and has never been involved in the use of animals in scientific or teaching activities, either in employment or beyond their undergraduate education (Category "D"). The Code recommends, but does not currently mandate the appointment of persons responsible for the routine care of animals within the institution (e.g. Animal Facility Managers) as AEC members. Invitation of persons with specific expertise to provide advice as required is also advised. Finally, to ensure a fair balance between the number of persons who are independent of the institution (Categories C and D) and those who may not be, it is stated that where an AEC has more than four members, Categories C and D should represent no less than one-third of the members.

The primary role of the AEC is to ensure that the use of animals is justified, provides for the welfare of those animals and incorporates the principles of Replacement, Reduction and Refinement. However, in reality, the Code gives the AEC a relatively large role in that it is responsible, on behalf of the institution, not only for the ethical review and approval of proposed projects, but also for overseeing and ensuring that all aspects of animal work (including acquisition, use, husbandry, housing, care, welfare, and fate) are acceptable, and for ongoing monitoring of these parameters and the implementation of the 3Rs. On the surface, this may appear to be an impossibly large task for the AEC, as a small group of persons, at least a third of whom are generally serving on a voluntary and casual basis; however in reality, by law, the onus of compliance and implementation of the 3Rs actually rest directly on the institution and on the persons involved directly in animal use and animal care. Nevertheless by entrusting AECs with the oversight and monitoring of animal use and care to the AEC, the Code better provides them with the necessary authority to assist the institutions in humane use by intervening, suspending and making recommendations for change as necessary.

Post-approval monitoring of animal use is an AEC responsibility that is detailed in Section 2.3 of the Australian Code. The AEC members monitor animals by:

(1) Performing inspections of animal facilities where animals are housed and laboratories where animal use occurs; (2) Inspecting animal monitoring records maintained by researchers and teachers; (3) Reviewing written annual, progress and completion reports on projects; (4) Considering reports from investigators on unexpected adverse events.

OVERSIGHT AND ETHICAL REVIEW PROCESS—NEW ZEALAND

The key principle governing the use of laboratory animals in RTT revolves around the concept of a CEC. Persons or institutions may apply to the Director-General for approval of a CEC. This code is effectively a government license to use animals for the purpose of RTT. Approval is personal to the "code holder" and is not transferable. A senior institutional official is the code holder. For example, the code holder for each NZ university is generally the Vice-Chancellor of that university.

The contents of a code are specified in the AWA[22]:

1. "Each code of ethical conduct must contain provisions that set out, in relation to the carrying out of the research, testing, or teaching to which the code relates, the policies to be adopted and the procedures to be followed,—
   a. by the code holder; and
   b. by an animal ethics committee appointed by the code holder.
2. The policies and procedures must—
   a. enable the animal ethics committee to carry out its functions effectively; and

    b. enable persons who are members of the animal ethics committee but who are not employed by the code holder to have an effective input into the working of the committee; and

    c. make provision for adequate monitoring of compliance with the conditions of project approvals to be carried out; and

    d. make provision for the code holder to collect the information and to maintain the records required by regulations made under this Act; and

    e. specify animal management practices and facilities that are such as to enable the purposes of this Part to be met adequately; and

    f. be such as to ensure that where any member of the animal ethics committee makes a complaint, that complaint may be dealt with fairly and promptly by the animal ethics committee or the code holder; and

    g. include, if necessary, the policies and procedures referred to in section 84(1)(b).

3. The provisions of each code of ethical conduct must—

    a. be consistent with this Act and with any standards or policies prescribed by regulations made under this Act."

NAEAC has a "Guide to the Preparation of Codes of Ethical Conduct".[33] At the time of writing there are approximately 35 CECs in force in NZ, all slightly different, but all in compliance with the main welfare objectives as defined by NAEAC. Central to the CEC is the requirement for the code holder to establish an AEC. Details of the AEC are provided below.

CECs are generally issued for up to 5 years. Approval of each code is publicly notified in the Gazette. Code holders must reapply upon expiry of the CEC.

The process of code application involves an "independent review" by an accredited reviewer appointed by the code holder, from a national panel of certified accredited reviewers. The terms of reference for the review are detailed in the AWA. The review process examines institutional compliance with the specific operational procedures detailed in the CEC. The independent review report is evaluated by NAEAC and the Ministry of Primary Industries.

Each code holder must establish an AEC. The functions and powers of the AEC are:

a. "to consider and determine on behalf of the code holder applications for the approval of projects;

b. to consider and determine, under section 84(1)(a), applications for the approval of projects;

c. to set, vary, and revoke conditions of project approvals;

d. to monitor compliance with conditions of project approvals;

e. to monitor animal management practices and facilities to ensure compliance with the terms of the code of ethical conduct;

f. to consider and determine applications for the renewal of project approvals;

g. to suspend or revoke, where necessary, project approvals;

h. to recommend to the code holder amendments to the code of ethical conduct.

(2) Each animal ethics committee has such powers as are reasonably necessary to enable it to carry out its functions."

In considering any application for the approval of a project and in setting, varying, or revoking conditions of the approval of a project, every animal ethics committee must have regard to such of the following matters as are relevant:

a. "the purposes of this Part; and
b. any matters that the committee is required to consider by regulations made under this Act; and
c. the scientific or educational objectives of the project; and
d. the harm to, or the distress felt by, the animals as a result of the manipulation, and the extent to which that harm or distress can be alleviated by any means (including, where the pain or distress cannot be held within reasonable levels, the abandonment of the manipulation or the humane destruction of animals); and
e. whether the design of the experiment or demonstration is such that it is reasonable to expect that the objectives of the experiment or demonstration will be met; and
f. the factors that have been taken into account in the choice of animal species; and
g. whether the number of animals to be used is the minimum necessary to ensure a meaningful interpretation of the findings and the statistical validity of the findings; and
h. whether adequate measures will be taken to ensure the general health and welfare of animals before, during, and after manipulation; and
i. whether suitably qualified persons will be engaged in supervising and undertaking the research, testing, or teaching; and
j. whether any duplication of an experiment is proposed and, if so, whether any such duplication will be undertaken only if the original experiment—
   – is flawed in a way that was not able to be predicted; or
   – needs to be duplicated for the purpose of confirming a result that was unexpected or has far-reaching implications; and
k. whether the same animals are to be used repeatedly in successive projects, and, if so, the cumulative effect of the successive projects on the welfare of the animals; and
l. whether there is a commitment to ensuring that findings of any experiment will be adequately used, promoted, or published; and
m. any other matters that the committee considers relevant."

"Membership of the AEC is defined as follows:

1. Each animal ethics committee is to consist of at least 4 members.
2. If the code holder is an organisation, the members of the animal ethics committee must be appointed by the chief executive of the organisation or his or her nominee.

3. One member must be—
   a. the code holder; or
   b. if the code holder is an organisation, a senior member of the organisation appointed by the chief executive to be a member of the committee.
4. Any senior member of an organisation who is appointed under subsection (3)(b) must be a person who is capable of evaluating—
   a. each proposal for a project; and
   b. the qualifications and skills of the proposer of a project; and
   c. the scientific value or the teaching value, as the case may require, of a project.
5. One member must be a veterinarian (not being a veterinarian who is an employee of, or is otherwise associated with, the code holder) appointed by the code holder on the nomination of the New Zealand Veterinary Association or a similar national body of veterinarians.
6. One member must be a person appointed by the code holder on the nomination of an approved organisation.
7. The person appointed under subsection (6) must not be—
   a. a person who is in the employ of, or is otherwise associated with, the code holder; or
   b. a person who is involved in the use of animals for research, testing, or teaching.
8. One member must be a person appointed by the code holder on the nomination of a territorial authority or regional council.
9. The person appointed under subsection (8) must not be—
   a. a person who is in the employ of, or is otherwise associated with, the code holder; or
   b. a person who is associated with the scientific community or an animal welfare agency."

REUSE: REQUIREMENTS—AUSTRALIA

Repeated use of individual animals is permitted, subject to requirements detailed in the *Australian Code for the care and use of animals for scientific purposes 8th edition* 2013 ("Australian Code"; recently revised).[1] Section 2.3.05 states that:

"When considering approval for the reuse of animals, the AEC must take into account:

1. the pain or distress, and any potential long-term or cumulative effects, caused by any previous activities and conditions;
2. the time allowed for recovery of the animals between activities;
3. whether an animal has recovered fully from the previous activities;
4. the pain and distress likely to be caused by the next and subsequent activities; and
5. the total time over which an animal will be used."

And Section 1.22 to 1.24: The Reduction benefits of reusing animals must also take into account the lifetime experience of the individual animal. "Reducing the number of animals used should not result in greater harm, including pain and distress, to the animals used."

REUSE: REQUIREMENTS—NEW ZEALAND

The AWA[22] requires the AEC in Section 100 (k) to determine if there is any proposed reuse of animals and make appropriate decisions regarding the welfare of animals. There are no formal controls on the number of reuse events to which an animal might be exposed.

SETTING FREE/REHOMING REQUIREMENTS—AUSTRALIA

Setting free or release of genetically altered, feral, domestic, exotic or pest animal species into the natural environment is often prohibited by state and territory legislation.[1] There are several concerns, including the welfare of the particular animal, the potential effects upon the natural environment or ecosystem, and biosecurity concerns relating to the risk of spread of disease from released animals to wild animal populations. Rehoming of suitable domestic animal species is a matter subject to AEC policy decision, or on a case-by-case basis. Native wildlife release requirements are detailed in Section 3.3.39 of the *Australian Code for the care and use of animals for scientific purposes 8th edition* 2013 ("Australian Code"; recently revised) but may also be subject to other legislation (e.g. Gene Technology Act 2000, Agricultural and Veterinary Chemicals Act 1994, Quarantine Regulations 2000, biosecurity, environmental pest animals). It is outside the scope of this chapter to discuss these, however references to relevant regulatory restrictions are detailed and listed in the Australian Code and related National Health and Medical Research Council guidelines, and Commonwealth, state and territory government websites.[1,12,14,17,19–21,25–27]

The Australian Code requires the capture, holding, transportation, handling and release of wildlife must be in accordance with the principles stated in Sections 3.3.33 to 3.3.46. Requirements for release of wildlife:

1. the duration of captivity must be minimized and consistent with the purpose and aims of the project. If animals are to be released they must be handled in a manner where the risk of disease transmission and environment disturbance is minimized. All possible steps must be taken to avoid animals becoming habituated to human activity;
2. release occurs at the site of capture, unless otherwise approved by the AEC;
3. the timing of the release coincides with the period of usual activity for the species. Animals are protected from injury and predation at the time of their release; and
4. animals that have been sedated or anaesthetized have recovered to full consciousness before their release.

SETTING FREE/REHOMING: REQUIREMENTS FOR NEW ZEALAND

There are no national standards on rehoming of animals. Institutions are free to develop their own policies.

OCCUPATIONAL HEALTH AND SAFETY—AUSTRALIA

Health and safety in the workplace is the shared responsibility of employers and employees, and is subject to state and territory occupational health and safety (OHS) legislation. A detailed description and comparison of the legislation in each jurisdiction is beyond the scope of this chapter, therefore the focus here is mainly on the requirements detailed in the Australian Code.[1]

The *Australian Code for the care and use of animals for scientific purposes 8th edition* 2013 ("Australian Code"; recently revised) requires the AEC to seek information from animal users concerning the potential OHS impact of their animal work upon other personnel and other animals in the animal facility, and also information concerning related discussions with the Institutional Biosafety and OHS Committees.[1] The purpose is not to duplicate the functions of other Committees, but rather for a better understanding of the impact of the experiment upon the animals and personnel, and to ensure that all other compliance matters have been addressed.

OCCUPATIONAL HEALTH AND SAFETY—NEW ZEALAND

The main Act for the management of health and safety in the workplace is the Health and Safety in Employment Act 1992 (HSE Act). The HSE Act was amended in 2002.[34] The principal objective of the HSE Act and amendment is to provide for the prevention of harm to employees at work. It achieves this through promoting health and safety management by employers, imposing duties on employers and others, and the provision of making regulations and approved codes of practice relating to particular significant hazards.

EDUCATION, TRAINING, AND COMPETENCE OF PERSONNEL—AUSTRALIA

The *Australian Code for the care and use of animals for scientific purposes 8th edition* 2013 ("Australian Code"; recently revised) requires scientific institutions, AECs, teachers and the animal users themselves to accept responsibility for ensuring education, professional training and competence of all persons involved with animal use.[1] Section 1.29 requires that all people caring for and using animals must be competent for the procedure they perform, or under the direct supervision of a person who is competent to perform the procedure. The Institution and the AEC is charged with the responsibility of determining whether investigators have the required education, training and competence to perform the techniques required for the particular activity. Teachers and supervisors must ensure that prior to using animals, the students receive appropriate instruction in their ethical and legal responsibilities as well as the appropriate methods of animal care and use (Sections 4.12 and 4.13).

Several states also require that researchers and members of AECs are provided with an understanding of the ethical and humane care and use of animals. Recognizing that the skills and knowledge of those involved are essential to achieve high standards of animal welfare and scientific outcomes, courses for people using animals in research and teaching have been developed.[35-37] In Victoria, a program to develop Standard Operating Procedures for registered training authorities for the use of animals in training is underway. Specific training for researchers and teachers is currently undertaken at the institutional level. Some training and education for AEC members have been identified as being provided by the Australian and New Zealand Council for the Care of Animals in Research and Teaching (ANZCCART) and the Animal Welfare Units in Queensland, NSW and Victoria.[21,35-37] An online training program in NSW is currently being revised and updated (see Animal Ethics Infolink website below for links to international courses and Caring for Animals in Research and Education course). Many universities and other research institutions require that new research, teaching and post-graduate students undertake an induction training program, which includes ethics and regulation of animals used for scientific purposes.

Publicly available online courses and recorded public seminars, webinars and lectures hosted by Australian institutions:

- NSW Animal Research Review Panel & NSW Dept. Primary Industries Animal Welfare Branch: Animal Ethics Infolink (http://www.animalethics.org.au/)[35]
- University of Melbourne (http://www.orei.unimelb.edu.au/content/online-resources)[38] and (https://www.coursera.org/#unimelb)[39]
- Animal Welfare Science Centre (http://www.animalwelfare.net.au/article/scientific-seminars)[40]
- Australian Animal Welfare Strategy (http://www.australiananimalwelfare.com.au/).[41]

Animal Care Personnel

The Australian Code requires that institutions should ensure that animal care staff are appropriately trained and instructed (Sections 2.1.2 and 2.1.5).[1] In many cases, these staff will have attended or be attending appropriate tertiary education institutions providing training in animal technology, veterinary nursing, and other animal care courses. It would be expected that these staff have had training so that they:

- understand the ethical principles that govern the care and uses of animals as detailed in the code;

- are familiar with the laws, guidelines and codes of practice which govern the acquisition, care, breeding, handling and use of animals in their workplace;
- and are able to provide and maintain animals in accordance with the relevant codes to meet species-specific needs and to ensure animals are maintained in an optimum state of wellbeing.

Education, Training, and Competency with Regard to Non-Human Primates

The National Health and Medical Research Council provides the principal funding, and the key policy and guidelines, for scientific use of non-human primates in Australia.[1,24,25] The *Policy on the use of non-human primates for scientific purposes 2003* states:

"To ensure the wellbeing of the animals and to facilitate their management, all staff must be well trained in handling methods and have a sound knowledge of the species in their charge. Training for new or inexperienced staff and investigators must be arranged in consultation with the management of the National Health and Medical Research Council National Non-human primate Breeding Colonies (NBCs). Training through the NBC will facilitate familiarity between the animal and the investigator and the transition of the non-human primates to a new environment.

Investigators and the team working with non-human primates should be familiar with all aspects of the care and use of non-human primates, before obtaining animals or commencing projects. When preparing an application to the AEC for a research project involving the use of non-human primates, investigators should consult with NBC management as a source of helpful information. Application to the AEC should document evidence of the competence of the investigator's team in dealing with the species."[25]

EDUCATION, TRAINING, AND COMPETENCE OF PERSONNEL—NEW ZEALAND

The AWA[22] requires the AEC to determine the qualifications of personnel using animals in research, testing and teaching; as detailed in Section 100 (i): "whether suitably qualified persons will be engaged in supervising and undertaking the research, testing, or teaching."

The specific details of how each institution determines this qualification and the level of training required to gain competency is left to the discretion of each institution.

To date there is no national training program in New Zealand for personnel working in research, testing, and teaching. Each institution provides for the training of their staff.

TRANSPORT—AUSTRALIA

Importation of animals into Australia poses a biosecurity risk, which is managed and administered by the Australian Government Department for Agriculture, Fisheries and Forestry and their Quarantine Inspection Service. Importation of exotic and wild animal species is subject to Commonwealth of Australia quarantine legislation and international treaty such as the Convention on International Trade in Endangered Species of Wild Fauna and Flora (CITES).[42,43]

International and domestic air transport of exotic and genetically modified animals is subject to Commonwealth, state, and territory government biosecurity regulations,[42] Office of the Gene Technology Regulator transportation guidelines for genetically modified organisms[28] and International Air Transport Association (IATA) Live Animal Regulations 2012 guidelines.[44] Domestic transport of domestic animals, native animals, and animal tissues within Australia are subject to state and territory regulatory requirements and restrictions.

The *Australian Code for the care and use of animals for scientific purposes 8th edition* 2013 ("Australian Code"; recently revised), Section 3.2[1] provides guidance on transport of animals used for scientific purposes, and stipulates that domestic animal and livestock transport must comply with the relevant Commonwealth, state, and territory animal transportation legislation and standards. These include the 2012 Australian Animal Welfare Standards and Guidelines for the Land transport of Livestock.[16] IATA guidelines apply to air transport of animals within Australia.

Specific NHMRC Importation Requirements for Non-Human Primates

The National Health and Medical Research Council *Policy on use of non-human primates for scientific purposes* (2003) states:

"Before importing non-human primates permission must be obtained from two authorities:

- Australian Quarantine Inspection Service of the Commonwealth Department of Agriculture, Fisheries and Forestry—Australia; and
- Environment Australia, Department of the Environment and Heritage.

If animals are to be imported for NHMRC-funded research, the AWC of the NHMRC must be notified after approval of the project by the relevant AEC(s) and before importation of the animals.

Before non-human primates are transported, the following must be considered:

- The requirement for AEC approval of projects before the animals are transported
- The personnel responsible for the animals from point of departure to destination including during any break in the journey

- Physical containment of the animals
- Twenty four hour contact details on the container in case of an emergency
- Whether animals are fit to travel. For example, pregnant animals in the last third of gestation should not be subjected to transport stress
- The likelihood of sudden or extreme changes in temperature during transport and how any such changes will be managed
- Whether animals will best be transported singly, in pairs or as a group
- The total time for transportation, which will influence the timing of feeding and watering
- For the importation of non-human primates, the Department of Agriculture, Fisheries and Forestry-Australia's conditions must be met
- For air travel, assurance that the animals will travel in a pressurised and temperature controlled compartment in compliance with the document Live Animals Regulations of the International Air Transport Association (IATA)."[25,42–44]

TRANSPORT—NEW ZEALAND

The Good Practice Guide for the Use of Animals in Research, Testing, and Teaching[23] provides general recommendations for the transport of animals. There is a Code of Recommendations and Minimum Standards for the Transport of Animals within New Zealand.[45]

The transfer of genetically modified animals requires authorization by the Ministry of Primary Industries.

HOUSING AND ENRICHMENT—AUSTRALIA

Care of all animals is addressed in state and territory animal welfare legislation,[2–9] and other state and territory animal welfare codes of practice. In general, the owner or person in charge of any animal must provide the animal with adequate and appropriate food, water, shelter, exercise, and veterinary attention, provide a suitable living environment and take all reasonable steps to prevent animal harm. These obligations apply equally to the owners of companion animals, livestock, or animals used for scientific purposes. In the case of domestic livestock, the states and territories have based their animal welfare codes of practice upon the Primary Industries Standing Committee Model Animal Welfare Codes of Practice, which provide guidance on animal housing, husbandry, transport, euthanasia, and other relevant issues.[15] The Model Codes are currently under review as part of the *Australian Animal Welfare Strategy* and will be replaced by regulated standards under the authority of state and territory legislation.[12,16,41]

Section 4 of the *Australian Code for the care and use of animals for scientific purposes 8th edition* 2013 ("Australian Code"; recently revised) discusses care of animals in production and holding facilities in scientific establishments.[1] This section provides an outline of the issues that need to be satisfactorily addressed, including: species-suitable accommodation (buildings and pens or cages); building maintenance and repair; cleaning and hygiene of buildings and animal pens or cages; vermin exclusion and control; water supply and drainage; environmental controls (temperature, humidity, ventilation, light, noise, ammonia, dust, odors); contingency planning for emergencies. Routine husbandry procedures on livestock that are not part of a project must comply with relevant legislation and animal welfare codes of practice.

The Australian Code does not prescribe cage, pen or engineering standards. Guidance on environmental enrichment strategies is provided in Part III Fact Sheet E from the National Health and Medical Research Council *Guidelines to promote the wellbeing of animals used for scientific purposes: The assessment and alleviation of pain and distress in research animals* (2008).[46]

In the state of Victoria, *The Prevention of Cruelty to Animals Act, 1986* mandates compliance with the *Code of Practice for the Housing and Care of Laboratory Mice, Rats, Guinea Pigs and Rabbits*.[47] This code applies to all institutions breeding or using laboratory rats, mice, guinea pigs, and rabbits for scientific purposes. The primary objective of this code is to establish minimum standards and promote more detailed recommendations for the housing and care of these species than is currently provided in the Australian Code. The particular aspects this Housing and Care Code include are nutrition, animal enclosures, climate and environment, behavior and environmental enrichment, hygiene, handling and basic procedures, health monitoring, transportation, euthanasia, and monitoring and record keeping. It specifies minimum standards for the housing, and recommends monitoring sheets, welfare assessment sheets, other record keeping sheets and a phenotype report for genetically modified animals.

In the state of New South Wales, similar standards apply. These are detailed in six species-specific Animal Research Review Panel *Guidelines* for the housing and care of dogs, rats, rabbits, guinea pigs, sheep, and mice in scientific establishments.[17]

HOUSING AND ENRICHMENT—NEW ZEALAND

The *Good Practice Guide for the use of animals in research, testing and teaching*[23] provides general recommendations for housing conditions and enrichment. New Zealand has elected to avoid the use of engineering standards for cage sizes. Instead, performance standards are in place.

Animal accommodation should be designed and managed to meet species-specific needs. Pens, cages, and containers should be designed, constructed, and maintained to ensure the comfort and wellbeing of the animals. Any variations to these requirements as part of a project must receive prior AEC approval. The following factors should be taken into account:

- species-specific behavioral requirements, including the availability and design of space to allow free movement and activity, sleeping, enclosed spaces, contact with others of the same species, and the opportunity to perform a species-specific behavioral repertoire;

- species-specific environmental requirements such as lighting, temperature, air quality, appropriate day/night cycles, and protection from excessive noise and vibrations;
- provision of single housing for animals when appropriate for the species and, if necessary, for the purpose of the experiment, e.g. during recovery from surgery or collection of samples;
- the need to provide ready access to food and water;
- the need to clean the pen, cage or container;
- protection from spread of pests and disease;
- requirements of the project; and
- the need to observe the animals readily.

Pens, cages and containers should:

- be constructed of durable, impervious materials;
- be kept clean;
- be secure and escape-proof;
- protect the animals from climatic extremes;
- not cause injury to the animals;
- be large enough to ensure the wellbeing of the animal or animals, with adequate space to allow them to stretch out when recumbent and to stand upright;
- be maintained in good repair.

Wire floor cages for rodents should not be used unless essential to the research protocol and then only for brief periods. Animals should have a solid resting area when housed on wire floors.

The population density of animals within cages, pens or containers and the placement of these in rooms should be such that acceptable social and environmental conditions for the species can be maintained. Where it is necessary to individually house animals of a species, which are normally kept in a social group, the conditions should be managed to minimize the impact of social isolation. Animals should be housed in these circumstances for the minimum time necessary.

Bedding, litter or other environmental provisions should be provided, if appropriate, to the species, and should be comfortable, absorbent, dust-free, nonpalatable, nontoxic, able to be sterilized if needed, and suitable for the particular research purpose. Pregnant animals must be provided with nesting materials where appropriate.

Enrichment and Environmental Complexity

Most animals used in research, testing, and teaching are housed in unnatural environments. Wherever possible such animals should be provided with an environment that can accommodate the behavioral and physiological needs of the species.

Almost all the species of animals used in research, testing, and teaching have well defined social structures and prefer to live in groups, although care must be taken to ensure that animals are socially compatible. Individual housing is stressful for social animals, and social isolation should be avoided whenever possible and limited to meet the specific research objectives as approved by an AEC. The effects of physical isolation should be minimized where possible by the use of noncontact communication, whether visual, auditory or olfactory. Judicious use of mirrors can also be helpful, as can an environment of increased complexity.

The living areas of the animals should be set up and provisioned with the means that will enable them to perform a behavioral repertoire appropriate to the species.

HUSBANDRY AND ENVIRONMENT—AUSTRALIA

Section 4 of the *Australian Code for the care and use of animals for scientific purposes 8th edition* 2013 ("Australian Code"; recently revised), discusses care of animals in production and holding facilities in scientific establishments.[1] The Victorian and New South Wales government regulators also provide minimum standards and recommendations which have been discussed above, and which are similar to those that apply in New Zealand. Additional detail on suggested enrichment is provided for each particular species in the National Health and Medical Research Council (NHMRC) *Guidelines to promote the wellbeing of animals used for scientific purposes: The assessment and alleviation of pain and distress in research animals* (*2008*) and the National Health and Medical Research Council, Victorian and New South Wales state animal housing and care guidelines listed in the General Framework section.[17,46,47]

Codes and guidelines for meeting the needs of domestic livestock species, captive native species, and exotic zoo animal species may apply in some situations. National and International guidelines for husbandry and environment of captive wildlife, zoo and aquatic species exist, however a detailed listing of these is outside the scope of this chapter. References to Commonwealth of Australia Model Codes for animal welfare and Australian animal welfare standards and guidelines are provided earlier in the General Framework section, and other references are provided in the Australian Code and at the Australian Animal Welfare Strategy website.[1,12,15,16,41]

HUSBANDRY AND ENVIRONMENT—NEW ZEALAND

The Good Practice Guide for the use of animals in research, testing and teaching[23] provides recommendations:

"Animals should receive appropriate, clean and nutritionally adequate food according to accepted requirements for the species. The food should be in sufficient quantity and of appropriate composition to maintain normal growth of immature animals or normal weight of adult animals and to provide for the requirements of pregnancy or lactation."

"Uneaten perishable food should be removed promptly unless contrary to the needs of the species. Where possible, alteration to dietary regimes should be gradual. When animals are fed in groups, there should be sufficient trough space or feeding points to avoid undue competition for food, especially if feed is restricted. Feeding space is determined by the size and number of animals that must eat at one time. Drinking water should be constantly and reliably available, and be clean, fresh and uncontaminated."

"Routine husbandry procedures should comply with any code of welfare for the species involved and must be performed by competent personnel. Variations to normal procedure as part of an experimental project must receive prior AEC approval. Procedures applied to the maintenance of breeding stock and supply of animals are viewed as routine husbandry and fall outside the definition of manipulation. When special breeding requirements are integral to a research or teaching project such as in the creation of a genetically modified animal, then procedures applicable to breeding must be regarded as a manipulation and should be included in the proposal to the AEC."

VETERINARY CARE—AUSTRALIA

Rather uniquely, compared to the rest of the world, there is currently no legislative requirement for a dedicated veterinary surgeon, who is laboratory animal science- or laboratory animal medicine-trained, to be appointed by institutions to assist with proactive implementation of initiatives consistent with the 3Rs principles in the use, care and welfare of the animals. Instead there is a more basic requirement in the *Australian Code for the care and use of animals for scientific purposes 8th edition* 2013 ("Australian Code"; recently revised), for institutions to ensure "availability and access to veterinary advice for the management and oversight of a program of veterinary care, quality management and project design to safeguard animal wellbeing."[1] However, the current Australian Code does advise institutions to consider appointing an officer with veterinary, or other appropriate qualifications who is authorized by the AEC to ensure that projects are proceeding in compliance with the Code and the decisions of the AEC. Furthermore, many institutions do appoint a veterinary officer or animal welfare officer to fulfill the duties and responsibilities often undertaken by veterinarians appointed as, for example, "designated veterinarians" in the European Union, "clinical or attending veterinarians" in the United States, "named veterinary surgeons" in the United Kingdom, and "animal welfare officers" in Canada, the Netherlands, etc.

VETERINARY CARE—NEW ZEALAND

The provision of veterinary care is not addressed directly. The AWA[22] makes indirect reference to the need for appropriate veterinary care in Section 9:

1. "requires owners of animals, and persons in charge of animals, to take all reasonable steps to ensure that the physical, health, and behavioural needs of the animals are met in accordance with both—
   a. good practice; and
   b. scientific knowledge; and

2. requires owners of ill or injured animals, and persons in charge of such animals, to ensure that the animals receive, where practicable, treatment that alleviates any unreasonable or unnecessary pain or distress from which the animals are suffering."

The Good Practice Guide[23] specifies responsibilities of Investigators as follows:

"People who use animals for scientific purposes have an obligation to treat the animals humanely and to consider their welfare as an essential factor when planning and conducting experiments."

"Investigators have direct and ultimate responsibility for all matters related to the welfare of the animals under their control, including the general husbandry and housing of those animals as well as the specific manipulations. They should act in accordance with their specific AEC approval. The responsibility of investigators extends over all facets of the care and use of animals in projects approved by the AEC, beginning when the animal is allocated to the approved project and ending with its fate at the end of the project. Investigators are responsible for the standard of animal care and use by all other persons involved in the project. They should ensure that the extent of supervision is compatible with the level of competence of each person and the responsibilities they are given."

"However, it is recognised that in many institutions the duty of managing routine animal husbandry is delegated to professional animal care personnel on a daily basis. Strategies must be in place for the facility manager to effectively communicate with the investigator regarding animal welfare and research concerns."

"Investigators have a legal and ethical responsibility to ensure that animals on a project are manipulated using medical and surgical techniques which are consistent with the principles of good practice and scientific knowledge in the discipline of laboratory animal veterinary medicine. Investigators should consult with veterinarians whenever unexpected adverse effects occur in order that standard veterinary care treatment regimes are immediately implemented. This responsibility parallels the public's duty of care to seek veterinary management of any sick animals in their charge."

CONDUCT OF EXPERIMENTAL PROCEDURES—AUSTRALIA

Planning to Maximize Well-Being and Minimize Pain and Distress in Animals

The *Australian Code for the care and use of animals for scientific purposes 8th edition* 2013 ("Australian Code"; recently revised), is based upon the general principles of prevention of cruelty, duty-of-care, and Russell and Burch's 3Rs principles of humane animal use (Replacement, Reduction, Refinement).[1] Investigators are required to treat animals with respect and to consider their welfare when planning and conducting scientific research, teaching and testing. Implementation of the 3Rs at the planning stage is a process that involves both the investigators and the Animal Ethics Committee, assisted by the structured animal use Proposal or Application documentation, and the AEC assessment process (Australian Code Sections 2 and 3). The Australian Code (Sections 2.4 and 3) places emphasis on the need for the animal user or investigator to anticipate the animal welfare impacts of all the scientific procedures that comprise a particular experiment, study or use, and to prepare a strategy that will promote animal wellbeing and limit animal pain and distress (Section 3). Ensuring technical skills and competence of all personnel involved in animal care and use is a responsibility of the scientific investigator or teacher. Both anticipated and unanticipated (or accidental) adverse impacts need to be incorporated in the strategy. Often the AEC requires an animal use flowchart or chart that outlines step-by-step what will happen (and when) to each experimental group of animals. A period of conditioning or acclimatization of the animal to the project environment, personnel and procedures prior to commencing the animal use is recommended to be included in the plan. Familiarity with the species-specific signs of pain and distress, and incorporation of these in a record of animal welfare monitoring observations is usually required, and is a most useful tool if tailored to the particular animal use proposed. The monitoring plan usually incorporates details of the frequency of animal welfare observations as determined by the animal use protocol and the animal's condition, and provides specific criteria for interventions that maximize wellbeing and minimize pain and distress. Interventions include withdrawal of the animal from the experiment, use of analgesia, provision of veterinary treatments, nursing and palliative care, and humane killing to limit pain and distress (euthanasia). Further details and suggestions concerning the importance of the planning process and ways in which wellbeing can be promoted are detailed in the National Health and Medical Research Council *Guidelines to promote the wellbeing of animals used for scientific purposes: The assessment and alleviation of pain and distress in research animals, Part II (2008).*[46]

Detecting and Limiting Pain and Distress

The Australian Code Section 3 (ii) states:

"Pain and distress may be difficult to evaluate in animals. Unless there is evidence to the contrary, it must assume that procedures and conditions that would cause pain and distress in humans cause pain and distress in animals. Decisions regarding the possible impact of procedures or conditions on an animal's wellbeing must be made in consideration of an animal's capacity to experience pain and distress." And Section 2.4.18 (x) states: "Use of pharmacological agents such as anaesthetics, analgesics and sedatives must be appropriate to the species, the individual animal (e.g age, physiological status) and the scientific aims, and must be consistent with current veterinary or medical practice. Anaesthesia must be used for procedures that are likely to cause pain of a kind and degree for which anaesthesia would normally be used in veterinary or medical practice."

Further details and suggestions concerning the importance of the detecting and limiting animal pain and distress are detailed in the National Health and

Medical Research Council *Guidelines to promote the wellbeing of animals used for scientific purposes: The assessment and alleviation of pain and distress in research animals, Parts II and III Fact sheets—Pain management: anaesthesia, analgesia and anxiolytics, and humane killing and euthanasia (2008)*.[46]

Monitoring, Reporting, and Reviewing Research and Teaching Use

Reporting annually to the AEC on progress and outcomes of animal use projects is a requirement of the Australian Code, and the responsibility of the scientific investigator or teacher. The investigator is also responsible for reporting promptly to the AEC any unexpected adverse events. The AEC is required to perform inspections of animal facilities and animal use projects and records to ensure that welfare of animals is monitored as agreed. The AEC reports at least annually to the governing body (or delegate) of the scientific institution. The AEC report includes a summary of its activities, and the animal use conducted by the institution.[1] The scientific institution reports on scientific use of animals to the government regulator, in accord with state and territory legislative requirements.[2–9]

Currently neither the Australian Code nor the state/territory legislation provides any guidelines on severity classifications of projects or procedures. However, in the state of Victoria, the "impact" (severity) of procedures is required to be specified as part of the annual use statistics-reporting requirement (required by the Regulations associated with the Victorian *Prevention of Cruelty to Animals Act*).[3] The categories used are as follows: observational, animal unconscious without recovery, minor conscious intervention no anesthesia, minor surgery with recovery, minor physiological challenge, surgery with recovery, physiological challenge and "death as an end point". Note that the latter lethality study is a special category, which requires special approval from the Minister of Agriculture.

CONDUCT OF EXPERIMENTAL PROCEDURES—NEW ZEALAND

The Good Practice Guide[23] provides guidance on the conduct of experiments by discussing limits on pain and distress, welfare monitoring strategies, study endpoints, humane endpoints, repeated used of animals, limits on duration of experiments, and restraint/handing strategies.

In 1994, NAEAC developed a system to record the number of animals manipulated within different scales of severity. This generated statistics collected in accordance with the Animal Welfare (Records and Statistics) Regulations 1999.[48]

In December 2008, the "severity" scale was replaced with a grading system that was designed to better reflect the overall estimate of the impact or invasiveness of each animal use. The five grades are:

Grade A "No impact or virtually no impact"
Grade B "Little impact"
Grade C "Moderate impact"

Grade D "High impact"

Grade E "Very high impact"

The Good Practice Guide[23] addresses the management of painful procedures in Appendix 1. The basis of the Guide's recommendations on pain management is the Swiss Academy of Medical Sciences Ethical Principles and Guidelines for Experiments on Animals,[49] which states in Section 4:

"Experiments on animals shall be carried out in accordance with the latest developments. Known prophylactic, diagnostic and therapeutic processes shall be taken into account and the scientific guidelines provided by international expert bodies shall be observed."

"If pain, suffering or stress are inevitable concomitants of an experiment, their duration and intensity must be limited to the minimum. To this end, the animals shall be monitored by specially trained personnel in accordance with predefined criteria and at predefined times and measures necessary to alleviate suffering shall be taken insofar as this is compatible with the objective of the experiment. The animal must be able to express its sensations and where possible avoid painful stimuli. Hence, the use of substances that induce paralysis without loss of consciousness and analgesic effects is unauthorized."

EUTHANASIA—AUSTRALIA

The *Australian Code for the care and use of animals for scientific purposes 8th edition* 2013 ("Australian Code"; recently revised), requires animals to be killed humanely in circumstances where animal welfare is compromised and animal pain and distress is unable to be alleviated promptly (Section 3.1).[1] The Australian Code Section 3.3.45 states:

"The method and procedures used for killing an animal must be humane and: (i) avoid pain or distress and produce rapid loss of consciousness until death occurs; (ii) be compatible with the scientific and eductational aims of the project or activity; (iii) be appropriate to the species, age, developmental stage and health of the animal; (iv) require minimum restraint of the animal; (v) be reliable, reproducible and irreversible; (vi) ensure that animals are killed in a quiet, clean environment away from other animals; (vii) ensure that death is established before disposal of the carcass, fetuses, embryos and fertilized eggs."

The local documents most commonly relied upon to provide guidance on methods of euthanasia are the Australian and New Zealand Council for the Care of Animals in Research and Teaching (ANZCCART) Guidelines on Euthanasia (2001, currently under revision)[50] and the National Health and Medical Research Council *Guidelines to promote the wellbeing of animals used for scientific purposes: The assessment and alleviation of pain and distress in research animals, Parts II and III Fact sheets—Pain management: anaesthesia, analgesia and anxiolytics, and Humane killing and euthanasia (2008).*[46] In the state of Victoria the mandatory *Code of Practice for the Housing and Care of Laboratory Mice, Rats, Guinea Pigs and Rabbits* contains guidelines on euthanasia methods in those species.[47] Likewise, the New South Wales the Animal Research Review Panel and other states' animal care guidelines also include suitable humane species-specific methodologies.[20,21,35] All of these guidelines and methods are similar to the *American and Veterinary Medical Association Guidelines for the euthanasia of animals: 2013 edition.*[51]

As mentioned earlier in the chapter, the *Australian Animal Welfare Strategy* promotes humane animal protocols for a wide range of wild and domestic species and circumstances,[41] and Regulated and Model Animal Welfare Codes include details of acceptable euthanasia techniques for domestic livestock species.[15,16]

EUTHANASIA—NEW ZEALAND

The Good Practice Guide[23] provides general guidance on the principles of euthanasia, and the Guide references the ANZCCART guidelines on euthanasia and the AVMA Guidelines for Euthanasia.[50,51]

EQUIPMENT AND FACILITIES—AUSTRALIA

As discussed earlier in this chapter, the *Australian Code for the care and use of animals for scientific purposes 8th edition* 2013 ("Australian Code"; recently revised), does not prescribe cage, pen or engineering standards, however in some cases Commonwealth, state and territory legislation and animal welfare Codes of Practice stipulate general standards or provide specific guidance.[1–9,14–17,19,25–29]

Regulatory requirements apply in circumstances where facilities are used for holding or transporting animals or biological materials subject to Australian Quarantine Inspection Service and Department of Agriculture, Fisheries and Forestry (DAFF) biosecurity, and/or Office of the Gene Technology Regulator Physical Containment (PC) conditions for genetically modified animals and organisms.[27,28,42]

As imported and genetically modified laboratory rodents are commonly used in biomedical research, many animal facilities are constructed or modified to accommodate animals requiring PC2 and Quarantine Approved Premise biocontainment. There are no specific regulations for more general animal research equipment or facilities except for requirements relating to building maintenance, hygiene and OHS (discussed earlier in this chapter), compliance with license requirements for scientific use of animals,[2–9] or to operate diagnostic radiography units, annual certification of autoclave chambers and so on.[1]

EQUIPMENT AND FACILITIES—NEW ZEALAND

There are no specific regulations for equipment or facilities except for compliance with license requirements to operate diagnostic radiography units, annual certification of autoclave chambers and so on.

REFERENCES

1. National Health and Medical Research Council 2004. *Australian Code of Practice for the care and use of animals for scientific purposes.* 8th ed. Canberra: Australian Government; 2013. (Internet). (cited 2013 Jul 26); NHMRC reference no. EA28. Available from: http:// www.nhmrc.gov.au/guidelines/publications/ea28.

2. New South Wales Government 2013. Animal research act 1985. (Internet). (cited 2013 Mar 23); Act no. 123 of 1985. Available from: http://www.austlii.edu.au/au/legis/nsw/consol_act/ ara1985134/.

3. Victorian Government 2012. Prevention of cruelty to animals act 1986. (Internet). (cited 2013 Mar 23); Act no. 46 of 1986. Available from: http://www.austlii.edu.au/au/legis/vic/consol_act/ poctaa1986360/.

4. Queensland Government 2012. Animal care and protection act 2001. (Internet). (cited 2013 Mar 23); Act no. 64 of 2001. Available from: http://www.legislation.qld.gov.au/LEGISLTN/ CURRENT/A/AnimalCaPrA01.pdf.

5. Australian Capital Territory 2012. Animal welfare act 1992. (Internet). (cited 2013 Mar 23); Act no. 45 of 1992. Available from: http://www.austlii.edu.au/au/legis/act/consol_act/ awa1992128/.

6. Northern Territory Government 2013. Animal welfare act 1999. (Internet). (cited 2013 Mar 23); Act no. 44 of 1999. Available from: http://www.austlii.edu.au/au/legis/nt/consol_act/awa128/.

7. Western Australian Government 2008. Animal welfare act 2002. (Internet). (cited 2013 Mar 23); Act no. 033 of 2002. Available from: http://www.slp.wa.gov.au/legislation/statutes.nsf/ main_mrtitle_50_currencies.html.

8. South Australian Government 2012. Animal welfare act 1985. (Internet). (cited 2013 Mar 23); Act no. 106 of 1985. Available from: http://www.austlii.edu.au/au/legis/sa/consol_act/ awa1985128/.

9. Tasmanian Government 2013. Animal welfare act 1993. (Internet). (cited 2013 Mar 23); Act no. 63 of 1993. Available from: http://www.austlii.edu.au/au/legis/tas/consol_act/awa1993128/.

10. National Consultative Committee on Animal Welfare 2012. National guidelines for animal welfare. (Internet). (cited 2013 Mar 23); Available from: http://www.daff.gov.au/animal-plant-health/welfare/nccaw/guidelines/research.

11. Council of Australian Governments' Standing Council on Primary Industries 2012. (Internet). (cited 2013 Mar 23); Available from: http://www.daff.gov.au/scopi/about_scpi.

12. Australian Government Department of Agriculture Food and Forestry 2012. Animal welfare in Australia. (Internet). (cited 2013 Mar 23); Available from: http://www.daff.gov.au/animal-plant-health/welfare.

13. National Health and Medical Research Council 2012. Animal research ethics. (Internet). (cited 2013 Mar 23); Available from: http://www.nhmrc.gov.au/health-ethics/animal-research-ethics.

14. National Health and Medical Research Council 2012. Animal research ethics publications. (Internet). (cited 2013 Mar 23); Available from: http://www.nhmrc.gov.au/health-ethics/ animal-research-ethics/animal-research-ethics-publications.

15. Primary Industries Standing Committee (PISC/SCARM). *Primary industries report series.* Collingwood Victoria: CSIRO Publishing. (Internet). 2013 (cited 2013 Mar 23); Available from http://www.publish.csiro.au/nid/22/sid/11.htm.

16. Animalwelfarestandards.net.au (Internet). Canberra: Animal Health Australia Australian animal welfare standards and guidelines 2013. (updated 2013 Mar 14; cited 2013 Mar 23); Available from: http://www.animalwelfarestandards.net.au/.

17. New South Wales Animal Research Review Panel 2013. Policies and guidelines. (Internet). (cited 2013 Mar 23); Available from: http://www.animalethics.org.au/policies-and-guidelines/.

18. New South Wales Department of Primary Industries 2013. Animal Welfare Branch fact sheet 21: Supply of dogs and cats for use in research. (Internet). (cited 2013 Mar 23); Available from: http://www.dpi.nsw.gov.au/agriculture/livestock/animal-welfare/research-teaching/factsheets/aw-fact21.

19. Victoria Department of Primary Industries 2013. Victorian codes of practice for animal welfare. (Internet). (cited 2013 Mar 23); Available from: http://www.dpi.vic.gov.au/agriculture/about-agriculture/legislation-regulation/animal-welfare-legislation/codes-of-practice-animal-welfare.

20. South Australia Department of Environment 2012. Water and Natural Resources. Policies and guidelines for wildlife research. (Internet). (cited 2013 Mar 23); Available from: http://www.environment.sa.gov.au/Plants_Animals/Animal_welfare/Policies_and_guidelines_for_wildlife_research.

21. ANZCCART (Internet) 2013. Adelaide: Australian and New Zealand Council for the Care of Animals in Research and Teaching Australia. (updated 2013 Mar 22; cited 2013 Mar 23); Available from: http://www.adelaide.edu.au/ANZCCART/resources.

22. New Zealand Animal Welfare Act (Internet) 1999. (cited 2013 Mar 25); Available from: http://www.legislation.govt.nz/act/public/1999/0142/latest/DLM49664.html.

23. New Zealand Good Practice Guide for the Use of Animals in Research 2010. Testing and Teaching (Internet). (cited 2013 Mar 25); Available from: http://www.biosecurity.govt.nz/files/regs/animal-welfare/pubs/naeac/guide-for-animals-use.pdf.

24. NHMRC (Internet) Canberra: National Health and Medical Research Council 2013. Use of non-human primates. (updated 2013 Jan 18; cited 2013 Mar 23); Available from: http://www.nhmrc.gov.au/health-ethics/animal-research-ethics/use-non-human-primates.

25. National Health and Medical Research Council 2003. *Policy on the use of non-human primates for scientific purposes*. Canberra: Australian Government. (Internet). (updated 2011 Apr 4; cited 2013 Mar 23); NHMRC reference no. EA14. Available from: http://www.nhmrc.gov.au/health-ethics/animal-research-ethics/policy-care-and-use-non-human-primates-scientific-purposes and http://www.nhmrc.gov.au/guidelines/publications/ea14.

26. Australian Government 2012. Gene technology act 2000—C2012C00172. (Internet). (cited 2013 Mar 23); Act no. 169 of 2000. Available from: http://www.comlaw.gov.au/Details/C2012C00172.

27. Australian Government 2011. Gene technology regulations 2001—F2011C00732. (Internet). (cited 2013 Mar 23); Statutory Rules 2001 no. 106. Available from: http://www.comlaw.gov.au/Details/F2011C00732.

28. Department of Health and Aging Office of the Gene Technology Regulator 2011. Guidelines for the transport, storage and disposal of GMOs 2011. (Internet). (cited 2013 Mar 23); Available from: http://www.ogtr.gov.au/internet/ogtr/publishing.nsf/Content/storageanddisp-3/$FILE/tsd-guidelines.pdf.

29. National Health and Medical Research Council Animal Welfare Committee 2007. *Guidelines for the generation, breeding, care and use of genetically modified and cloned animals for scientific purposes*. Canberra: Australian Government. (Internet). (cited 2013 Mar 23); NHMRC reference no. EA17. Available from: http://www.nhmrc.gov.au/_files_nhmrc/publications/attachments/ea17.pdf.

30. New Zealand Hazardous Substances and New Organisms Act 1996, 1996. (Internet). (cited 2013 Mar 25). Available from: http://www.legislation.govt.nz/act/public/1996/0030/latest/DLM381222.html.

31. New Zealand Biosecurity Act 1993, 1993. (Internet). (cited 2013 Mar 25). Available from: http://www.biosecurity.govt.nz/biosec/pol/bio-act.

32. New Zealand Biosecurity Authority Standard 154.03.03 Containment Facilities for Vertebrate Laboratory Animals 2003. (Internet). (cited 2013 Mar 25); Available from: http://www.epa.govt.nz/Publications/154-03-03-MAF-ERMA-Std-2002.pdf.

33. NAEZC Guide to the Preparation of Codes of Ethical Conduct 2012. (Internet). (cited 2013 Mar 25); Available at: http://www.mpi.govt.nz/Default.aspx?TabId=126&id=1331.

34. New Zealand Health and Safety in Employment Act 1992, 1992. (Internet). (cited 2013 Mar 25); Available at: http://www.osh.govt.nz/law/hse.shtml.

35. New South Wales Animal Research Review Panel and New South Wales Department of Primary Industries Animal Welfare Branch: Animal Ethics Infolink 2013. (Internet). (cited 2013 Mar 23); Available from: http://www.animalethics.org.au/.

36. Victoria Department of Primary Industries 2013. Using Animals for Research and/or Teaching; and Introductory AEC training. (Internet). (updated 2013 Jan 22; cited 2013 Mar 23); Available from: http://www.dpi.vic.gov.au/agriculture/innovation-and-research/animals-used-research-and-teaching/.

37. Queensland Department of Agriculture, Fisheries and Forestry 2012. Animal ethics training. (Internet). (updated 2012 Sept 05; cited 2013 Mar 23); Available from: http://www.daff.qld.gov.au/27_9917.htm.

38. University of Melbourne Office for research ethics and integrity 2012. Online resources. (Internet). (cited 2013 Mar 23); Available from: http://www.orei.unimelb.edu.au/content/online-resources.

39. Corsera.org 2013, University of Melbourne. (Internet). (cited 2013 Mar 23); Available from: https://www.coursera.org/#unimelb.

40. Animal Welfare Science Centre 2013. Scientific seminars. (Internet). (cited 2013 Mar 23); Available from: http://www.animalwelfare.net.au/article/scientific-seminars.

41. Australian Government 2013. Australian Animal Welfare Strategy. (Internet). (updated 2013 Mar 19; cited 2013 Mar 23); Available from: http://www.australiananimalwelfare.com.au/.

42. Australian Government Department of Agriculture, Fisheries and Forestry 2012. Biosecurity in Australia. (Internet). (updated 2012 Dec 06; cited 2013 Mar 23); Available from: http://www.daff.gov.au/aqis.

43. CITES.org (Internet) 2013. Geneva; Convention on International Trade in Endangered Species of Wild Fauna and Flora (CITES) (cited 2013 Mar 23); Available from: http://www.cites.org/.

44. IATA.org (Internet) 2013. International Air Transportation Association. Live animal regulations manual 2012. (cited 2013 Mar 23); Available from: http://www.iata.org/.

45. New Zealand Code of Animal Welfare (Transport within New Zealand) Code of Welfare 2011, 2011. (Internet). (cited 2013 Mar 25); Available at: http://www.biosecurity.govt.nz/files/regs/animal-welfare/req/codes/transport-within-nz/transport-code-of-welfare.pdf.

46. National Health and Medical Research Council 2008. *Guidelines to promote the wellbeing of animals used for scientific purposes: the assessment and alleviation of pain and distress in research animals.* Canberra: Australian Government. (Internet). (updated 2012 Feb 20; cited 2013 Mar 23); NHMRC reference no. EA17. Available from: http://www.nhmrc.gov.au/guidelines/publications/ea18.

47. Victoria Department of Primary Industries 2010. Code of practice for the housing and care of laboratory mice, rats, guinea pigs and rabbits. (Internet). (updated 2010 Sep 27; cited 2013 Mar 23); Available from: http://www.dpi.vic.gov.au/agriculture/about-agriculture/legislation-regulation/animal-welfare-legislation/codes-of-practice-animal-welfare/care-of-laboratory-mice-rats-guinea-pigs-rabbits/code-of-practice-for-the-housing-and-care-of-laboratory-animals.

48. New Zealand Animal Welfare (Records and Statistics) Regulations 1999. (Internet). (cited 2013 Mar 25); Available at: http://www.legislation.govt.nz/regulation/public/1999/0392/latest/DLM1045.html.

49. Swiss Academy of Medical Sciences 2005. Ethical Principles and Guidelines for Experiments on Animals. (Internet). (cited 2013 Mar 25); Available at: http://www.bvet.admin.ch/themen/tierschutz/index.html?lang=en&download=.

50. Australian and New Zealand Council for the Care of Animals in Research and Teaching. (Internet). (updated 2013 Mar 23; cited 2013 Mar 23). In: Reilly JS, editor. *Euthanasia of animal for scientific purposes*. 2nd ed. Adelaide: ANZCCART; 2001. Available at: http://www.adelaide.edu.au/ANZCCART/publications/Euthanasia.pdf.

51. American and Veterinary Medical Association 2013. *Guidelines for the euthanasia of animals: 2013 edition*. Schaumberg, IL: AVMA. (Internet). (cited 2013 Mar 23); Available at: https://www.avma.org/KB/Policies/Pages/Euthanasia-Guidelines.aspx.

Index

Note: Page numbers with "*f*" denote figures; "*t*" tables.

Printed and bound by CPI Group (UK) Ltd, Croydon, CR0 4YY
03/10/2024
01040410-0011